DISORDERED AND COMPLEX SYSTEMS

Related Titles from AIP Conference Proceedings

548 Fundamental Issues of Nonlinear Laser Dynamics
Edited by Bernd Krauskopf and Daan Lenstra, December 2000, 1-56396-977-7

524 Nonlinear Acoustics at the Turn of the Millennium: ISNA 15; 15th International Symposium on Nonlinear Acoustics
Edited by Werner Lauterborn and Thomas Kurz, July 2000, 1-56396-945-9

517 Computing Anticipatory Systems: CASYS'99– Third International Conference
Edited by Daniel M. Dubois, June 2000, 1-56396-933-5

511 Unsolved Problems of Noise and Fluctuations: UPoN'99: Second International Conference
Edited by Derek Abbott and Lazlo B. Kish, March 2000, 1-56396-826-6

502 Stochastic and Chaotic Dynamics in the Lakes: STOCHAOS
Edited by David S. Broomhead, Elena A. Luchinskaya, Peter V. E. McClintock, and Tom Mullin, February 2000, 1-56396-915-7

501 Stochastic Dynamics and Pattern Formation in Biological and Complex Systems: The APCTP Conference
Edited by Seunghwan Kim, Kyoung J. Lee, and Wokyung Sung, January 2000, 1-56396-914-9

469 Slow Dynamics in Complex Systems: Eighth Tohwa University International Symposium
Edited by Michio Tokuyama and Irwin Oppenheim, April 1999, 1-56396-811-8

465 Computing Anticipatory Systems (CASYS-98): Second International Conference
Edited by Daniel M. Dubois, March 1999, 1-56396-863-0

To learn more about these titles, or the AIP Conference Proceedings Series, please visit the webpage **http://www.aip.org/catalog/aboutconf.html**

DISORDERED AND COMPLEX SYSTEMS

London, United Kingdom 10–14 July 2000

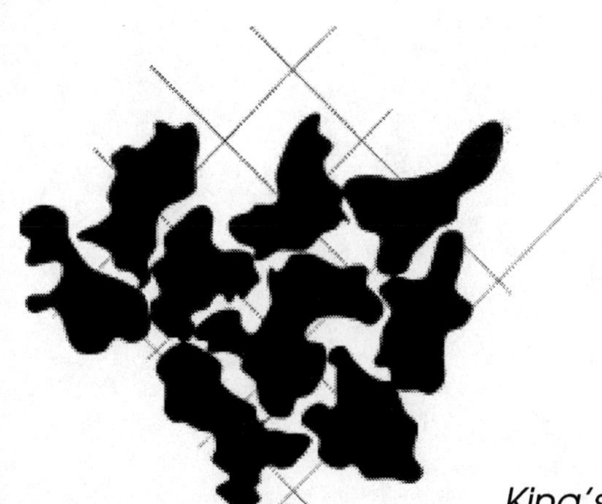

EDITORS
P. Sollich
A. C. C. Coolen
L. P. Hughston
R. F. Streater
King's College, London, United Kingdom

Melville, New York, 2001
AIP CONFERENCE PROCEEDINGS ■ VOLUME 553

Editors:

P. Sollich
A. C. C. Coolen
L. P. Hughston
R. F. Streater

Department of Mathematics
King's College London
Strand
London WC2R 2LS
UNITED KINGDOM

E-mail: peter.sollich@kcl.ac.uk
tcoolen@mth.kcl.ac.uk
lane.hughston@kcl.ac.uk
ray.streater@kcl.ac.uk

The figure on the cover and title page, taken from the article by M. Nicodemi in this volume, represents schematically the arrangement of grains in a granular medium. Granular materials such as sand constitute concrete examples of disordered and complex systems in the world around us.

L.C. Catalog Card No. 00-111913
ISBN 1-56396-983-1
ISSN 0094-243X
Printed in the United States of America

CONTENTS

*Italicized name indicates the author who presented the paper.

PART II: INFORMATION GEOMETRY

PART III: QUANTUM DYNAMICS AND QUANTUM CHAOS

*Italicized name indicates the author who presented the paper.

PART IV: REACTION-DIFFUSION EQUATIONS

PART V: NEW DIRECTIONS IN MATHEMATICAL FINANCE

*Italicized name indicates the author who presented the paper.

*Italicized name indicates the author who presented the paper.

PREFACE

From 10 to 14 July 2000, an international conference on *Disordered and Complex Systems* was held at King's College London; this was a satellite meeting of the XIIIth International Congress on Mathematical Physics, held the following week at Imperial College, London. The aim of the meeting was to take a deliberately broad view of the meaning of the terms "disorder and complexity", thus bringing together researchers working in a number of different areas and creating opportunities for possible synergies. The programme was structured around five areas: Glassy systems and neural networks, information geometry, quantum dynamics and quantum chaos, reaction-diffusion equations, and mathematical finance. While these topics may sound disparate, a substantial number of common threads appeared: Information geometry, for example, is being applied to subjects such as neural network learning, quantum mechanics and interest rate dynamics; the statistical mechanics of glassy systems is helping to understand the behaviour of large systems of economic agents; and Lévy statistics occur in glasses, in anomalous diffusion, and in asset price returns. In keeping with the aim of the conference, it is hoped that the reader will find in this book many other surprising connections between seemingly unrelated areas.

All speakers at the conference, as well as a selection of poster presenters, were invited to contribute a paper to the proceedings. The articles are grouped into five parts under the headings listed above; within each part, papers are arranged in alphabetical order by presenting author. Where there are several co-authors, the name of the presenting author is marked with an asterisk in the table of contents.

Part one of the proceedings is concerned with *Glassy systems and neural networks.* Here, the methods of the statistical mechanics of disordered systems are to the fore, and the papers in this section demonstrate the impressive sweep of topics to which these methods can be applied. There are, first of all, the structural glasses, of which window glass is the archetype; Barkai provides evidence for the occurrence of Lévy statistics in such systems. Cavagna studies a simple model for structural glasses and extracts information about its "energy landscape". Ritort focuses on fluctuations around a non-equilibrium state, and in particular on whether they can be understood by extending the concept of temperature to non-equilibrium situations. The kinetic spin model studied by Garrahan is more abstract—the "caging in" of atoms by their neighbours in a glass is represented by simple kinetic constraints—but nonetheless captures important features of glassy dynamics. Nicodemi applies similar ideas to granular materials, where "jamming" effects have recently been recognized as closely analogous to a glass transition. Goldbart studies yet another related class of materials, the vulcanized rubbers. These are in fact easier to deal with in an equilibrium framework than glasses, since the chemical crosslinks between different strands of a vulcanized polymer are permanent, rather than slowly evolving like the "cages" in a glass. Van Mourik, Nishimori and Parisi explore the theory of spin glasses, where glassiness arises from a random distribution of magnetic impurities (spins) in a host material. Many of the methods of the statistical mechanics of disordered systems—such as replica theory—were originally developed for these systems. Van Mourik advances a new interpretation of the Parisi scheme for the analysis of replica symmetry breaking, while Nishimori's work puts rigorous constraints on the regions of the phase diagram where such symmetry breaking can actually appear. Parisi, finally, discusses exciting links between equilibrium and non-equilibrium behaviour via fluctuation-dissipation relations and effective temperatures.

In neural network models, glassy effects often arise from the storage of conflicting patterns; Bollé and Skantzos address the resulting dynamical features, revealing links with information theory and random-field systems. Buhot and Kozłowski focus on capacity calculations, where one asks how many patterns can be stored before interference destroys any possible recall. Saad broadens the picture out to error-correcting codes; the decoding problem has similarities to the task of finding the ground state of a spin glass. Giardina analyses a model of population dynamics in a glass-like landscape, where energy barriers can be surmounted by growth of the population. Sherrington, finally, demonstrates that statistical physics also has a part to play in the study of large systems of interacting economic agents.

Part two of this book deals with *Information geometry*. In the classical version of this theory, the set of probability measures on a finite space has a natural structure as a Riemannian manifold (with the Fisher metric), and also has two dual flat affine structures, those of Amari. Pistone, Combe and Nencka explain some progress that has been made in extending these ideas to infinite sample spaces, using ideas from the theory of Orlicz spaces. Fukumizu shows how these extensions are used in the theory of neural networks to cope with singularities. The quantum version of information geometry has many different good choices for the metric. Jenčová analyses the family of metrics associated with Hasegawa's quantum α-entropy, and one metric is singled out by Streater & Grasselli as the only one for which the entropy and free energy are dual. Gibilisco & Isola construct a whole family of affine structures, on the tangent bundle with its dual cotangent bundle. Grasselli shows that the free energy is an analytic function on a manifold of relatively bounded perturbations of a given state; these ideas can be used to describe dissipative dynamics in quantum theory. Finally, a relation between the quantum and classical theories via the semi-classical limit is described by Ghikas.

In part three, on *Quantum dynamics and quantum chaos*, Hasegawa describes how the statistical mechanics of random matrices leads to an understanding of the physics of the metal-insulator transition, while Fyodorov extends the subject to the spectrum of non-Hermitian random matrices; such matrices occur naturally in quantum chaotic systems. De Cock introduces a new way of analysing the properties of a sequence of vectors, using the matrix of scalar products. Aneva discusses a geometric definition of the operator pq proposed by Berry as having a spectrum related to the Riemann zeros. Shukla suggests a universal method, based on Calogero-Moser dynamics, which seems to unify the treatment of many apparently different problems in chaos. Tomiya describes a numerical study of the distribution of modes in a quartic oscillator system, and finds that the transition to a Gaussian distribution is not at all sharp. The quantum L^p spaces of Majewski arise in the dynamics of infinite dissipative systems, while Lőrinczi's deep analysis tackles the infrared problem in a non-trivial quantum field theory. Cartas illustrates how topological bifurcations can occur as parameters controlling the dynamics of a model are varied.

Part four of the book contains papers on *Reaction-diffusion equations.* These are parabolic partial differential equations with a dissipative Lyapunov function that is identified with the entropy or free energy, and have been generalized to situations where the temperature is not assumed to be constant everywhere. Biler, Karch and Nadzieja take such a system and add the mutual energy, either Newtonian gravity or the Coulomb energy, to obtain an integro-differential system. They study existence, uniqueness and blow-up of the solutions; the methods involve versions of Leray-Schauder theory, and also make good use of scaling properties of the equations. Liskevich provides Gaussian estimates for general parabolic operators, which are useful also for nonlinear problems, while Archincheev shows that Lévy processes are behind some nonlinearities in model dynamics.

Part five, entitled *New Directions in Mathematical Finance*, includes eleven papers on interrelated topics concerning the behaviour of financial markets. Concepts arising in statistical physics, such as information manifolds and the scaling laws of critical phenomena, are put to use in various ways along side the established body of finance theory based on random price dynamics. For example, one of the central problems in modern finance is to understand the origins of the fat-tailed distributions of asset returns. This can be approached either by modelling the relevant processes directly, or by looking at the market as a whole and studying the interactions of agents. The resolution of this problem may have profound consequences for the risk management of derivatives. Several papers here address aspects of this problem (Balland; Bingham & Kiesel; Bolan, Hurd, Pivato & Seco; Iori; Shaw; and Woo). There are interesting overlaps with the mathematics of natural catastrophes and self-organised critical phenomena, with deep connections to the theory of Lévy processes. The great bond market crisis of 1998 provides a nice example of a "risk avalanche". This was triggered by a default on a sovereign debt, an event that can be modelled by a point process (as described in the article by Beumee). Two further papers here, by Brody & Hughston and by Webber, develop geometric ideas in the theory of interest rates with links to statistical physics. Finally, another key problem, closely related to modelling the optimal behaviour of market agents, is the more general task of managerial decision making, which is the topic of the theory of "real options", discussed in articles by Leppard and by Hughston & Zervos.

Neither the conference nor this book would have been possible without the help of a great many people. We thank the members of the scientific advisory committee (R Alicki, P Biler, G Pistone, D Petz and D Sherrington) for refereeing and selecting among the abstracts submitted for presentation. We are grateful to C Hunter and M Zervos, who assisted in the organization of the mathematical finance programme. Thanks are also due to S Glass, A Lyles and H Morton, who handled the local arrangements for the conference. Finally, we acknowledge generous financial support from the London Mathematical Society and King's College London; we are grateful also to S B Streater, of Forbidden Technologies, for a grant that assisted with the costs of this publication.

P Sollich, A C C Coolen, L P Hughston, R F Streater
20 October 2000

PART I:
GLASSY SYSTEMS AND NEURAL NETWORKS

Lévy Statistics for Single Molecule Spectroscopy in Low Temperature Glass

Eli Barkai*[1], Robert J. Silbey* and Gert Zumofen†

* *Department of Chemistry,*[2] *Massachusetts Institute of Technology, Cambridge, MA 02139.*
† *Physical Chemistry Laboratory, Swiss Federal Institute of Technology, ETH-Z, CH-8092 Zürich.*

Abstract. We investigate line shapes of single molecules in low temperature glass based on the standard tunneling model. Using the Kubo–Anderson line shape theory, and numerical simulations, we show that the cumulants of the lines are distributed according to Lévy stable laws. Thus the generalized central limit theorem is applicable to this problem.

INTRODUCTION

Single molecule spectroscopy (SMS) has made it possible to measure spectral properties of individual molecules embedded in a condensed phase. In the last decade the field has expanded at a breathtaking pace [1]. SMS eliminates ensemble averaging and hence has revealed many new phenomena. The method is used in many fields, including but not limited to biology, quantum dot spectroscopy and low temperature glass. SMS is used to answer fundamental questions as well as for many practical applications. We are interested in general aspects of theory of SMS in the condensed phase and in particular SMS in low temperature glass.

Line shapes of chemically identical single molecules in glassy materials vary from one molecule to the other since each molecule is in a unique environment. By investigating properties of line shapes of single molecules in a disordered condensed phase, one can learn about the dynamical and disorder properties of the host medium. Since each line shape is a random function one must consider the statistical properties of the lines. In this way SMS in a low temperature glass, has begun to reveal the nature of glass on a microscopic local level. An important question in this direction is whether the standard tunneling model of low temperature glass is valid [2]. We know for certain that this phenomenological model works well for many glassy materials (i.e., universality) when macroscopic measurements are

---

1) Email: barkai@mit.edu
2) This work was supported by the NSF.

CP553, *Disordered and Complex Systems*, edited by P. Sollich, et al.

considered (e.g., heat capacity), however on the more microscopic level we do not have much experimental (or theoretical [3,4]) proof that the model works well.

Recent experiments in Bordeaux have measured line shapes and spectral trails of single molecules (tenylene) in a polymer glass [5]. Spectroscopy of 70 molecules was investigated and 22 molecules exhibited behavior that seems not compatible with the standard model predictions. While the number of molecules investigated is not sufficient to determine whether the standard tunneling model works well, the experiment is a clear step forward towards a direct verification of the model. In order to understand the SMS in glass we assume the validity of the standard tunneling model and give our predictions based on it. This will help in the analysis of the statistical data from the experiments.

We use the approach of Geva and Skinner [6] who modeled single molecule line shapes in low temperature glass based on the sudden jump approach of Anderson and Kubo [7] and the standard tunneling model for the glass. The model assumes that the low temperature excitations of the glass, the two level systems (TLS), are distributed randomly and uniformally in the medium. These TLSs interact with the molecule via *long range interactions* (e.g., the dipole interaction) while the TLS-TLS interaction is neglected. Brown and Silbey [8] showed that the neglect of the TLS-TLS interaction for this type of problem is reasonable, at least for the parameter set we consider in this work. This is related to the low density of TLSs in the standard tunneling model (i.e., also most of the macroscopic measurements are compatible with the assumption of non interacting TLSs).

The line of each single molecule can be characterized by its cumulants κ_j, $(j = 1, 2, \cdots)$ which will vary from one molecule to the other. Following our recent publication [9] we show that the probability densities $P(\kappa_1)$, $P(\kappa_2)$ are symmetrical and one sided Lévy stable laws respectively, while higher order cumulants are described by Lévy statistics only in the *slow modulation limit.* Using numerical simulation we show that the slow modulation limit holds for a parameter set relevant to experiment.

Lévy stable probability densities, $L_{\gamma,\eta}(x)$, serve as a natural generalization of the normal Gaussian density [10]. Khintchine and Lévy found that stable characteristic functions, $\hat{L}_{\gamma,\eta}(k)$, are of the form

$$\ln\left[\hat{L}_{\gamma,\eta}\left(k\right)\right] = i\mu k - z_\gamma |k|^\gamma \left[1 - i\eta \frac{k}{|k|} \tan\left(\frac{\pi\gamma}{2}\right)\right] \tag{1}$$

for $0 < \gamma \leq 2$ (for the case $\eta \neq 0$, $\gamma = 1$ see [10]). Four parameters are needed for a full description of a stable law. The constant γ is called the characteristic exponent, the parameter μ is a location parameter which is unimportant in the present case, $z_\gamma > 0$ is a scale parameter and $-1 \leq \eta \leq 1$ is the index of symmetry. When $\eta = 0$ $(\eta = \pm 1)$ the stable density $L_{\gamma,\eta}(x)$ is symmetrical (one sided). Today the term Lévy behavior is often used in a wider sense than that implied by the functions in Eq. (1) and is used to describe behavior with broad tails. Examples for applications of Lévy statistics are sub recoil laser cooling [11] and the characterization of geometrical properties of random fractals [12].

Lévy statistics is known to describe several long range interaction systems in diverse fields such as astronomy [10], turbulence [13] and spin glass [14]. Other statistical aspects of spectroscopy, in systems with long range interactions, belong to the Lévy universality class (see [15,16] and references therein), this was usually not noticed. For example according to the theory of Stoneham [17] normalized inhomogeneous lines are Lévy stable densities [16].

MODEL

We assume a single molecule coupled to non identical independent TLSs at distances $\mathbf{r}$ in dimension d. Each TLS is characterized by its asymmetry variable A and tunneling element J. The energy of the TLS is $E = \sqrt{A^2 + J^2}$. The TLSs are coupled to phonons or other thermal excitations such that the state of the TLS changes with time. The state of the n th TLS is described by an occupation parameter, $\xi_n(t)$, that is equal 0 or 1 if the TLS is in its ground or excited state respectively. The probability for finding the TLS in its upper $\xi = 1$ state, p, is given by the standard Boltzmann form $p \equiv 1/\{1+\exp[E/(k_b T)]\}$. The transitions between the ground and excited state are described by the up and down transition rates K_u, K_d, which are related to each other by the standard detailed balance condition. The transition rate of the TLS is $K \equiv K_u + K_d = cJ^2 E \coth\left[E/(2k_b T)\right]$ and c is the TLS phonon coupling constant.

The excitation of the n th TLS shifts the single molecule's transition frequency by ν_n. Thus, the single molecule's transition frequency is

$$\omega(t) = \omega_0 + \sum_{n=1}^{N_{\text{act}}} \xi_n(t)\, \nu_n, \tag{2}$$

where N_{act} is the number of active TLSs in the system (see details below) and ω_0 is the bare transition frequency that differs from one molecule to the other depending on the local static disorder. We consider a wide class of frequency perturbations

$$\nu = 2\pi\alpha\Psi(\Omega)\, f(A, J)\, \frac{1}{r^\delta}, \tag{3}$$

α is a coupling constant with units $\left[\text{Hz nm}^\delta\right]$, $\Psi(\Omega)$ is a dimensionless function of order unity, Ω is a vector of angles determined by the orientations of the TLS and molecule (in some simple cases Ω depends on polar angles only), $f(A, J) \geq 0$ is a dimensionless function of the internal degrees of freedom of the fluctuating TLS, δ is the interaction exponent. For the glass-single molecule system under investigation $\delta = 3$ (dipole interaction), $d = 3$ and $f(A, J) = A/E$. The line shape of the single molecule is given by the complex Laplace transform of the relaxation function

$$I(\omega) = \frac{1}{\pi}\text{Re}\left[\int_0^\infty dt e^{i\omega t} \Pi_{n=1}^{N_{\text{act}}} \Phi_n(t)\right]. \tag{4}$$

The relaxation function for a single TLS is $\Phi_n(t) = \overline{\exp(i\nu \int_0^t \xi_n(t)\mathrm{d}t)}$, where the over bar denotes an average with respect to the stochastic process. It was evaluated [18] based on methods developed in [7]. For a bath of TLSs the line shape, Eq. (4), is a formidable function of the random TLS parameters (r, Ω, A, J) as well as the system parameters (α, T, etc) (see details in [9]). The standard tunneling model assumes a particular distribution of the TLSs parameters A, J, hence we can generate (on a computer) the glass environment and then using Eq. (4) and the fast Fourier transform, find the line shape of a molecule. The details of this procedure and further discussion are given in [6,9,16]. Typically one uses 1000 TLSs to mimic the glass. For typical set of parameters, the structured line of a single molecule is very different from any other, a feature also observed in experiment.

RESULTS

In the fast modulation limit $K \gg |\nu|$, one finds a simple behavior: all lines are Lorentzian with half width $\tilde{\Gamma} = \sum_n^{N_{act}} p_n(1-p_n)\nu_n^2/K_n$ which varies from one molecule to the other. One can show that the probability density function $P(\tilde{\Gamma})$ is under certain conditions a one sided Lévy stable density [16], however as we show below for glass systems the fast modulation limit is not valid (i.e., the lines are not simple Lorentzians).

In the slow modulation limit $K \ll |\nu|$ one finds $\Phi(t) = 1 - p + pe^{-i\nu t}$ implying that the line shape of a molecule coupled to a single TLS is composed of two delta peaks, the line shape of a molecule coupled to two TLSs is composed of four delta peaks, etc (splitting).

The spectral line is characterized by its cumulants κ_j $(j = 1, 2, \cdots)$ that vary from one molecule to the other, and we investigate the cumulant probability density $P(\kappa_j)$. It is interesting to recall that theory and experiment of line shape moments (not single molecule lines) is discussed in Chapter 4 in [19]. We have derived the cumulants of the single molecule line shape, and the first four cumulants are $\kappa_1 = \sum_n p_n \nu_n$, $\kappa_2 = \sum_n p_n (1-p_n) \nu_n^2$, $\kappa_3 = \sum_n \left[p_n(1-p_n)(2p_n - 1)\nu_n^3 + ip_n(1-p_n)K_n\nu_n^2\right]$ and

$$\kappa_4 = \sum_n \left\{ p_n(-1+p_n)\left[K_n^2 + \nu_n^2\left(-1 + 6p_n - 6p_n^2\right)\right]\nu_n^2 - \right.$$

$$\left. 2iK_n(-1+p_n)p_n(-1+2p_n)\nu_n^3 \right\}. \tag{5}$$

The exact calculation of the cumulants was made possible using symbolic programming without which the calculation becomes cumbersome. We observe that cumulants of order $j \leq 2$ are real while cumulants of order $j > 2$ are complex. The summation, $\sum_n$, in Eq. (5), is over the active TLSs, namely those TLSs which flip on a cutoff time scale τ (i.e., $K_n > 1/\tau$). Usually τ is taken to be the time of observation t_o, in this case the line shape can become unstable provided that a TLS in the vicinity of the molecule flips only a few times on a the time scale t_o.

For such a molecule the line shape theory does not work well. A way out of this is to follow the spectral trail of the molecule and then determine if it is coupled to TLSs whose flipping time rate is $K_n > 1/\tau$ and choose $\tau \ll t_o$. Thus only a sub ensemble of molecule (or TLSs) is considered and our results that follow depend on the choice of τ. We consider the slow modulation limit, soon to be justified, which means that we consider the case $K_n \ll \nu_n$. To investigate this limit we set $K_n = 0$ in Eq. (5), then all the cumulants are real and are rewritten as $\kappa_j = \sum_n H_{jn} \nu_n^j$, where H_j are functions of p only and $H_1 = p$, $H_2 = p(1-p)$, $H_3 = p(1-p)(2p-1)$ etc. Note that for κ_1 and κ_2 no approximation is made.

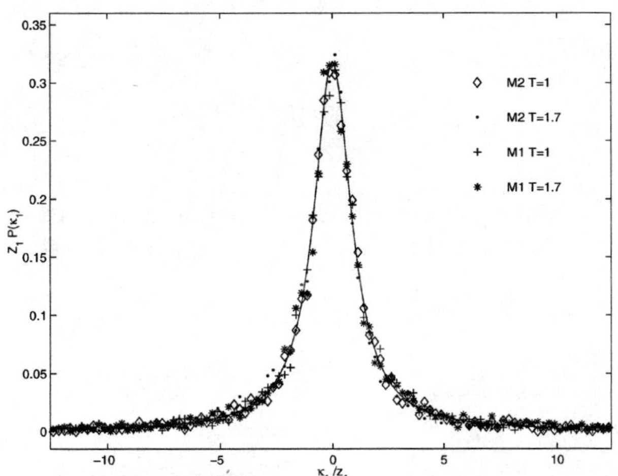

FIGURE 1. Scaled probability density of first cumulant $P(\kappa_1)z_1$ versus κ_1/z_1. The theory is plotted as a solid curve, it predicts a Cauchy density $P(\kappa_1) = L_{1,0}(\kappa_1)$ with a scaling parameter z_1 which varies from one set of data to the other. The symbols are the simulation results. From Ref. [9]. Copyright (2000) by the American Physical Society.

Let $\langle \cdot \rangle_{r\Omega AJ}$ denote an averaging over the random TLS parameters. Using the slow modulation limit assumption, the characteristic function of the j cumulant can be written in a form

$$\langle \exp\left(ik\kappa_j\right) \rangle_{r\Omega AJ} = \exp\left[-\rho_{\text{eff}} \left\langle \int d\Omega \int_0^\infty \frac{d(r^d)}{d} \left(1 - e^{ikB_j/r^{\delta j}}\right) \right\rangle_{AJ}\right], \tag{6}$$

ρ_{eff} is the density of the active TLS and $B_j = (2\pi\alpha)^j \Psi^j(\Omega) f^j(A, J) H_j$. To derive Eq. (6) we have used the assumption of independent TLSs uniformly distributed in the system. For odd j cumulants we find

$$\langle \exp\left(ik\kappa_j\right) \rangle_{r\Omega AJ} = \hat{L}_{\gamma,0}(k), \tag{7}$$

with characteristic exponent $\gamma = d/(\delta j)$ and the scale parameter

$$z_\gamma = \rho_{\text{eff}} (2\pi\alpha)^{\frac{d}{\delta}} \left\langle f^{d/\delta}(A,J)\, |H_j|^\gamma \right\rangle_{AJ} c_\gamma \int d\Omega |\Psi^j(\Omega)|^\gamma \tag{8}$$

with $c_\gamma = \cos(\gamma\pi/2)\,\Gamma(1-\gamma)$, $c_1 = \pi/2$. Eq. (7) shows that odd cumulants are described by symmetrical Lévy stable density, i.e., $P(\kappa_j) = L_{\gamma,0}(\kappa_j)$. Two conditions must be satisfied for such a behavior, $0 < \gamma < 2$ and $\int d\Omega \sin[\Psi^j(\Omega)] = 0$. The latter condition gives the symmetry condition, $\eta = 0$, which means that negative and positive contributions to κ_j are equally probable. For even cumulants and $0 < \gamma < 1$ we find

$$\langle \exp(ik\kappa_j) \rangle_{r\Omega AJ} = \hat{L}_{\gamma,\eta}(k) \tag{9}$$

with a scale parameter Eq. (8) and with Lévy index of symmetry

$$\eta = \frac{\left\langle f^{j\gamma}(A,J)\, |H_j|^\gamma \frac{H_j}{|H_j|} \right\rangle_{AJ}}{\left\langle f^{j\gamma}(A,J)\, |H_j|^\gamma \right\rangle_{AJ}}. \tag{10}$$

Eq. (9) implies that even cumulants are distributed according to $P(\kappa_j) = L_{\gamma,\eta}(\kappa_j)$. The characteristic exponent γ depends only on the general features of the model (namely on d and δ). In contrast, the Lévy index of symmetry η and the scaling parameter z_γ depend on the details of the model (e.g., on $f(A,J)$) and on system parameters (T, τ, etc).

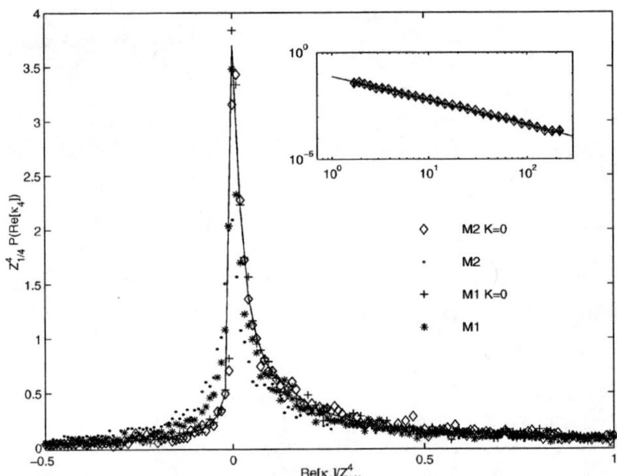

FIGURE 2. The solid curve is $L_{1/4,\eta}[\text{Re}(\kappa_4)]$ with an index of asymmetry $\eta = 0.6104$. The symbols are simulation results; in the inset we show the power law tail of the scaled probability density. From Ref. [9]. Copyright (2000) by the American Physical Society.

In Fig. (1) we show scaled $P(\kappa_1)$ for different temperatures T and for two models of the angular dependence $\psi(\Omega)$ M1 and M2. For details see [9]. The first cumulant exhibits the Lévy behavior and $P(\kappa_1)$ is the Cauchy density since

$d = \delta$). In simulations we reject TLS from a sphere of radius 1 [nm] centered at the molecule, showing that the result in the figure is not sensitive to finite small cutoff. As mentioned κ_1 does not depend on the dynamical properties of the model (i.e., on K) hence this result is not limited to the slow modulation limit.

The question remains if a slow modulation limit is valid for the standard tunneling model parameters. Consider the distribution of $\mathrm{Re}[\kappa_4]$, which in the slow modulation limit is distributed according to $L_{1/4,\eta}(\mathrm{Re}[\kappa_4])$, Eq. (9). In Fig. 2 we compare between numerical results and the prediction Eq. (9) (we use $T = 1.7$K). We also show simulation results in which all rates are set to zero ($K = 0$). We find that the deviation between simulation and theory is small so the assumption of slow modulation limit is justified. This implies that it is the static (not dynamic) features of the TLSs that are encoded in the distribution of line shapes.

Permission from the American Physical Society to reprint copyrighted material from Ref. [9] is gratefully acknowledged.

REFERENCES

1. W. E. Moerner and M. Orrit, *Science* **283**, 5408 (1999).
2. P. W. Anderson, B. I. Halperin, C. M. Varma, *Philos. Mag.* **25**, 1 (1971). W. A. Philips, *J. Low. Temp. Phys.* **7**, 351 (1972).
3. A. Heuer and R. J. Silbey *Phys. Rev. Lett.* **70**, 3911 (1993).
4. R. Kühn and U. Horstmann, *Adv. in Solid State Physics* **38**, 425 (1999).
5. A. M. Boiron, P. Tamarat, B. Lounis, R. Brown and M. Orrit, *Chem. Phys.* **247**, 119 (1999).
6. E. Geva and J. L. Skinner, *J. Phys. Chem. B* **101**, 8920 (1997).
7. P. W. Anderson, *J. Phys. Soc. Jpn.* **9**, 316 (1954). R. Kubo, *ibid* **6**, 935 (1954).
8. F. L. H. Brown and R. J. Silbey, *J. Chem. Phys.* **108**, 7434 (1998).
9. E. Barkai, R. Silbey and G. Zumofen, *Phys. Rev. Lett.* **84**, 5339 (2000).
10. W. Feller, *An introduction to probability Theory and Its Applications* Vol. 2 (John Wiley and Sons, 1970).
11. B. Saubamea, M. Leduc, C. Cohen-Tannoudji, *Phys. Rev. Lett.* **83**, 3696 (1999).
12. J. P. Hovi, A. Aharony, D. Stauffer and B. B. Mandelbrot, *Phys. Rev. Lett.* **77**, 877, (1996).
13. B. N. Kuvshinov, T. J Schep, *Phys. Rev. Lett.* **84**, 650 (2000); I. A. Min, I. Mezic, A. Leonard, *Phys. of Fluid* **8**, 1169 (1996); H. Takayashu *Prog. Theor. Phys.* **72**, 471 (1984).
14. Y. Furukawa, Y. Nakai and N. Kunitomi, *J. of the Phys. Soc. Japan* **62**, 306 (1993).
15. G. Zumofen and J. Klafter, *Chem. Phys. Lett.* **219**, 303 (1994).
16. E. Barkai, R. Silbey and G. Zumofen, *J. Chem. Phys.* (accepted).
17. A. M. Stoneham, *Rev. Mod. Phys.* **41**, 82 (1969).
18. P. D. Reilly and J. L. Skinner, *J. Chem. Phys.* **101**, 959 (1994).
19. A. Abragam, *The Principle of Nuclear Magnetism* (Clarendon, Oxford 1961).
20. Feller mentions that Holtsmark was an astronomer, in fact he was a spectroscopist. Chandrasekhar, who used Holtsmark's work was a well known astrophysicist.

Information Content of Neural Networks with Self-Control and Variable Activity

D. Bollé*, S.I. Amari†, D.R.C. Dominguez Carreta§ and G. Massolo‡

*Instituut voor Theoretische Fysica, K.U.Leuven, B-3001 Leuven, Belgium
†RIKEN Frontier Research Program, Wako-shi, Saitama 351-01, Japan
§ESCET, Universidad Rey Juan Carlos, Mostoles, 28933 Madrid, Spain
‡ Università degli Studi dell'Insubria, Facoltá di Scienze, I-22100 Como, Italy

Abstract. A self-control mechanism for the dynamics of neural networks with variable activity is discussed using a recursive scheme for the time evolution of the local field. It is based upon the introduction of a self-adapting time-dependent threshold as a function of both the neural and pattern activity in the network. This mechanism leads to an improvement of the information content of the network as well as an increase of the storage capacity and the basins of attraction. Different architectures are considered and the results are compared with numerical simulations.

INTRODUCTION

It is well known by now that efficient neural network modelling requires *autonomous functioning* independent of external constraints or control mechanisms. For fixed-point retrieval by an attractor associative memory this requirement is mainly expressed by the *robustness* of its learning and retrieval capabilities *against external noise*, *against malfunctioning of some of the connections* and so on. This assures *large basins of attraction* for the memorized patterns.

A problem is that these basins of attraction become smaller when the loading capacity gets bigger. This especially occurs in sparsely coded (=low activity) models ([1] and references therein). Therefore, some *control of the activity of the neurons* is necessary.

An important question is then whether the basins of attraction can be *optimized without imposing external constraints*, keeping the simplicity of the (given) architecture of the network model.

A solution to this problem has been advocated in [2] (see also [3]) for binary networks with extremely diluted architectures and extended in [4] to models for sequential patterns and in [5] to layered feedforward architectures. It consists out of introducing a *complete self-control* mechanism in the dynamics through a time-dependent threshold chosen as a function of the noise and the pattern activity.

CP553, *Disordered and Complex Systems*, edited by P. Sollich, et al.

In this contribution we look at this self-control mechanism for extremely diluted networks with three-state neurons (see also [6]) and extend it to fully connected architectures. The latter is highly non-trivial because of the feedback correlations in the dynamics. Full details will be published elsewhere [7].

A SPIN-GLASS MODEL

Consider a network of N three-state neurons $\sigma_{i,t} \in \{0, \pm 1\}, i = 1, \ldots, N$ *updated synchronously* at a discrete time step t

$$\sigma_{i,t+1} = F_{\theta_t}(h_{i,t}), \quad h_{i,t} = \sum_{j(\neq i)}^{N} J_{ij}\sigma_{j,t}. \tag{1}$$

In the above $h_{i,t}$ is the local field and $F_{\theta_t}(x)$ the transfer function given by

$$F_{\theta_t}(x) = \begin{cases} \text{sign}(x) & \text{if } |x| > \theta_t \\ 0 & \text{if } |x| < \theta_t \end{cases} \tag{2}$$

The couplings J_{ij} in (1) are a function of the memorized patterns $\xi_i^\mu \in \{0, \pm 1\}, \mu = 1, \ldots, p = \alpha N$ (α the loading capacity)

$$J_{ij} = \frac{1}{Na} \sum_{\mu=1}^{p} \xi_i^\mu \xi_j^\mu \tag{3}$$

with $a = \langle |\xi_i^\mu|^2 \rangle$ the pattern *activity*. (The brackets $\langle \cdots \rangle$ denote the average over the ξ^μ.)

The patterns ξ_i^μ are IIDRV and chosen according to the probability distribution

$$p(\xi_i^\mu) = a\delta(|\xi_i^\mu|^2 - 1) + (1 - a)\delta(\xi_i^\mu). \tag{4}$$

The relevant order parameters are the retrieval overlap and the neural activity

$$m_{N,t}^\mu \equiv \frac{1}{aN} \sum_i \xi_i^\mu \sigma_{i,t}, \qquad q_{N,t} \equiv \frac{1}{N} \sum_i |\sigma_{i,t}|^2. \tag{5}$$

Furthermore, the activity-overlap [6] determining the overlap between the active neurons and the active parts of a memorized pattern is important for calculating the mutual information function

$$n_{N,t}^\mu \equiv \frac{1}{aN} \sum_i^N |\sigma_{i,t}|^2 |\xi_i^\mu|^2. \tag{6}$$

DYNAMICS

The synchronous dynamics of this network depends on the architecture (extremely diluted, layered feedforward, fully connected) and may be difficult to solve because of strong feedback correlations.

A recursive dynamical scheme has been developed [8] for the distribution of the local field at an arbitrary time step using a probabilistic signal-to-noise analysis. Suppose that the initial configuration of the network $\{\sigma_{i,0}\}$, is a collection of IIDRV with mean $\langle\sigma_{i,0}\rangle = 0$, variance $\langle(\sigma_{i,0})^2\rangle = q_0$, and correlated with only one stored pattern, say $\{\xi_i^1\}$, through an initial overlap m_0 . Then, in general, the local field $h_{i,t}$ in the thermodynamic limit consists out of a *signal* from the retrieved pattern, a *normally distributed noise* part and a *discrete noise* part both coming from the other $(p-1)$ patterns, viz.

$$h_{i,t} = \xi_i^1 m_t^1 + \sqrt{\alpha a D_t}\,\mathcal{N}(0,1) + B_{i,t} \tag{7}$$

$$D_t = \mathrm{Var}\left[\lim_{N\to\infty}\frac{1}{a\sqrt{N}}\sum_i \xi_i^\nu \sigma_{i,t}\right], \nu > 1\,, \quad B_{i,t} = \sum_{t'=0}^{t-1} \alpha \left[\prod_{s=t'}^{t-1}\chi_s\right]\sigma_{i,t'} \tag{8}$$

with $\mathcal{N}(0,1)$ denoting a Gaussian random variable with mean zero and variance 1 and where χ is the susceptibility

$$\chi_t = \frac{1}{\sqrt{\alpha a D_t}}\left\langle\!\!\left\langle \int \mathcal{D}z\; z\, F_{\theta_t}\left(\xi^1 m_t^1 + \sqrt{\alpha a D_t}\, z\right)\right\rangle\!\!\right\rangle \tag{9}$$

with $\mathcal{D}$ the Gaussian measure.

The idea of self-control [2] is to let the network itself counter the noise terms in the local field at each time step by introducing the following time-dependent threshold in the dynamics (1)-(2)

$$\theta_t = \sqrt{-2\ln(a)\alpha a D_t}. \tag{10}$$

On the basis of the recursion relation for the quantity D_t (see [8]) one can argue that for extremely diluted and layered feedforward networks $D_t = q_t/a$ [2], [5], while for the fully connected network $\sqrt{\alpha a D_t} = a\sqrt{2/\pi} + \sqrt{\alpha q_t}$ [7].

Then the following results have been obtained. Firstly, for the extremely diluted and layered feedforward models a closed set of coupled equations for the order parameters m_t^μ and q_t has been derived. Secondly, for the fully connected model recursion relations can be written down determining the full time evolution of these order parameters. For the explicit form of these relations we refer to [7].

THE MUTUAL INFORMATION

The retrieval quality of this network is measured by the *mutual information function* (see, e.g., [9]). This concept is standard in information theory and measures the average amount of information that can be received by the user by observing the signal at the output of a channel. For the retrieval process of memorized patterns considered here at each time step the process can be regarded as a channel with input ξ_i^μ and output $\sigma_{i,t}$. The mutual information is then defined by

$$I(\sigma_{i,t};\xi_i^\mu)= S(\sigma_{i,t}) - \langle S(\sigma_{i,t}|\xi_i^\mu)\rangle_{\xi_i^\mu} \tag{11}$$

with $S(\sigma_{i,t})$ and $S(\sigma_{i,t}|\xi_i)$ the entropy and conditional entropy of the output

$$S(\sigma_{i,t}) = -\sum_\sigma p(\sigma_{i,t})\ln[p(\sigma_{i,t})]\,, \quad S(\sigma_{i,t}|\xi_i^\mu) = -\sum_\sigma p(\sigma_{i,t}|\xi_i^\mu)\ln[p(\sigma_{i,t}|\xi_i^\mu)]\,. \tag{12}$$

Here $p(\sigma_{i,t})$ denotes the probability distribution for the neurons at time t and $p(\sigma_{i,t}|\xi_i^\mu)$ indicates the conditional probability that the i-th neuron is in a state $\sigma_{i,t}$ at time t given that the i-th site of the stored pattern to be retrieved is ξ_i^μ. One can argue [2] that it is a better measure than the Hamming distance, especially for small a. The mutual information can then be calculated analytically in terms of the order parameters m_t^μ, q_t and the parameter n_t [2], [7].

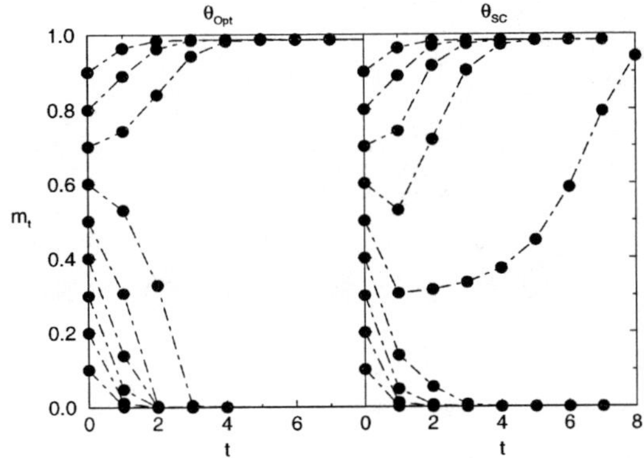

FIGURE 1. The evolution of the overlap m_t for several initial values m_0, with $q_0 = 0.01 = a$ and $\alpha = 3$ for the diluted model with and without self-control.

NUMERICAL RESULTS

We have solved the time evolution of the threshold dynamics and studied the behaviour of the networks for variable pattern activities a. A self-control time-dependent threshold (10) is compared with a fixed time-independent threshold selected manually for every loading α in order to maximize the information content [6], [7]. Furthermore, the results are compared with numerical simulations. For a detailed description we refer to [5]- [7]. Here we list the main features.

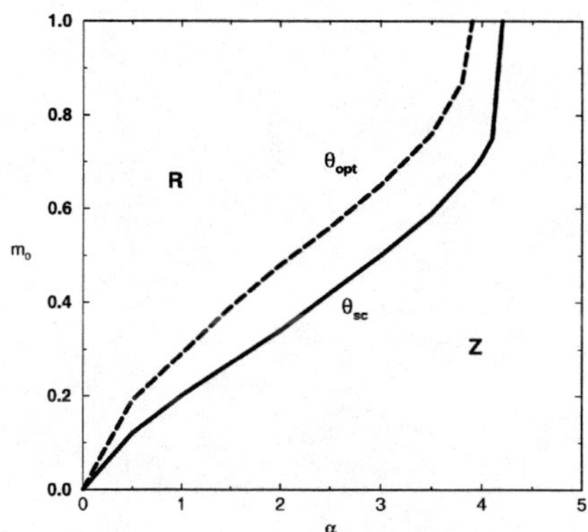

FIGURE 2. The basin of attraction as a function of α for $a = 0.01$ and initial $q_0 = a$ for the diluted model with (right) and without (left) self control.

Self-control forces more of the overlap trajectories to go to the retrieval attractor $m_t = 1$ (Fig. 1) such that the basins of attractions and the loading capacity are larger (Fig. 2). The information content of the network is much higher (Fig. 3).

CONCLUSIONS

In this work we have discussed self-control dynamics for neural networks with different architectures by introducing an adaptive time-dependent threshold. This leads to a large improvement of the quality of retrieval. The relevant quantity to measure this is the mutual information function. The information content as well as the basin of attraction and the loading capacity are optimised through the self-control mechanism. These ideas may work for general attractor dynamics.

REFERENCES

1. Okada M., *Neural Networks* **9**, 1429 (1996).
2. Dominguez D.R.C., and Bollé D., *Phys. Rev. Lett.* **80**, 2961 (1998).
3. Grosskinsky S., *J. Phys. A: Math. Gen.* **32**, 2983 (1999).
4. Kitano K., and Aoyagi T., *J. Phys. A: Math. Gen.* **31**, L613 (1998).
5. Bollé D., and Massolo G., *J. Phys. A: Math. Gen.* **33**, 2597 (2000).
6. Bollé D., Dominguez D.R.C., and Amari S.I., *Neural Networks* **13**, 455 (2000).
7. Bollé D., and Dominguez Carreta D.R.C., cond-mat/9912101 to appear in *Physica A*.
8. Bollé D., Jongen G., and Shim G.M., *J. Stat. Phys* **91**, 125 (1998).
9. Deco G., and Obradovic D., *An information-theoretic approach to neural computing*, New York: Springer-Verlag, 1997.

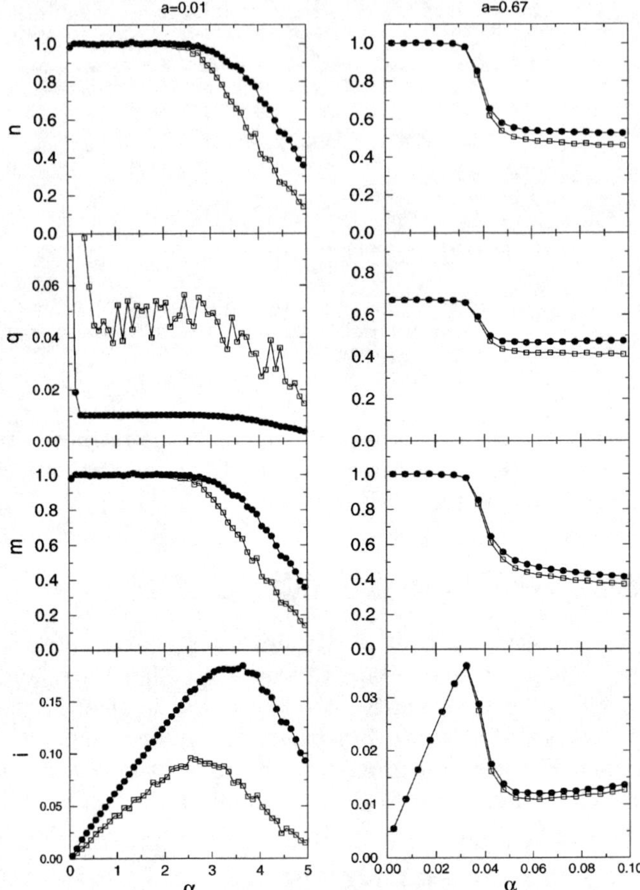

FIGURE 3. Simulations for $N = 10^4$ neurons of the information $i = \alpha I$, the overlap m, the neural activity q and the activity-overlap n as a function of α for the fully connected self-control model (filled circles) and a fixed threshold model (squares) after five time steps.

Storage Capacity of the Tilinglike Learning Algorithm

Arnaud Buhot* and Mirta B. Gordon†

*Theoretical Physics, University of Oxford, 1 Keble Road, Oxford OX1 3NP, UK.
†SPSMS, DRFMC, CEA Grenoble, 17 av. des Martyrs, 38054 Grenoble Cedex 9, France.
Also at Centre National de la Recherche Scientifique

Abstract. The storage capacity of an incremental learning algorithm for the parity machine, the Tilinglike Learning Algorithm, is analytically determined in the limit of a large number of hidden perceptrons. Different learning rules for the simple perceptron are investigated. The usual Gardner-Derrida rule leads to a storage capacity close to the upper bound, which is independent of the learning algorithm considered.

INTRODUCTION

The storage capacity is one of the most important characteristics of a neural network. It is the maximal number of random input-output patterns per input component that a network is able to classify correctly with probability one. This quantity is independent of the algorithm used to learn the weights of the network; it only depends on its architecture. We will refer to it as the *architecture storage capacity* α_c^{arch} in the following, to distinguish it from the algorithm-dependent one, hereafter called *algorithm storage capacity*, α_c^{alg}.

The simplest neural network, the perceptron, has the inputs directly connected to the output. Geometrical arguments [1] and a statistical mechanics calculation [2] determined that $\alpha_c^{arch} = 2$. Several perceptron learning algorithms, like the Adatron [3,4], are known to achieve such storage capacity.

The capacity can be increased using networks with more complicated architectures. The next one in order of increasing complexity is the extensively studied [5] monolayer perceptron (MLP), which has k "hidden" perceptrons connected to the output unit. As an MLP can store any function of its inputs, provided that the number of hidden units is adequate, it is not worth considering more complex architectures for the storage problem.

Given an input pattern, the hidden unit states define a k-dimensional vector, the pattern's *internal representation* (IR). The network's output is a function of the IR. In the following, we consider binary neurons, with states ± 1, and focus on the parity machine, whose output is the product of the k components of the IR.

CP553, *Disordered and Complex Systems*, edited by P. Sollich, et al.

The main problem when training MLPs is that the IRs of the input patterns are unknown. It has been proposed [6] to build the hidden layer using a constructive procedure, called Tilinglike Learning Algorithm (TLA), in which the hidden perceptrons are included one after the other and trained to correct the learning errors of the preceding unit. As each unit can at least correct one error, convergence is ensured [6]. It is straightforward to show that the TLA generates a parity machine [6], but the number of included hidden units depends crucially on the performance of the perceptron learning algorithm used to train them.

Geometric arguments [7] and a statistical mechanics replica calculation [8], showed that the architecture storage capacity of the parity machine in the limit of a large number of hidden units k is $\alpha_c^{arch}(k) \simeq k \ln k / \ln 2 \simeq 1.44\, k \ln k$. However, it was not clear whether this storage capacity could be actually achieved with a learning algorithm. In this paper, we show that the Tilinglike Learning Algorithm (TLA) can reach a storage capacity close to the architecture storage capacity, provided that the hidden perceptrons are trained with an appropriate learning algorithm. We first describe more precisely the setting and the TLA. The analytical expression of the algorithm storage capacity in the limit of large k is then determined, and we show that the learning algorithm used to train the hidden units must satisfy a stringent condition for the TLA to converge with a finite number of hidden units. Finally, we discuss why these conditions rule out some perceptron learning algorithms, like the Adatron, and present our conclusions.

THE TILINGLIKE LEARNING ALGORITHM

Let us assume a training set $\mathcal{L}_\alpha = \{\mathbf{x}_\mu, \tau_\mu\}_{\mu=1,\cdots,P}$ of $P = \alpha N$ input-output patterns. The inputs $\mathbf{x}_\mu$ are random Gaussian N-dimensional vectors with zero mean and unit variance in each direction. The corresponding outputs $\tau_\mu = \pm 1$ are the learning targets. Their values τ are randomly selected with probability:

$$P(\tau; \varepsilon) = \varepsilon\, \delta(\tau + 1) + (1 - \varepsilon)\, \delta(\tau - 1), \tag{1}$$

with $\varepsilon = 1/2$. The rôle of the bias ε introduced in (1) will become clear in the following. The probability of the targets τ_μ is unbiased.

The TLA constructs the parity machine by including successive perceptrons in the hidden layer. Each unit is connected to the input $\mathbf{x}$ through weights $\{\mathbf{J}, \theta\} = (J_1, \ldots, J_N, \theta)$ where θ is a threshold. The inputs are classified through $\sigma = \text{sign}(\mathbf{J} \cdot \mathbf{x} - \theta)$. Thus, a perceptron linearly separates the input space with a hyperplane orthogonal to $\mathbf{J}$ (we assume $\mathbf{J} \cdot \mathbf{J} = 1$) at a distance θ to the origin. The weights and the threshold are learned through the minimization of a cost function:

$$E\left(\{\mathbf{J}, \theta\}; \mathcal{L}_\alpha\right) = \sum_{\mu=1}^{P} V\left(\tau_\mu(\mathbf{J} \cdot \mathbf{x}_\mu - \theta)\right). \tag{2}$$

where the potential $V(\lambda)$ is the contribution of each pattern to the cost, and $|\lambda|$ is the distance of the pattern to the hyperplane.

Within the TLA heuristics, the first perceptron $k = 1$ is trained to learn targets $\tau_\mu^1 = \tau_\mu$. After learning, its weights are $\{\mathbf{J}^1, \theta^1\}$; its training error is $\varepsilon_t^1 = (1/P)\sum_\mu \Theta(-\sigma_\mu^1 \tau_\mu^1)$ where $\Theta(x)$ is the Heaviside function and σ_μ^1 the perceptron's output for pattern μ. If $\varepsilon_t^1 = 0$, the training set is correctly classified; the TLA stops with only one simple perceptron. Otherwise, a new perceptron is introduced. The successive perceptrons i are trained to learn training sets $\mathcal{L}_\alpha(i) = \{\mathbf{x}_\mu, \tau_\mu^i\}$ with targets $\tau_\mu^i = \tau_\mu^{i-1}\sigma_\mu^{i-1}$, that is, $\tau_\mu^i = 1$ if the pattern μ is correctly classified by the previous perceptron and $\tau_\mu^i = -1$ otherwise. If the perceptron learning algorithm is correctly chosen it can be shown that the successive training errors ε_t^i are strictly decreasing [6,9]. Thus, the TLA procedure necessarily converges to a MLP with k units, where the kth perceptron is the first one to meet the condition $\varepsilon_t^k = 0$. Then, the product $\sigma_\mu = \sigma_\mu^1 \cdots \sigma_\mu^k = \tau_\mu$ gives the correct output for all patterns of the training set $\mathcal{L}_\alpha$.

STORAGE CAPACITY

The algorithm storage capacity of the TLA, $\alpha_c^{alg}(k)$, is simply the inverse function of $k(\alpha)$, the average number of perceptrons typically included by the TLA when the training set has a size $\alpha = P/N$. In order to determine $k(\alpha)$, consider the ith hidden unit: The probability of its targets τ_μ^i depends on the training error ε_t^{i-1} of the previous perceptron. Although there exist some correlations between the outputs τ_μ^i, due to the correlations in the weights of the successive perceptrons, in the limit of large training sets ($\alpha \to \infty$) they may be neglected [10]. Thus, we may assume that the targets τ_μ^i are independently drawn with probability (1), with a bias ε_t^{i-1}. Then, the successive training errors satisfy a simple recursive relation [10–12]:

$$\varepsilon_t^i = \mathcal{E}_t\left(\alpha, \varepsilon_t^{i-1}\right) \tag{3}$$

where $\mathcal{E}_t(\alpha, \varepsilon_t^{i-1})$ is the training error of a simple perceptron trained with a training set of size α and biased targets τ_μ^i drawn with a probability $P(\tau_\mu^i; \varepsilon_t^{i-1})$ given by (1). The number k of perceptrons necessary to correctly classify the initial training set satisfies [11,12]:

$$\circ_k f_\alpha(1/2) = \underbrace{f_\alpha \circ \cdots \circ f_\alpha}_{k \text{ times}} (1/2) = 0 \tag{4}$$

where $f_\alpha(\varepsilon)$ stands for $\mathcal{E}_t(\alpha, \varepsilon)$ and the symbol $\circ$ for the composition of functions. In the limit of a large α, the training error $\mathcal{E}_t(\alpha, \varepsilon)$ is close to ε and the number of simple perceptrons $k(\alpha)$ is large. It is thus possible to use the continuum limit:

$$\varepsilon_t^{i+1} - \varepsilon_t^i \simeq \frac{1}{k}\frac{d\varepsilon}{dx} = \mathcal{E}_t(\alpha, \varepsilon) - \varepsilon \tag{5}$$

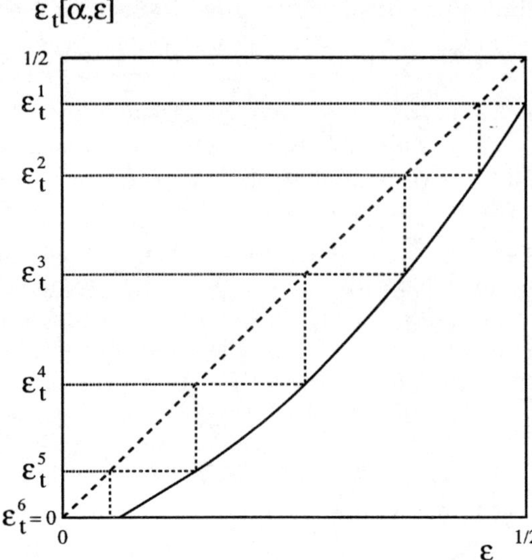

FIGURE 1. Successive training errors ε_t^i. The full curve corresponds to $\mathcal{E}_t(\alpha,\varepsilon)$. In this case six perceptrons are necessary for the convergence of the Tiling-like Learning Algorithm. From Ref. [11]. Copyright (2000) by IOP Publishing Limited.

with $x = i/k$. After integration of this differential equation we obtain [11,12] the typical number of hidden units introduced by the TLA:

$$k(\alpha) = \int_0^{1/2} \frac{d\varepsilon}{\varepsilon - \mathcal{E}_t(\alpha,\varepsilon)}. \tag{6}$$

$k(\alpha)$ depends on the specific cost function (2) used to train the perceptrons through $\mathcal{E}_t(\alpha,\varepsilon)$, which is the training error of a simple perceptron learning a training set with biased targets.

RESULTS FOR DIFFERENT LEARNING POTENTIALS

In this section we determine the perceptron's training error $\mathcal{E}_t(\alpha,\varepsilon)$ for biased training sets, in the thermodynamic limit ($N \to \infty$ with $\alpha = P/N$ fixed). For different possible choices of the potential V in (2), we deduce the number of hidden units, $k(\alpha; V)$ and the algorithm storage capacity of the TLA. Only the main results are presented here; the interested reader can find the details in [11,12].

Adatron cost function

The Adatron potential is $V(\lambda) = (\kappa - \lambda)^2\Theta(\kappa - \lambda)$, where κ is a positive parameter called stability. All the patterns with negative λ and those with $\lambda > 0$ but closer than κ to the hyperplane (which are correctly classified) contribute to the cost. The training error $\mathcal{E}_t(\alpha,\varepsilon)$ is obtained through a replica calculation assuming replica symmetry (RS), which can be shown to hold for all α. It turns out that for fixed κ, in the limit of large α, $\mathcal{E}_t(\alpha, \varepsilon) > \varepsilon$. As a consequence, the constraint that the successive training errors are strictly decreasing, necessary for the convergence of the TLA, is not satisfied. This problem can be circumvented at the price of considering κ as a free parameter, and minimizing the training error with respect to it. In that case, $\mathcal{E}_t(\alpha, \varepsilon) < \varepsilon$ and the algorithm storage capacity is $\alpha_c^{alg}(k, Adatron) \simeq 4.55 \ln k$ in the limit of large α.

Gardner-Derrida cost function

The potential of the Gardner-Derrida (GD) cost function [13] is $V(\lambda) = \Theta(\kappa - \lambda)$. The hypothesis of RS is incorrect for this potential, and the obtained value of $\mathcal{E}_t$ is a lower bound on the true training error. Consequently, the replica calculation only allows one to determine an upper bound on $\alpha_c^{alg}(k)$. If $\kappa = 0$, the cost is nothing else but the number of misclassified patterns. It gives the lowest bound on $\mathcal{E}_t$. In the limit of large training set size α, we obtain:

$$\mathcal{E}_t(\alpha, \varepsilon) \simeq \varepsilon - \frac{2}{a^2(\varepsilon)} \frac{\ln \alpha}{\alpha}, \tag{7}$$

where $a(\varepsilon)$ satisfies $\varepsilon = [e^a(1-a) - 1]/[2(\cosh a - a \sinh a - 1)]$. This leads to $k(\alpha, GD) \simeq 0.475\, \alpha / \ln \alpha$ and $\alpha_c^{alg}(k, GD) \simeq 2.11\, k \ln k$, larger than $\alpha_c^{arch}(k)$, probably due to the failure of the RS hypothesis.

In order to obtain a lower bound on $\alpha_c^{alg}(k)$ we used the Kuhn-Tucker cavity method [14,11,12], which gives an upper bound on $\mathcal{E}_t$. As a result of both calculations, we can bound $\alpha_c^{alg}(k, GD)$:

$$0.924\, k \leq \alpha_c^{alg}(k, GD) \leq 2.11\, k \ln k. \tag{8}$$

On view of this result, we expect that the algorithm storage capacity of the TLA behaves like $k(\ln k)^\nu$ with $0 \leq \nu \leq 1$. A calculation with one step of replica symmetry breaking would give an estimate of the exponent ν. Since the RS solution gives a better approximation of the training error than the Kuhn-Tucker cavity method, we expect the exponent ν to be close to 1, leading to an algorithm storage capacity close to the architecture's capacity. This result shows that the TLA may build a nearly optimal network.

In order to improve the robustness against noise in the data, it is generally useful to impose some finite stability κ to the patterns. The corresponding GD potential is $V(\lambda) = \Theta(\kappa - \lambda)$. As with $\kappa = 0$, also here the RS solution is unstable. The bounds on $\alpha_c^{alg}(k, \kappa, GD)$ deduced from the results for $\mathcal{E}_t$ obtained with the RS hypothesis and the Kuhn-Tucker cavity method [11,12] give:

$$\frac{k}{\kappa\sqrt{2 \ln k}} \leq \alpha_c^{alg}(k, \kappa, GD) \leq \frac{k}{2\kappa^2}. \tag{9}$$

Strikingly, imposing a finite stability κ has an important effect on the algorithm storage capacity, which in this case behaves as $k(\ln k)^\nu$ with $-1/2 \leq \nu \leq 0$. The prefactors of the two bounds on $\alpha_c^{alg}(k, \kappa, GD)$ are κ-dependent and they both diverge for $\kappa \to 0$. The exponent ν is independent of κ for finite κ but differs from the one corresponding to $\kappa = 0$.

CONCLUSION

We have determined analytically the storage capacity of the Tilinglike Learning Algorithm for the parity machine, a constructive procedure generating a monolayer perceptron of binary hidden units. A training set of input-output examples is used to determine the number of hidden units, which are introduced one after the other. These are simple perceptrons that have to learn their weights using increasingly biased target distributions.

We have shown that the storage capacity of the TLA depends crucially on the learning errors of the successively introduced perceptrons. The properties of the algorithm used to train the latter thus have dramatic consequences on the size of the hidden layer generated by the TLA, and may even hinder the convergence to a finite size network. This arises, in particular, if the perceptrons are trained with the Adatron algorithm unless the stability is adapted to the successive targets' biases.

The smallest network is obtained using the Gardner-Derrida cost function with vanishing stability, which corresponds to minimizing the training error. Based on the results obtained within the replica symmetry hypothesis, and those using the Kuhn-Tucker cavity method, we expect a supra-linear storage capacity $\alpha_c^{alg}(k, GD) \simeq k(\ln k)^\nu$ with $\nu > 0$, very close to the theoretical capacity corresponding to the architecture considered.

The work of AB is supported by the Marie Curie Fellowship HPMF-CT-1999-00328.

REFERENCES

1. Cover, T. M., *IEEE Trans. Electron. Comput.*, **14**, 326 (1965).
2. Gardner, E., *Europhys. Lett.*, **4**, 481 (1987), *J. Phys. A: Math. Gen.*, **21**, 257 (1988).
3. Abbott, L. F., and Kepler, T. B., *J. Phys. A: Math. Gen.*, **22**, 2031 (1989).
4. Anlauf, J. K., and Biehl, M., *Europhys. Lett.*, **10**, 687 (1989).
5. Watkin, T. H. L., Rau, A., and Biehl, M., *Rev. Mod. Phys.*, **65**, 499 (1993).
6. Biehl, M., and Opper, M., *Phys. Rev. A*, **44**, 6888 (1991).
7. Mitchinson, G. J., and Durbin, R. M., *Biol. Cybern.*, **60**, 345 (1989).
8. Xiong, Y. S., and Kwon, C., and Oh, J.-H., *J. Phys. A: Math. Gen.*, **31**, 7043 (1998).
9. Gordon, M. B., in *Proc. ICNN'96*, IEEE Publications, 1996, p. 381.
10. West, A. H. L., and Saad, D., *J. Phys. A: Math. Gen.*, **31**, 7043 (1998).
11. Buhot, A., and Gordon, M. B., *J. Phys. A: Math. Gen.*, **33**, 1713 (2000).
12. Buhot, A., *PhD. thesis* Université de Grenoble (Unpublished) (1999).
13. Gardner, E., and Derrida, B., *J. Phys. A: Math. Gen.*, **21**, 271 (1988).
14. Gerl, F., and Krey, U., *J. Physique I*, **7**, 303 (1997).

Saddles on the Potential Energy Landscape of a Lennard-Jones Liquid

Kurt Broderix*, Kamal K. Bhattacharya*, Andrea Cavagna†, Annette Zippelius* and Irene Giardina**

* *Institut für Theoretische Physik, Universität Göttingen, D-37073 Göttingen, Germany*
† *Department of Physics and Astronomy, The University, Manchester, M13 9PL, UK*
** *Service de Physique Thèorique, CEA Saclay, 91191 Gif-sur-Yvette, France*

Abstract. By means of molecular dynamics simulations, we study the stationary points of the potential energy in a Lennard-Jones liquid, giving a purely geometric characterization of the energy landscape of the system. We find a linear relation between the degree of instability of the stationary points and their potential energy, and we locate the energy where the instability vanishes. This threshold energy marks the border between saddle-dominated and minima-dominated regions of the energy landscape. The temperature where the potential energy of the Stillinger-Weber minima becomes equal to the threshold energy turns out to be very close to the mode-coupling transition temperature T_c.

The low temperature dynamics of supercooled liquids and glasses is often put in relation with the geometric properties of the potential energy landscape of these systems. In particular, the presence of a large number of inequivalent glassy minima has stimulated many studies in the past [1–5]. More recently, the study of mean-field models of spin-glasses has strengthened the persuasion that the dynamical behaviour of glassy systems is deeply connected to the topology of the energy landscape [6]. Moreover, it has been shown that spin-glass systems exhibiting one-step replica symmetry breaking (1RSB) have many dynamical properties in common with fragile structural glasses [7], suggesting that 1RSB mean-field spin-glasses and fragile glasses may have a similar energy landscape.

In this context, crucial questions are: How to characterize the energy landscape of a glassy system? How to quantify the similarity of the energy landscape of fragile glasses and 1RSB spin-glasses? Although utterly relevant, the structure of minima of the potential energy is not enough: even at low temperatures, when activation is the only mechanism of diffusion in liquids, overcoming a barrier implies crossing a *saddle* of the potential energy. Furthermore, at higher temperatures the system spends more time around saddles than minima, hence the structure of unstable stationary points is important for understanding the crossover from a non-activated to an activated dynamics upon cooling [8,9]. The statistical properties of the stationary points of the potential energy are of course independent of the temperature, as also should be a thorough description of the energy landscape itself.

CP553, *Disordered and Complex Systems*, edited by P. Sollich, et al.

Here we will focus on the purely geometric properties of the energy landscape of a glassy system, by studying the statistical properties of *all* the stationary points of its potential energy, be they minima or saddles. We classify them according to number of unstable directions, (or *index* K), potential energy and smallest eigenvalue of the Hessian matrix. We show that in this way it is possible to quantitatively compare the energy landscape of different systems, pointing out similarities and differences. Moreover, we discover a threshold energy below which it is highly unlikely to find saddles and the relaxation requires activation. We thus establish a connection between this threshold energy and the critical temperature T_c of mode-coupling theory (MCT) [10]. Finally, by means of our results, we support some recent speculations on the role of saddles in supercooled liquids [9].

The system under consideration is a binary mixture of Lennard–Jones (LJ) particles [11] (for details see [12]). Throughout this study we present results for systems with $N = 60$ particles. In order to explore the stationary points of the potential energy, we use the following method: We equilibrate a configuration at a given temperature T using a standard molecular dynamics (MD) simulation technique. To locate a saddle close to the equilibrium configuration we then perform a quench on a pseudo-potential energy landscape $W(x)$ given by the modulus square of the force, $W(x) = \vec{\nabla}U(x) \cdot \vec{\nabla}U(x)$, where $U(x)$ is the original potential energy [12,13]. All *absolute* minima of $W(x)$ are stationary points of $U(x)$, hence every saddle of $U(x)$ has a well defined basin of attraction. The *local* minima of $W(x)$, however, do not correspond to zeros of the real force. These points are frequently sampled, but they can easily be distinguished from the absolute minima and are excluded from our analysis. However, this means that the method does not associate to all the configurations a nearby saddle and therefore it cannot be used to build a natural dynamics of the relevant saddles, equivalent to the Stillinger-Weber (SW) one for minima [2]. Furthermore, the relation between the initial MD equilibrium configuration at temperature T and the final stationary point found by this algorithm, is not straightforward. We prefer to perform here a purely geometric analysis of the stationary points, independent of the way we have sampled them [14].

Given a stationary point, we compute its index density $k = K/(3N)$ and its potential energy density $u = U/N$. In Fig.1 we show the results obtained by sampling saddles at two different values of the temperature. This plot clearly suggests that there is an underlying curve $k(u)$ independent of the temperature, which encodes a purely geometric feature of the landscape. By sampling stationary points at different values of T we are simply exploring different portions of the same geometric curve: temperature acts as a light spot needed to unveil the underlying function $k(u)$. In Fig.2 we show the average index density as a function of the energy density. What is most striking of this plot is how well defined the function $k(u)$ is: due to its geometric nature there are no thermal fluctuations. This curve shows that if we cut the potential energy landscape with a plane of constant energy density $u = u_0$, the stationary points on this plane (or within a narrow shell around this plane) will be dominated by saddles with index density $k(u_0)$. Furthermore,

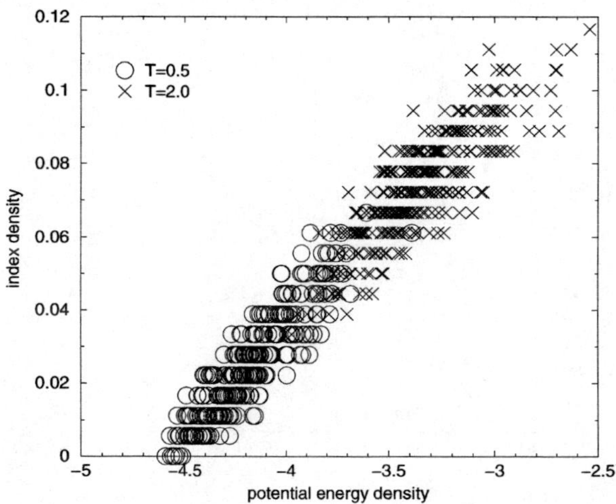

FIGURE 1. Index density as a function of the potential energy density: sampling of stationary points at two different temperatures, $T = 0.5$ and $T = 2.0$.

$k(u)$ is to a very good approximation a linear function in the explored regime of u. This implies that the curve extrapolates to zero at a well defined energy, which we call the *threshold energy* u_{th}, in analogy with 1RSB spin-glasses. The linear interpolation of all the data and the linear interpolation of the last four points give the same estimate for the threshold, that is $u_{th} = -4.55$.

The threshold energy marks the border between the saddles-dominated portion of the energy landscape and the minima-dominated one. A crucial point is that u_{th} is *above* the energy of the deepest glassy minima found with the SW method [2], that is $u_0 = -4.65$: there is a finite energy density interval, $u \in [u_0, u_{th}]$, where minima are entropically dominant over saddles, as it happens in 1RSB spin-glasses [15]. In those systems, however, $k(u)$ is *not* a linear function and $k'(u)$ vanishes at the threshold. This difference may be related to the mean-field nature of 1RSB models as opposed to real liquids. Indeed, in [9] the slope of $k(u)$ has been connected to the energy barriers in the system by the relation $\Delta U \sim 1/k'(u)$: in the mean-field case we expect barriers among minima to diverge, implying $k'(u_{th}) = 0$ (as found in 1RSB models), while this cannot be true in finite dimensional systems.

To further compare the energy landscape of LJ and 1RSB systems, we consider in Fig.3 the lowest eigenvalue λ_0 of the Hessian in a stationary point, as a function of its potential energy density. As expected, $\lambda_0 \to 0$ for $u \to u_{th}$, implying that the potential energy landscape at the threshold has some flat directions, i.e. it is *marginal* [6]. What is somewhat surprising is that $\lambda_0(u)$ is approximately a linear function of the energy, $\lambda_0 \sim (u_{th} - u)$, exactly as in 1RSB spin-glasses [17]. We conclude that both, the index density $k(u)$ and the smallest eigenvalue $\lambda_0(u)$,

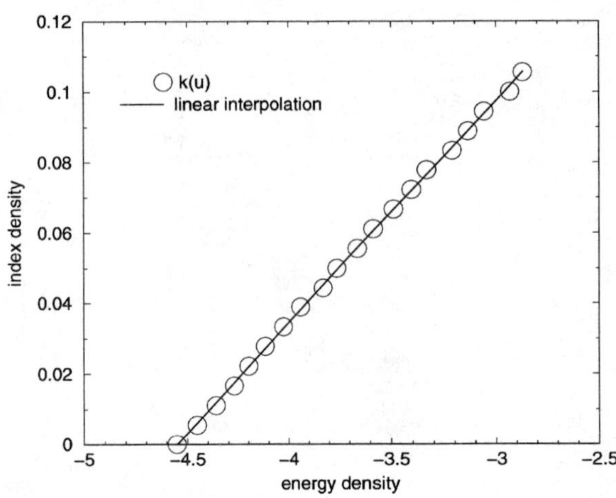

FIGURE 2. Index density as a function of the potential energy density. Average over all the data obtained by sampling at $T \in [0.3, 2.0]$. The full line is a linear fit of the data.

provide a quantitative measure and allow a direct comparison of the properties of the energy landscape for two very different systems, LJ liquids and 1RSB mean-field spin-glasses.

Our next task is to relate the threshold potential energy u_{th} to the dynamical behaviour of the supercooled liquid in the proximity of the glass transition. In Fig.4 we plot the average energy density $u_{SW}(T)$ of the SW minima as a function of the temperature of the initial MD trajectory [3–5], in comparison to $\delta(T) \equiv \langle U(x)/N \rangle(T) - 3/2T$, i.e. the difference between the average potential energy density (from MD simulations) and the vibrational energy in the harmonic approximation. For a harmonic potential $\delta(T)$ is just the energy of the minimum of the well. We find that $\delta(T) \sim u_{SW}(T)$ for $T \leq 1.2$ [16]. This is the range of temperatures which is dominated by the energy landscape and the timescales for the two processes of relaxation - vibrations inside a minimum and hopping between different minima - start to separate. Close to the glass transition (depending on the cooling rate) the system falls out of equilibrium, as indicated by the saturation of both quantities, $u_{SW}(T)$ and $\delta(T)$. As we can see, the extrapolation of the equilibrium part of these curves reaches the threshold energy at the MCT transition temperature T_c. The index density vanishes at the threshold, so that below this energy minima are the entropically dominant stationary points. Of course, there are minima also above the threshold, but they are *not* dominant, while saddles are. The MCT T_c therefore corresponds to the temperature below which minima visited by the dynamics become entropically dominant, while saddles become statistically irrelevant.

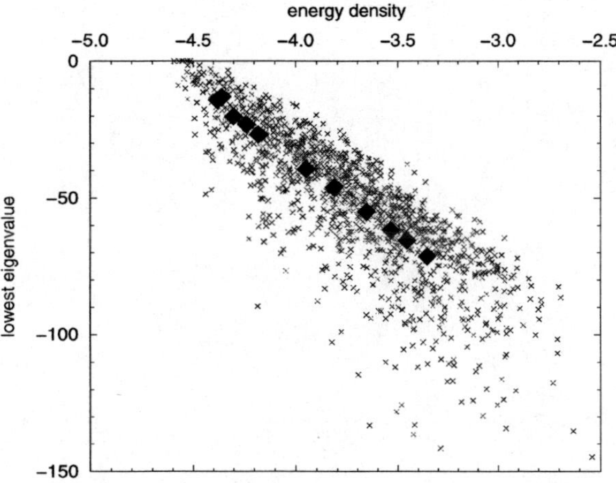

FIGURE 3. Lowest eigenvalue of the Hessian as a function of the energy density. Full diamonds are the average of all the data obtained at the same values of T as in Fig.2.

To conclude, we consider the potential energy barriers between different minima in our system. According to [9] we can obtain an estimate of the barriers from the slope of $k(u)$, as $\Delta U = 1/[3k'(u_{th})] \sim 5.0$, which has the right order of magnitude (see, for example, the data of [18]). Following the scenario of [9], we must locate the geometric crossover temperature T_B, below which saddles no longer contribute to diffusion: as we have seen above, this temperature must be identified in this system with T_c, giving $T_B \sim 0.44$. On the other hand, the other crossover temperature T_A, ruling activation, is fixed by the barrier size, $T_A \sim \Delta U \sim 5.0$. Hence, for this system we have $T_B < T_A$: at T_B barriers are already quite large as compared to T, and following [9] this implies that the system is fragile. This result is consistent with the common classification of LJ liquids as fragile and therefore supports the description of fragile vs strong liquids behaviour given in [9].

REFERENCES

1. M. Goldstein, *J. Chem. Phys* **51**, 3728 (1969).
2. F.H. Stillinger and T.A. Weber, *Phys. Rev.* A **25**, 978 (1982).
3. H. Jonsson and H.C. Andersen, *Phys. Rev. Lett.* **60**, 2295 (1988);
4. S. Sastry, P.G. Debenedetti and F. Stillinger, *Nature*, **393**, 554 (1998).
5. K. K. Bhattacharya, K. Broderix, R. Kree and A. Zippelius, *Europhys. Lett.* **47**, 449 (1999)
6. See, for example, J.-P. Bouchaud, L.F. Cugliandolo, J. Kurchan and M. Mézard, in *Spin Glasses and Random Fields*, Singapore: World Scientific Publishing, 1998, editor A.P. Young; pp. 161-223, and references therein.

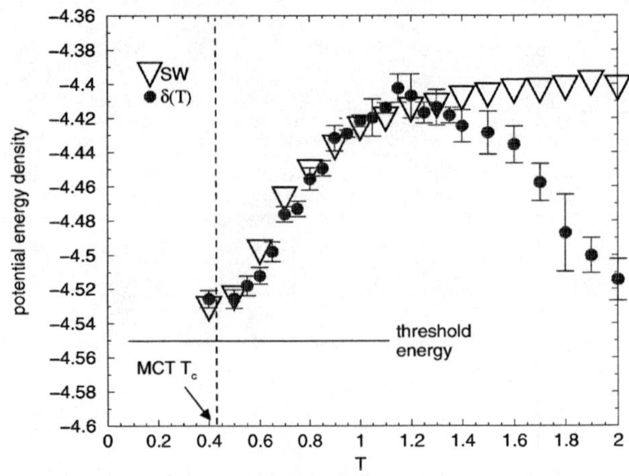

FIGURE 4. Triangles represent the energy of the Stillinger-Weber minima u_{sw} as a function of the temperature of the initial MD trajectory. Circles represent the quantity $\delta(T)$. The MCT transition is at $T_c \sim 0.44$.

7. W. Kob and J.-L. Barrat, *Phys. Rev. Lett.* **78**, 4581 (1997).
8. B. Madan and T. Keyes, *J. Chem. Phys.* **98**, 3342 (1992).
9. A. Cavagna, preprint cond-mat/9910244 (1999).
10. U. Bengtzelius, W. Goetze and A. Sjolander, *J. Phys. Chem.* **17**, 5915 (1984); E. Leutheusser, *Phys. Rev.* A **29** 2765 (1984).
11. W. Kob and H. C. Andersen, *Phys. Rev. Lett.* **73**, 1376 (1994); *Phys. Rev. E* **51**, 4626 (1995)
12. K. Broderix, K.K. Bhattacharya, A. Cavagna, A. Zippelius and I. Giardina, preprint cond-mat/0007258 (2000).
13. L. Angelani, R. Di Leonardo, G. Ruocco, A. Scala and F. Sciortino, preprint cond-mat/0007241 (2000).
14. A quench on the pseudo-potential $W(x)$ is likely to decrease on average the real energy $U(x)$ as a consequence of the fact that the Hessian has more positive than negative eigenvalues. The increment of the energy during a gradient descent step, $\vec{dx} = -\vec{\nabla}W\,dt$, is given by: $dU = \vec{\nabla}U \cdot \vec{dx} = -\vec{\nabla}U \cdot \vec{\nabla}W\,dt = -\vec{\nabla}U \cdot \mathcal{H} \cdot \vec{\nabla}U\,dt$, where $\mathcal{H}$ is the Hessian of U, which at low energies has only a small fraction of negative eigenvalues. Thus, if $\mathcal{H}$ and $\vec{\nabla}U$ are weakly correlated, we do expect the quadratic form $\vec{\nabla}U \cdot \mathcal{H} \cdot \vec{\nabla}U$ to be positive on average, and therefore dU to be negative.
15. A. Cavagna, I. Giardina and G. Parisi, *Phys. Rev.* B **57**, 11251 (1998).
16. See also S. Sastry, *J. Phys.: Condens. Matter* **12**, 6515 (2000).
17. J. Kurchan, G. Parisi and M.A. Virasoro, *J. Phys. I France* **3**, 1819 (1993).
18. T.B. Schrøder, S. Sastry, J.C. Dyre and S. Glotzer, *J. Chem. Phys.* **112**, 9834 (2000).

Glassy Dynamics of a Two-Dimensional Non-Disordered Spin Model

Juan P. Garrahan[1] and M. E. J. Newman[2]

[1] *Theoretical Physics, University of Oxford, 1 Keble Road, Oxford OX1 3NP, United Kingdom*
[2] *Santa Fe Institute, 1399 Hyde Park Road, Santa Fe, NM 87501, U.S.A.*

Abstract. We study the low temperature dynamics of a two dimensional spin system, which displays glassiness despite the absence of disorder or frustration, and is an explicit realization of the "hierarchically constrained dynamics" scenario for glassy systems. We solve exactly the statics of the model, and study in detail the dynamical behaviour of one-time and two-time quantities.

INTRODUCTION

Understanding the nature of the low-temperature dynamics of glasses and other strongly interacting many body systems remains one of the outstanding open problems in condensed matter and statistical physics [1,2].

An natural approach to modeling glassy systems is the "hierarchically constrained dynamics" of Palmer *et al.* [3], which postulates that it should be possible to describe glassy dynamics in terms of hierarchies of degrees of freedom, from fast to slow, independent of the presence of disorder or even frustration. These hierarchies would be weakly interacting in the energetic sense, but their dynamics would be constrained, the faster modes constraining the slower ones. It is known that the presence of kinetic constraints in the dynamics can directly induce glassiness, like in the kinetically constrained Ising chains [4–6] and the Backgammon model [7]. However, these models have somewhat *ad hoc* dynamics, and represent only one side of the scenario discussed in Ref. [3].

In this paper, we study in detail a model introduced in [8], which is a two-dimensional Ising model with uniform short-range ferromagnetic interactions among triplets of spins. Despite the absence of either disorder or frustration, this model displays glassy behaviour at low temperatures. The cause of this behaviour is the presence of energy barriers which grow logarithmically with the size of correlated regions. The model has a dual description in terms either of strongly interacting spins subject to simple single-spin-flip dynamics, or of free "defects" subject to a constrained dynamics. It is thus an explicit realization of the constrained dynamics scenario of Palmer *et al.* [3].

CP553, *Disordered and Complex Systems*, edited by P. Sollich, et al.

THE MODEL AND ITS STATICS

The model introduced in Ref. [8] consists of N Ising spins $\sigma = \pm 1$ on a triangular lattice with uniform short-range three-spin ferromagnetic interactions between nearest-neighbours only in groups of three lying at the vertices of a downward-pointing triangle on the lattice. The Hamiltonian for the model is

$$H = \tfrac{1}{2} J \sum_{mn} \sigma_{mn}\, \sigma_{m,n+1}\, \sigma_{m-1,n+1} + \tfrac{1}{2} N J, \tag{1}$$

where the indices m and n run along the unit vectors of the lattice $\vec{a}_1 \equiv \hat{x}$ and $\vec{a}_2 \equiv \frac{1}{2}(\hat{x}+\sqrt{3}\hat{y})$. The model can also be formulated using the defect variables $\tau_{mn} \equiv \sigma_{mn}\, \sigma_{m,n+1}\, \sigma_{m-1,n+1}$, in terms of which the Hamiltonian is $H = \frac{1}{2} J \sum_{mn} \tau_{mn} + \frac{1}{2} N J$. On lattices of size a power of two in at least one direction, with periodic boundary conditions, there is a one-to-one correspondence between spin and defect configurations, and hence the partition function is given by $Z = (2\, e^{-\frac{1}{2}\beta} \cosh \frac{1}{2}\beta)^N$, where N is the number of spins, and we set $J = 1$ from here on. The equilibrium energy density is then $\varepsilon_{\rm eq} = \frac{1}{2}(1 + \langle \tau \rangle) = \frac{1}{2}(1 - \tanh \frac{1}{2}\beta)$. We can write the spins as functions of the defects

$$\sigma_{mn} = - \prod_{\substack{n \le l \\ m-l \le k \le m}} (-\tau_{kl})^{\binom{l-n}{m-k}}. \tag{2}$$

and with this result we can compute all equilibrium correlation functions of the spins. For the magnetization we obtain

$$\langle \sigma_{mn} \rangle = \begin{cases} -1 & \text{for } T = 0 \\ 0 & \text{for } T > 0, \end{cases} \tag{3}$$

implying that the system has a $T = 0$ static phase transition. The first non-zero correlations are those for triplets of spins at the vertices of inverted equilateral triangles of side 2^k,

$$C_k^{(3)} = \langle \sigma_{mn}\, \sigma_{m,n+2^k}\, \sigma_{m-2^k,n+2^k} \rangle = -(\tanh \tfrac{1}{2}\beta)^{3^k}. \tag{4}$$

Notice that $C_{k+1}^{(3)}(\tanh \frac{1}{2}\beta) = C_k^{(3)}(\tanh^3 \frac{1}{2}\beta)$, similar to the scaling relation seen in the one-dimensional Ising model. At low temperatures the system is in a scaling region, and using Eq. (4) we find a correlation length of $\xi = (\log \coth \frac{1}{2}\beta)^{-\log 2/\log 3}$.

DYNAMICS

It was found in [8] that the model studied here shows glassy behaviour under a single-spin-flip dynamics: it was shown that following a quench from $T = \infty$ the system was unable to equilibrate in finite time at low enough temperature, and the system also fell out of equilibrium for exponential cooling with a variety of cooling rates. The reason for this glassiness is that single-spin flips correspond to flips of the defect variables on triples of sites forming upward-pointing triangles on the lattice, and sets of such flips can be combined to flip upward-pointing triangles of side 2^k for any integer k. Any isolated such triangle with $k > 0$ is locally stable; in order to remove it we have to cross an energy barrier of height k. Thus as the system relaxes it has to cross energy barriers which grow logarithmically with the size of the equilibrated regions. This means that at low temperatures the flipping of an isolated defect is heavily suppressed, since such a flip requires the presence of another neighbouring defect: the defects are non-interacting in the Hamiltonian, but their low temperature dynamics is effectively constrained.

From the observation that barriers grow logarithmically with linear size it is straightforward to estimate equilibration time. From the Arrhenius formula and the equilibrium density of defects we obtain $t_{\rm eq} \sim \exp(1/2\,T^2 \log 2)$. This form is similar to that obtained in Ref. [6] for the asymmetrically constrained Ising chain (ACIC) [5]. It has no finite-temperature singularity, consistent with the absence of a finite-temperature phase transition. The behaviour of the relaxation, however, is "fragile" [1].

One time quantities

In Fig. 1a we show numerical results for the evolution in time of the energy density after a quench from $T = \infty$ to a variety of final temperatures. The energy displays "plateaus", which are more pronounced the lower the temperature, corresponding to the system becoming trapped in locally stable configurations. As in the case of the ACIC, the natural scaling time variable is $\nu = T \log t$ [6], see inset of Fig. 1a. At low temperatures, the relaxation can also be regarded as an anomalous coarsening process. The coarsening dynamics for T close to zero can be studied approximately using the generating function method of Sollich and Evans [6], by effectively mapping the problem into one dimension. In Fig. 1b we show that this crude approximation gives reasonable results: we plot the average linear length of the domains $d(t) = 1/\sqrt{\varepsilon(t)}$ as a function of $\nu = T \log t$ from our simulations, and the "staircase" function to which they should tend at $T \to 0$ as predicted from the theory. In equilibrium, the first non-zero correlations are those for triplets of spins at the vertices of downward pointing triangles of linear size 2^k. This is also the case in the out-of-equilibrium regime following a quench. In Fig. 1c we show the absolute value of the 3-spin correlations $|C_k^{(3)}|$ as a function of time after a quench, for various values of k. The behaviour of these curves illustrates the nature of the stages by which relaxation takes place.

Two-time quantities

In Fig. 2a we show the local two-time spin autocorrelation function, $C(t, t_w) = N^{-1} \sum_{mn} \langle \sigma_{mn}(t) \sigma_{mn}(t_w) \rangle$, for three different values of the waiting time t_w, as a

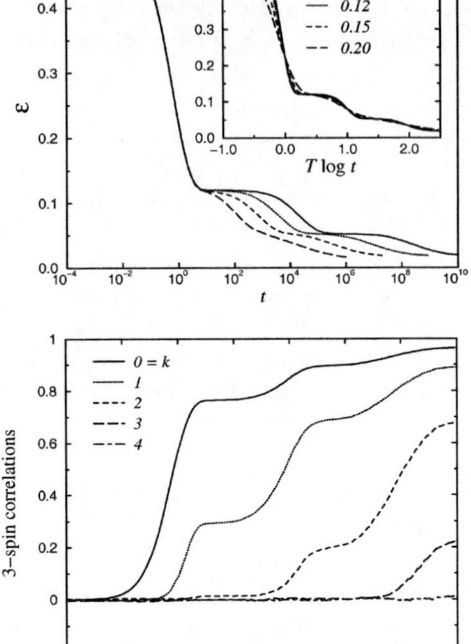

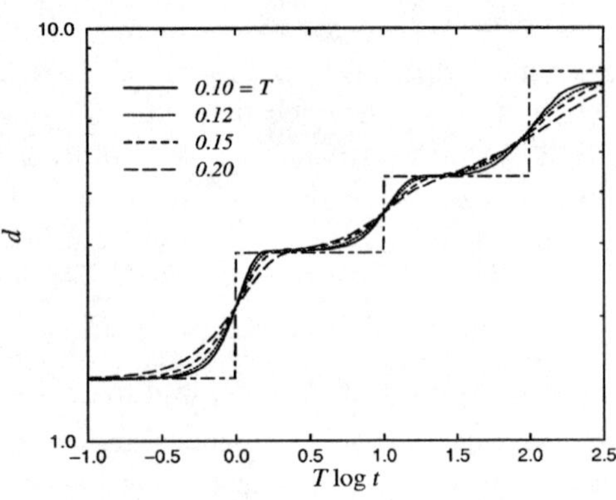

FIGURE 1. (a, top left) Energy density $\varepsilon =$ as a function of time following a quench from $T = \infty$ to $T = 0.10$, 0.12, 0.15 and 0.20, from Monte Carlo simulations for a system of 256×256. Inset: same as a function of rescaled time $\nu = T \log t$. (b, top right) Average distance between defects d as a function of rescaled time $\nu = T \log(t)$. The dot-dashed line corresponds to the $T = 0$ coarsening approximation. (c, left) Three-spin correlation functions $C_k^{(3)}$ as a function of time, for linear sizes $2^{k=0,1,2,3,4}$.

function of the rescaled time difference $\nu = T \log \tau$, with $\tau = t - t_w$. The correlation functions have a clear dependence on the waiting time, but do not obey a simple aging form, scaling with t/t_w [2]. This is because the length-scale associated with the relaxation of the spins does not grow as a simple power of time (see Figs. 1a and 1c). In general, if equilibration is associated with the growth of a length-scale $l(t)$, then the two-time correlation functions should scale with $l(t)/l(t_w)$ [2]. In our model, however, we have not been able to find a suitable length-scale such that the scaling with $l(t)/l(t_w)$ holds.

Since all off-diagonal spin-spin correlations vanish at all times, to study the response function associated with the local autocorrelation function we need only to measure the response to a uniform field. We have performed simulations in which the Hamiltonian was perturbed with $\Delta H(t) = h_0 \theta(t - t_w) \sum_{mn} \sigma_{mn}$ and have measured the integrated response $\chi(t, t_w) = \frac{1}{N h_0} \sum_{nm} \langle \sigma_{nm}(t) \rangle_h$. For $h_0 = 0.02$ we are well within the linear regime. In Fig. 2b we show the measured response for $T = 0.12$ and $\nu_w = T \log t_w = 0$ and 1. Notice that the response is not a monotonically increasing function of time, but shows cusps at times corresponding to the jumps between plateaus. For higher temperatures the cusps are less pronounced and eventually disappear (see inset).

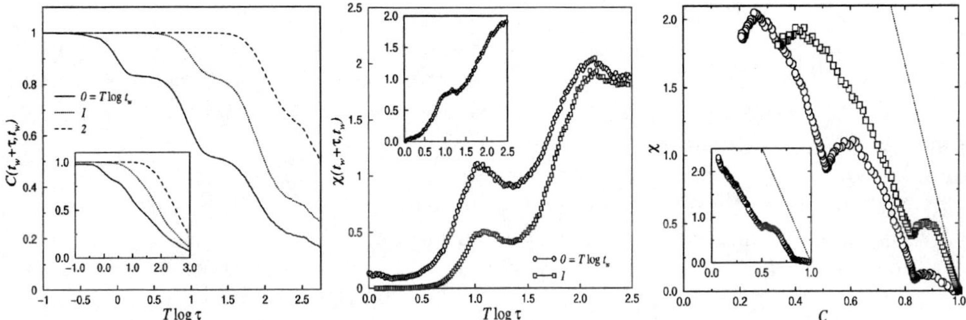

FIGURE 2. (a) Local spin-spin correlation functions as a function of the scaled time difference $T \log \tau$, following a quench to $T = 0.12$. Inset: same for $T = 0.20$. (b) Integrated response function vs. rescaled time difference, for $T = 0.12$ and $T \log t_w = 0$ and 1. Inset: same for $T = 0.20$. (c) Fluctuation-dissipation plot for $T = 0.12$ and $T \log t_w = 0$ and 1. The dotted line is the slope $-1/T$. In the inset we show the same for $T = 0.20$ and $T \log t_w = 0$.

Configurational entropy and fluctuation–dissipation relations

A configuration of our model is a local energy minimum if and only if no two defects occupy adjacent sites on the lattice. A nice feature of the model is that the set of local minima is isomorphic to the set of allowed states of Baxter's hard-hexagon model (HHM) [9]. The partition function per site of the HHM, $\kappa(z) \equiv \lim_{N\to\infty}[Z_N(z)]^{1/N}$, where z is the fugacity, is known exactly in the thermodynamic limit [9]. The entropy density of the local minima, known as configurational entropy, can be then obtained parametrically from $\varepsilon(z) = \frac{\partial \log \kappa}{\partial \log z}$ and $S_c[\varepsilon(z)] = \log \kappa(z) - \varepsilon(z) \log z$. At low defect densities, $S_c(\varepsilon)$ reduces to $S_c(\varepsilon) = -\varepsilon \log \varepsilon + \mathrm{O}(\varepsilon)$, which is the general form for a low concentration of non-interacting point defects in an ordered structure [10]. The configurational entropies of the disordered Ising chain, constrained Ising chains, and the Backgammon model all have this asymptotic form [11,12].

In Fig. 2c we show the fluctuation-dissipation plot for the spin response and correlation functions. The unusual non-monotonic shape is a consequence of the non-monotonicity of the response function. The curves are composed of a sequence of segments, each associated with one of the plateaus. The first part of each segment has a shape similar to that found in Ref. [13] for the one-dimensional Ising model. An important question is whether the configurational entropy plays any role in the out-of-equilibrium dynamics. A plausible explanation is that the initial part of each segment corresponds to quasi-stationary thermal excitations of fast modes, while the latter part corresponds to slower large-scale rearrangements before the transition to the next plateau takes place. It is noteworthy that the final slopes of each segment are roughly equal to $1/S_c'(\varepsilon_k)$, with values approximately 1/2, 1/3, and 1/4 for the first three plateaus.

CONCLUSIONS

We have studied the low temperature behaviour of a glassy two-dimensional spin model with uniform ferromagnetic short-range three-spin interactions, which is an explicit realization of the hierarchical constrained dynamics scenario of Palmer *et al.* [3]. We have solved exactly the statics and studied in detail the dynamics after a quench to low temperatures. The remarkable features are the supra-Arrhenius equilibration time and the existence of quasi-stationary stages or "plateaus" in the dynamics due to the presence of energy barriers which grow with the logarithm of the size of correlated regions. At low temperatures, the response functions have the unusual property of being non-monotonic: they display humps at exactly those times at which the system jumps between plateaus. This behaviour has also been observed in other models at times well within the activated regime [14,11,15]. An important open question is whether this is a generic feature of the out-of-equilibrium dynamics of activated processes.

ACKNOWLEDGEMENTS

We thank J.-P. Bouchaud, P. Goldbart, J. Kurchan, F. Ritort, D. Sherrington, and P. Sollich for useful discussions, and the Santa Fe Institute and Theoretical Physics, Oxford, for their kind hospitality. This work was funded in part by EC Grant No. ARG/B7-3011/94/27 and EPSRC Grant No. GR/M04426.

REFERENCES

1. C. A. Angell, Science **267**, 1924 (1995).
2. J.-P. Bouchaud, L. F. Cugliandolo, J. Kurchan and M. Mézard, in *Spin-glasses and random fields,* A. P. Young (ed.), World Scientific, Singapore (1997).
3. R. G. Palmer, D. L. Stein, E. Abraham, and P. W. Anderson, Phys. Rev. Lett. **53**, 958 (1984).
4. G. H. Fredrickson and H. C. Andersen, Phys. Rev. Lett. **53**, 1244 (1984).
5. J. Jäckle and S. Eisinger, Z. Phys. B **84**, 115 (1991).
6. P. Sollich and M. R. Evans, Phys. Rev. Lett. **83**, 3238 (1999).
7. F. Ritort, Phys. Rev. Lett. **75**, 1190 (1995).
8. M. E. J. Newman and C. Moore, Phys. Rev. E **60**, 5068 (1999).
9. R. J. Baxter, *Exactly Solved Models in Statistical Mechanics,* Academic Press, London (1982).
10. F. H. Stillinger, J. Chem. Phys. **88**, 7818 (1988).
11. A. Crisanti, F. Ritort, A. Rocco, and M. Sellitto, cond-mat/0006045.
12. G. Biroli and R. Monasson, cond-mat/9912061.
13. C. Godreche and J. M. Luck, J. Phys. A **33**, 1151 (2000).
14. M. Nicodemi, Phys. Rev. Lett. **82**, 3734 (1999).
15. L. Davison and D. Sherrington, cond-mat/0008039.

Proliferation Assisted Barrier Crossing and Population Dynamics

Irene Giardina*, Jean-Philippe Bouchaud†, Marc Mézard **

*Service de Physique Théorique et †Service de Physique de l'Etat Condensé, Commissariat à l'Energie Atomique, Orme des Merisiers, 91191 Gif-sur-Yvette CEDEX, France.
**Laboratoire de Physique Théorique et Modèles Statistiques Université Paris Sud, Bat. 100, 91 405 Orsay CEDEX, France

Abstract. We investigate a model of population dynamics in a random force field, where two competing mechanisms of barrier slowing down and proliferation induced super-diffusion are present simultaneously. A one-loop RG analysis close to the critical dimension $d_c = 2$ shows that the random force fixed point is unstable in the presence of a small proliferation term, and flows towards an uncontrolled strong coupling regime. Numerical results in $d = 1$ suggest that a new intermediate diffusion behaviour appears, consistent with the RG analysis. We introduce the idea of proliferation assisted barrier crossing and give a Flory like argument to understand qualitatively this non trivial diffusive behaviour.

The statistical properties of random walks can be dramatically altered by the presence of disorder, leading to non standard diffusion behaviour. In some cases the effect of disorder may lead to sub-diffusion. For example, this is what happens for random walks in a random potential: the walkers may remain trapped in deep potential wells for long times, and the typical travelled distance thus grows more slowly than the square-root of time [1]. In other cases the disorder may lead to super-diffusion. For example this happens for random walks in a rotational random force field: the walkers can be convected far away by long streamlines and their motion becomes faster than diffusive [1]. Another interesting mechanism of super-diffusion is *random proliferation*, that is when each random walker can either die or give birth to new random walkers at a rate which is random, both in time and space. Random walkers that have by chance travelled a distance much greater than the square-root of time, may reproduce and represent an appreciable fraction of the whole population, leading to a motion of the center of mass faster than diffusive.

In this paper we investigate the case where two mechanisms leading to sub-diffusion and super-diffusion are present simultaneously [2]. We consider a model of population dynamics, where the individuals live in a random environment and are subjected to a random growth rate. The random environment is modelled as a delta correlated static random force field, and represents the sub-diffusive mechanism.

CP553, *Disordered and Complex Systems*, edited by P. Sollich, et al.

The random growth rate is modelled by a Gaussian time and space dependent noise, and represents the 'proliferation' super-diffusive mechanism. The questions we want to address concern the diffusion properties of this 'mixed' model. As we shall see, the presence of the two opposite, competing mechanisms generates a new intermediate behaviour between super-diffusion and sub-diffusion. Applications of this problem can be found in many different contexts, whenever a population dynamics is involved. Interesting examples are in biology (i.e. bacteria dynamics in a random substrate, see also [3]), and economics (individuals acting via trading and speculation [4]).

We consider the following equation for the local population density $P(\vec{x},t)$ in d dimensions:

$$\frac{\partial P(\vec{x},t)}{\partial t} = \nu_0 \Delta P(\vec{x},t) - \vec{\nabla}\cdot(\vec{F}(\vec{x})P) + \eta(\vec{x},t)P(\vec{x},t), \tag{1}$$

where ν_0 is the bare diffusion constant, $\vec{F}(\vec{x})$ a space dependent static Gaussian random force such that $\langle F_\mu(\vec{x})F_\nu(\vec{x}')\rangle_F = 2\sigma_F^2\delta_{\mu,\nu}\delta^d(\vec{x}-\vec{x}')$, and $\eta(\vec{x},t)$ a Gaussian random growth rate, with $\langle\eta(\vec{x},t)\eta(\vec{x}',t')\rangle_\eta = 2\sigma_\eta^2\delta(t-t')\delta^d(\vec{x}-\vec{x}')$. Due to the last term, the total population $Z(t) = \int d\vec{x}P(\vec{x},t)$ is not conserved.

The quantities of interest, which describe how the population spreads in time are, for example, the average center of mass motion,

$$x_{cm}^2(t) = \langle\Big(\frac{1}{Z}\int \vec{x}P(\vec{x},t)\; d\vec{x}\Big)^2\rangle_{F,\eta} \tag{2}$$

and other higher moments which can be defined in a analogous way.

It is useful to briefly review the two cases where only one kind of disorder is present:

- Setting $\eta = 0$ in equation (1), we recover the problem of a random walk in a Random Force field (RF). This problem has been much studied in the past years via Renormalization Group techniques and its diffusion properties are well known. One finds that for d lower than the marginal dimension of the problem $d_c = 2$, $x_{cm}(t)$ grows as t_F^ν with $\nu_F < 1/2$, and the system behaves sub-diffusively [5,6,1].

- Setting $F = 0$, our model describes a purely Random Proliferation (RP) dynamics, and equation (1) represents the well known KPZ (or Directed Polymer) problem. For this problem, too, the marginal dimension is $d_c = 2$. RG calculations for the KPZ problem provides the exact result $\nu_\eta = 2/3$ in $d = 1$ [7], while for $3/2 < d < 2$ the diffusion exponent cannot be computed via RG (there is no accessible fixed point, see [7]), but is found numerically to be greater than $1/2$ thus corresponding to a super-diffusive behaviour.

For the mixed model with $F \neq 0$ and $\eta \neq 0$ we have performed an RG analysis for $d = d_c - \epsilon$, where $d_c = 2$ is the marginal dimension of the combined problem. The

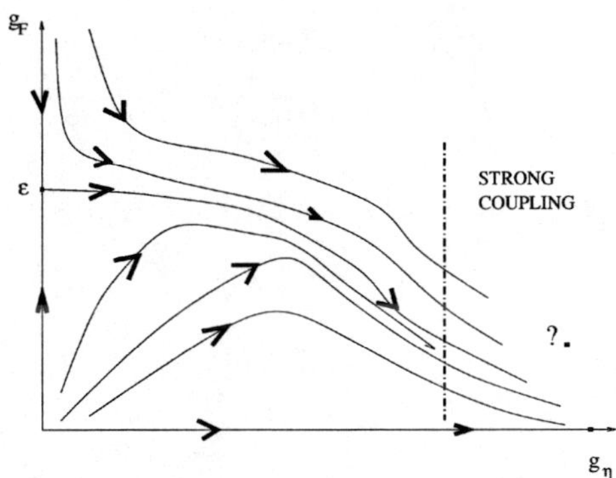

FIGURE 1. One loop RG flow in the g_η, g_F plane. As soon as g_η is non zero, it flows towards the strong coupling region.

resulting RG flows is shown in Fig. 1. $g_F = \sigma_F^2/(2\pi)\nu_0^2$ is the coupling constant related to the Random Force problem, while $g_\eta = \sigma_\eta^2\lambda^2/(2\pi)\nu_0^3$ is the coupling constant related to the KPZ part. As we can see in the figure the sub-diffusive Random Force fixed point $g_\eta = 0$, $g_F = \epsilon$ is unstable in the presence of a small 'proliferation' term g_η. Unfortunately, at one loop, g_η flows towards the strong coupling region, as is the case in the standard KPZ case $g_F = 0$.

In order to obtain some information about this strong coupling behaviour, we have performed some numerical simulations in one dimension evolving a space and time discretized version of Eq. (1) [2]. Starting from a localized packet $P(x = ia, t = 0) = \delta_{i,0}$, we have found that as soon as both coupling constants g_F^0 and g_η^0 are non zero, the exponent ν describing the diffusion of the center of mass $x_{cm}(t)$ at large times is found to be close to the value $\nu^* = 1/2$ (see Fig. 2). The ratio g_F^0/g_η^0 affects only the short time transient behaviour, which is either Random Force like or DP/KPZ like. In the RG language, this suggests that a non trivial *attractive* fixed point g_F^*, g_η^* appears. This is compatible with the flow diagram of Fig. 1 [8], although this new fixed point is out of reach at the one-loop level.

The value of $\nu^* = 1/2$ is equal to the value obtained in the case of free-diffusion. However the motion of the packet for a given environment is very far from simple diffusion, as the analysis of the width of the packet shows. Indeed, both the Random Force problem and the DP/KPZ problem are characterized by a strong localization of the packet for a given sample [9–13] and this feature survives in the combined problem.

We would like to understand, at least qualitatively, the observed non trivial diffusion behaviour, and in particular the value $\nu^* = 1/2$ of the diffusion exponent. In order to do this, it is again useful to look separately at the two competing mechanisms, and then try to guess how they interact:

- In $d = 1$, the Random Force problem simplifies in that the force $F(x)$ is

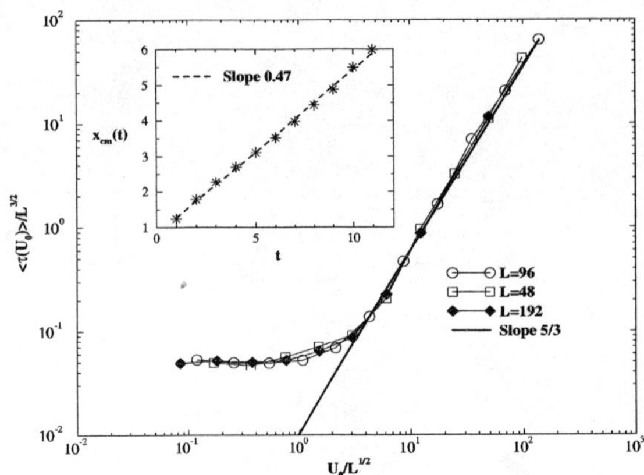

FIGURE 2. Average barrier crossing time $\langle\tau\rangle$, rescaled by $L^{3/2}$, as a function of the barrier height U_0 rescaled by $\sqrt{L}$, for different sizes L, and in log-log coordinates. Inset: Behaviour of the average center of mass $x_{cm}(t)$ as a function of time, for $\sigma_\eta/\sigma_F = 0.125$.

always derived from a potential: $U(x) = \int_0^x dyF(y)$. The problem becomes the well known Sinai model [9], and many results are available on its diffusion properties [14,11]. Qualitatively we can imagine that the particles feel an effective potential $U_{eff}(x) \sim \langle U(x)^2\rangle^{1/2} = \sigma_F\sqrt{x}$ which is a random walk in space. Thus the typical barrier encountered at time t grows as $x(t)^{\frac{1}{2}}$, and a naive activation argument leads to $x_{cm}(t) \propto \log^2(t)$, which is indeed the correct result [9].

- The DP/KPZ problem also simplifies in $d = 1$, since a fluctuation-dissipation theorem holds (which is not true in higher dimensions). In this case it is possible to obtain the stationary distribution of the effective free-energy $h(x,t) = \log P(x,t)$, which turns out to be a random walk in space, as in the Sinai problem. (It is precisely this random walk behaviour in space which is responsible, in the two individual cases as well as in the combined problem, for the numerically observed localization properties of the packet; see [2].) Besides, as previously mentioned, the diffusion exponent can be computed exactly [15] and is equal to 2/3.

In the combined problem Sinai+KPZ, there exists a competition between the slowing down induced by the ever-growing Sinai barriers and the speeding up of the population spreading allowed by the multiplicative growth term η. Understanding the behaviour of the system when subjected to these two mechanisms, means first of all understanding how the presence of a proliferation effect can help in overcoming a barrier. For this reason, we start considering the simpler case of a unique barrier of height U_0, which develops on scale L. For definiteness, we have solved numerically the equation (1) on the interval $[0, L]$ with $F(x) = -U_0/L \sin(4\pi x/L)$. The initial condition is localized in the first well, and the crossing time τ is defined as the average time after which the relative weight of the population in the second well is half of that in the first well. For $\eta \equiv 0$, one finds the classical Arrhenius law: $\log \tau = U_0/\nu_0$. On the other hand, when $U_0 = 0$ the time for the particles to reach a distance L is given by the DP/KPZ scaling, i.e. $L \propto \tau^{2/3}$. When $\eta \neq 0$ and $U_0 \neq 0$, the behaviour of τ as a function of U_0 for different values of L is shown in Fig. 2. The result can be expressed as: $\tau \propto L^{3/2} f(U_0/\sqrt{L})$, with $f(y \to 0) = 1$ and $f(y \to \infty) \propto y^b$, and b very close to 5/3. The scaling of τ with L can easily be understood: the influence of the external potential U_0 becomes substantial when it becomes of the order of the effective KPZ potential $h(L, t) = \log P(L, t)$, which grows as as $\sqrt{L}$. Therefore, we find that the exponential increase of the crossing time with the barrier height is replaced by a power-law increase in the presence of the random growth term η. One can call this effect proliferation assisted barrier crossing.

It is possible to give a Flory like argument to estimate the exponent b. To do so, we will try to estimate the contribution to the probability of jumping the barrier coming from free diffusion (the only mechanism which is present for $\eta = 0$), and the contribution coming from the random growth rate:
i) The probability that a particle reaches the top of the barrier x^* by pure diffusion is simply given by $\exp(-U_0/\nu_0)$;
ii) The random growth term, on the other hand, provides a certain proliferation 'gain' factor which we will call $\exp \mathcal{G}(x^*, t)$. To estimate this factor it is convenient to write the population density $P(x, t)$ in the Directed Polymer formalism as a path integral [15],

$$P(x,t) = \int_{x_0,t_0}^{x,t} \mathcal{D}x \exp\{-\int_0^t ds[\frac{(\nabla_s x)^2}{4\nu_0} + \eta(x(s), s)]\} \tag{3}$$

In this expression, the gradient, elastic part is the one related to free-diffusion and can be seen as the entropy of the random walks, while the second one is due to the random growth rate. We want to estimate the typical contribution due to this second part. If the path leading from the initial point x_0 to x^* was unique, one would simply have $\mathcal{G}(x^*, t) = \int dt' \eta(x_C(t'), t')$, which typically behaves as $\sigma_\eta \sqrt{t}$. In fact, many paths contribute to $\mathcal{G}(x^*, t)$. This leads to a kind of preaveraging effect of the random growth term η over the width $w(t)$ of the paths. Therefore:

$$\mathcal{G}(x^*, t) \sim \sigma_\eta \left(\int_0^t \frac{dt'}{w(t')^d} \right)^{1/2}. \tag{4}$$

Since most paths leading to x^* spend their time in the thermally accessible region of the well, one can estimate $w(t')$ as $w = L/\sqrt{U_0}$.

The time τ is such that the proliferation factor must compensate the barrier, and equating the contributions i) and ii) we get $\tau \propto U_0^{3/2}$ (for $d = 1$). This simple argument therefore leads to $b = 3/2$, not very far from the numerical value $b \simeq 5/3$.

Actually, one can apply this argument to the unconfined case $U_0 = 0$, where the detrimental factor is now the entropy of the random walk $\exp(-x^{*2}/t)$. Using self-consistently $w(t') = x^*(t')$, the compensation argument now leads to $x^* \propto t^{3/(4+d)}$, which is precisely the Flory result for the DP/KPZ problem [16]. The value $b = 3/2$ can therefore be seen as a Flory value for this problem.

Once the simple case of a single barrier has been understood, it is straightforward to generalize the previous Flory argument to the full Sinai+KPZ problem. In this case the barrier is not fixed but grows as $\sigma_F\sqrt{x^*}$ (x^* now being the typical distance travelled at time t). The self consistent compensation argument now leads to $\sigma_F\sqrt{x^*} \sim \sigma_\eta\sqrt{t/x^*}$, or $x^* \sim (\sigma_\eta/\sigma_F)\sqrt{t}$. The $\sqrt{t}$ behaviour is close to the numerical results shown in the inset of Fig. 2. However, the dependence of x_{cm} on σ_F is found numerically to be weaker than the $1/\sigma_F$ behaviour predicted by this simple argument, and closer to $1/\sigma_F^{3/4}$.

REFERENCES

1. J. P. Bouchaud, A. Georges, Phys. Rep. **195**, 127 (1990).
2. I. Giardina, J. P. Bouchaud and M. Mezard, *Population dynamics in a random environment*, preprint cond-mat/0005187.
3. K. Dahmen, D. Nelson, N. Shnerb, cond-mat/9903276.
4. J. P. Bouchaud, M. Mézard, cond-mat/0002374, Physica A **282**, 536 (2000).
5. J. M. Luck, Nucl. Phys. **B 225**, 169 (1983).
6. D. S. Fisher, Phys. Rev. **A 30**, 960 (1984).
7. E. Frey, U. Tauber, Phys. Rev. **E 50**, 1024 (1994), E. Frey, U. Tauber, T. Hwa, Phys. Rev. **E 53**, 4424 (1996).
8. One can actually argue, using the Directed Polymer representation, that the DP/KPZ strong coupling fixed point is also unstable against a small static random force $F(x)$.
9. Ya G. Sinai, Theory Probab. Appl. **27**, 247–58 (1982).
10. A. Golosov, Comm. Math. Phys. **92**, 491 (1984).
11. D. S. Fisher, P. Le Doussal, C. Monthus, Phys. Rev. **E 59**, 4795 (1999).
12. M. Mézard, J. Physique (Paris) 51, 1831 (1990).
13. D. S. Fisher, D. Huse, Phys. Rev. **B 43**, 10728 (1991), T. Hwa, D. S. Fisher, Phys. Rev. **B 49**, 3136 (1994).
14. J.P. Bouchaud, A. Comtet, A. Georges and P. Le Doussal, Annals of Physics **201**, 285 (1990).
15. T. Halpin-Healey and Y. C. Zhang, Phys. Rep. **254**, 217 (1995).
16. M. Mézard, G. Parisi, J. Physique I **1**, 809 (1991). T. Garel and H. Orland, Phys. Rev. **B 55**, 226-30 (1997).

Vulcanization and the Random Solid State it Yields: A Statistical Mechanical Perspective[1]

Paul M. Goldbart and Weiqun Peng

Department of Physics, University of Illinois at Urbana-Champaign, 1110 West Green Street, Urbana, Illinois 61801-3080, U.S.A.

Abstract. In many interesting physical settings, such as the vulcanization of rubber, the introduction of permanent random constraints between the constituents of a homogeneous fluid can cause a phase transition, known as the vulcanization transition, to a random solid state. In this random solid state, particles are permanently but randomly localized, and a rigidity to shear deformations emerges. Owing to the permanence of the random constraints, this phase transition is an equilibrium transition, which confers on it a simplicity (at least relative to the conventional glass transition) in the sense that it is amenable to established techniques of equilibrium statistical mechanics.

The aims of this paper are two-fold: First, the briefest sketch will be given of aspects of the mean-field theory of the vulcanization transition and the emergent random solid state. (These aspects have been reviewed elsewhere rather recently.) Second, a more detailed account will be given of various new results on the vulcanization transition beyond mean-field theory.

INTRODUCTION

The vulcanization transition (VT) is an equilibrium phase transition from a liquid state of matter to an amorphous solid state. This transition occurs when a sufficiently large density of permanent random constraints (e.g. chemical crosslinks)—the quenched randomness—is introduced to connect the constituents (e.g. linear, flexible macromolecules), whose spatial coordinates constitute the thermally fluctuating—the annealed—variables of the system.

A rather detailed description of the VT has emerged over the past few years within the context of a mean-field approximation [1–5]. At the heart of this description is a carefully crafted order parameter together with a minimal model

1) Work supported by the U.S. National Science Foundation via grant DMR99-75187. The contributions of several collaborators, including N. D. Goldenfeld, A. Zippelius, H. E. Castillo and K. A. Shakhnovich are gratefully acknowledged.

CP553, *Disordered and Complex Systems*, edited by P. Sollich, et al.

(i.e. a Landau-Wilson effective free energy) that governs the equilibrium behavior of the order parameter.

In what follows, we shall begin by outlining the basic ingredients (order parameter, minimal model, etc.) of the theory of the VT. Then, as it has already been summarized several times (see, e.g., Refs. [1,5]), we shall very briefly sketch the mean-field level picture of the VT. We shall then turn to our main purpose: to discuss recent results concerning critical fluctuations near the VT [6]. The reader may notice that, in contrast with the oral version of this paper, the present version focuses more on recent developments concerning critical fluctuations and less on mean-field aspects of the theory of vulcanized matter.

MODELING THE VULCANIZATION TRANSITION

Order parameter. The appropriate order parameter for the VT, capable *inter alia* of distinguishing between the liquid and amorphous solid states, is the following function of the $n+1$ wave vectors $\{\mathbf{k}^0, \mathbf{k}^1, \mathbf{k}^2, \cdots, \mathbf{k}^n\}$:

$$\Big[\frac{1}{N}\sum_{j=1}^{N}\int_0^1 ds\, \langle e^{i\mathbf{k}^0\cdot\mathbf{c}_j(s)}\rangle_\chi \cdots \langle e^{i\mathbf{k}^n\cdot\mathbf{c}_j(s)}\rangle_\chi\Big]. \tag{1}$$

Here, N is the total number of macromolecules in the system, $\mathbf{c}_j(s)$ (with $j = 1,\ldots,N$ and $0 \le s \le 1$) is the position in d-dimensional space of the monomer at fractional arc-length s along the jth macromolecule, $\langle\cdots\rangle_\chi$ denotes a thermal average for a particular realization χ of the quenched disorder (i.e. the crosslinking), and $[\cdots]$ represents a suitable averaging over this quenched disorder. This form of order parameter is motivated by the order parameter employed in the context of spin glasses: the repeated thermal average detects the existence of static *random* density fluctuations (see, e.g., Ref. [1]). It is worth emphasizing that the disorder resides in the specification of what monomers are crosslinked together: the resulting constraints do not *explicitly* break the translational symmetry of the system, although this symmetry does become *spontaneously* broken at the VT, albeit in a hidden way, detectable only via this refined order parameter.

Minimal model. This describes the relative free-energy cost of configurations of the order-parameter field Ω, the expectation value of which yields the order parameter (1). It can be ascertained by (i) invoking the replica trick to incorporate the consequences of the permanent random constraints, and (ii) making the Deam and Edwards assumption [7] about the statistics of constraint realizations (i.e. that the statistics of the disorder is determined by the instantaneous correlations of the unconstrained system). Thus we are led to the need to work with the $n \to 0$ limit of systems of $n+1$ (rather than the usual n) replicas, the additional replica, labeled by $\alpha = 0$, describing the constraint distribution.

The minimal model can be determined within the spirit of Landau approach to continuous phase transitions, i.e. from the transformation properties of the order parameter field Ω and the symmetries of the effective free energy, along with the assumptions of the analyticity of the effective free energy and the continuity of the transition [2]. (It can also be derived from a variety of semi-microscopic models.) Either scheme leads to the following minimal model, which takes the form of a cubic field theory involving the order-parameter field $\Omega(\hat{k})$ which lives on $(n+1)$-fold replicated d-dimensional space [8]:

$$f \propto \lim_{n\to 0} n^{-1} \ln[Z^n], \qquad [Z^n] \propto \int \mathcal{D}^{\dagger}\Omega \, \exp(-\mathcal{S}_n), \tag{2a}$$

$$\mathcal{S}_n = N \sum_{\hat{k}\in \mathrm{HRS}} \left(-a\tau + \frac{b}{2}|\hat{k}|^2 \right) |\Omega(\hat{k})|^2 - Ng \sum_{\hat{k}_1,\hat{k}_2,\hat{k}_3 \in \mathrm{HRS}} \Omega(\hat{k}_1)\,\Omega(\hat{k}_2)\,\Omega(\hat{k}_3)\,\delta_{\hat{k}_1+\hat{k}_2+\hat{k}_3,\hat{0}}, \tag{2b}$$

where the reduced control parameter τ measures the density of random constraints. Averages weighted by $\exp(-\mathcal{S}_n)$ will be denoted $\langle \cdots \rangle^{S}$. The coefficients a, b and g are phenomenological coefficients which encode the microscopic details of the system. We use the symbol $\hat{k}$ to denote the replicated wave vector $\{\mathbf{k}^0, \mathbf{k}^1, \ldots, \mathbf{k}^n\}$, and define the extended scalar product $\hat{k}\cdot\hat{c}$ by $\mathbf{k}^0\cdot\mathbf{c}^0 + \mathbf{k}^1\cdot\mathbf{c}^1 + \cdots + \mathbf{k}^n\cdot\mathbf{c}^n$. The symbol $\hat{k} \in$ HRS indicates that a summation over replicated wave vectors is restricted to those wave vectors containing *at least two nonzero component-vectors* $\mathbf{k}^{\alpha}$. (We say that such wave vectors lie in the higher-replica-sector, i.e. the HRS.) This HRS condition reflects the fact that no crystalline order (or any other kind of macroscopic inhomogeneity) appears in the vicinity of the VT; it determines the symmetry of the effective free energy and, thus, plays a crucial role.

CONCLUSIONS AT THE MEAN-FIELD LEVEL

Let us give the briefest sketch of the mean-field theory of the VT and the emergent random solid state.

(i) The order parameter encodes a great deal of statistical information about the microstructure of the random solid state, including the fraction of monomers that are localized and the statistical distribution of localization lengths (i.e. RMS position fluctuations of the localized monomers). The localized fraction grows continuously and linearly with the excess crosslink density (i.e. the amount by which the crosslink density exceeds its critical value). For all near-critical crosslink densities, the distribution of localization lengths has a *universal mean-field scaling form*, so that the distributions, when suitably plotted, should collapse on to a single master-curve determined once and for all by the theory. As discussed, e.g., in Refs. [2,5], these features are all borne out by extensive molecular dynamics computer simulations of three-dimensional, off-lattice, interacting, macromolecular systems, due to Barsky and Plischke [9].

(ii) By explicitly bounding the eigenvalues of the appropriate Hessian matrix from below, the field-theory saddle-point corresponding to the random solid state has been shown to be stable with respect to all small fluctuations (up to the expected

Goldstone mode associated with the breaking of translational symmetry) [3].
(iii) By examining the increase in free energy associated with changing the shape of the system at fixed volume, the elastic properties of the random solid state have been determined. In particular, it has been shown that at the mean-field level the shear modulus becomes nonzero as the random solid state is entered, and grows continuously, with the third power of the excess crosslink density [4].

VULCANIZATION BEYOND MEAN-FIELD THEORY

We now turn to our main purpose: to outline certain recent results concerning critical fluctuations near the VT. A detailed account can be found in Ref. [6].
Ginzburg criterion. Let us begin by estimating the range of reduced crosslink densities $\delta\tau$ within which the effects of order-parameter fluctuations are relatively strong. This is accomplished by following the conventional strategy of computing the loop expansion for the 2-point vertex function to one-loop order. (In the present setting the loop expansion amounts to an expansion in the inverse monomer density.) This shows that the upper critical dimension for the VT is six. To determine the physical content of the Ginzburg criterion, we invoke the values of the coefficients appropriate for the semi-microscopic model of randomly crosslinked macromolecular systems (see, e.g., Ref. [1]) and, hence, find the following form of the Ginzburg criterion: for $d < 6$, fluctuations cannot be neglected for values of τ satisfying

$$|\tau| \lesssim (L/\ell)^{-\frac{d-2}{6-d}} \left(\varphi/g^2\right)^{-\frac{2}{6-d}}. \tag{3}$$

Here, L is the arc-length of each macromolecule, ℓ the persistence length, and $\varphi \equiv (N/V)(L/\ell)\ell^d$ is the volume fraction. Equation (3) indicates that the fluctuation-dominated regime is narrower for longer macromolecules and for higher densities (provided $2 < d < 6$). Such dependence on the degree of polymerization L/ℓ is precisely that argued for by de Gennes on the basis of a percolation picture [10].
Expanding around six dimensions. Let us now mention the results that emerge from the $\varepsilon(\equiv 6-d)$ expansion calculation. The procedure is standard, but turns out to be rather non-trivial to implement, owing to the constraint that the critically fluctuating fields reside in the higher replica sector only.

As reported in detail in Ref. [6], we consider the order parameter correlator $\langle\Omega(\hat{k})\,\Omega(-\hat{k})\rangle^S$, which obeys the homogeneity relation $\langle\Omega(\hat{k})\,\Omega(-\hat{k})\rangle^S \sim |\hat{k}|^{-2+\eta}\, g(|\hat{k}|\,|\tau|^{-\nu})$ [in which $g(x) \sim x^{2-\eta}$ for $x \to +0$ and approaches a constant value for large x]. For this correlator we find that: (i) For ε negative (i.e. for $d > 6$) there is only the *Gaussian* fixed point, for which the exponents take on their classical values: $\nu^{-1} = 2$ and $\eta = 0$. (ii) For ε positive (i.e. $d < 6$), in addition to the Gaussian fixed point, a new fixed point—the *Wilson-Fisher* fixed point—emerges, and controls the critical behavior. The critical exponents are (to first order in ε) $\nu^{-1} = 2 - (5\varepsilon/21)$ and $\eta = -\varepsilon/21$.

A standard scaling argument [albeit with the subtlety associated with that fact that the theory does not contain $\Omega(\hat{0})$] leads us to the critical exponent β for the gel fraction. We find that for $d > 6$ we have $\beta = 1$, and for $d < 6$ we have $\beta = \nu(d+2-\eta)/2 = 1-\varepsilon/7$. In fact, under the (not unreasonable) assumption that in the random solid state the scale of the fluctuation correlation length is the same as that of the localization length, we can go further and propose a general scaling hypothesis for the (singular part of) $\langle\Omega(\hat{k})\rangle^S$, namely $\langle\Omega(\hat{k})\rangle^S \propto \tau^\beta\, w(\hat{k}^2\tau^{-2\nu})$, which is supported by the mean-field result [2].

Both above and below six dimensions, the critical exponents are identical (at least to first order in ε) to those in percolation theory, as computed via the Potts field theory [12,13]. In order to be sure that identical exponents arising in both theories describe physically analogous quantities, we discuss, in next section, the physical content of the VT order-parameter correlator along with its relationship with the correlation function of percolation theory.

ORDER-PARAMETER CORRELATOR AND ITS PHYSICAL SIGNIFICANCE

In order to explore the physical content of the order parameter correlator [14], we examine the construction $C_{\mathbf{t}}(\mathbf{r}-\mathbf{r}') \equiv$

$$\left[\frac{V}{N}\sum_{j,j'=1}^{N}\int_0^1 ds\,ds'\langle\delta^{(d)}(\mathbf{r}-\mathbf{c}_j(s))\,\delta^{(d)}(\mathbf{r}'-\mathbf{c}_{j'}(s'))\rangle\langle e^{-i\mathbf{t}\cdot(\mathbf{c}_j(s)-\mathbf{r})}\,e^{i\mathbf{t}\cdot(\mathbf{c}_{j'}(s')-\mathbf{r}')}\rangle\right] \quad (4a)$$

$$=\frac{N}{V}\sum_{\mathbf{k}}\mathrm{e}^{i(\mathbf{k}+\mathbf{t})\cdot(\mathbf{r}-\mathbf{r}')}\left[\frac{1}{N^2}\sum_{j,j'=1}^{N}\int_0^1 ds\,ds'\,\langle\mathrm{e}^{-i\mathbf{k}\cdot(\mathbf{c}_j(s)-\mathbf{c}_{j'}(s'))}\rangle_\chi\langle\mathrm{e}^{-i\mathbf{t}\cdot(\mathbf{c}_j(s)-\mathbf{c}_{j'}(s'))}\rangle_\chi\right]. \quad (4b)$$

(The two versions are simply related by Fourier representations of the delta functions.) Note that, in the second line, the expression involving the square brackets is precisely the microscopic prototype of a typical order-parameter correlator $\langle\Omega(\mathbf{0},\mathbf{k},\mathbf{t},\mathbf{0},\ldots,\mathbf{0})^*\,\Omega(\mathbf{0},\mathbf{k},\mathbf{t},\mathbf{0},\ldots,\mathbf{0})\rangle^S$. What about the meaning of $C_{\mathbf{t}}(\mathbf{r}-\mathbf{r}')$? The first expectation value in Eq. (4a) accounts for the likelihood that monomers (j,s) and (j',s') will respectively be found around $\mathbf{r}$ and $\mathbf{r}'$; the second describes the correlation between the respective fluctuations of monomer (j,s) about $\mathbf{r}$ and monomer (j',s') about $\mathbf{r}'$.

In the case that $\mathbf{t}=\mathbf{0}$, the quantity $C_{\mathbf{t}}(\mathbf{r}-\mathbf{r}')$ is simply (V/N times) the real-space density-density correlator, and is not of central relevance at the VT. The small-$\mathbf{t}$ limit of $C_{\mathbf{t}}(\mathbf{r}-\mathbf{r}')$ addresses the connectedness of clusters formed from crosslinked macromolecules, and can be shown [6] to be almost exactly the pair-connectedness function defined in (the on-lattice version of) percolation theory [13]. In the case of general $\mathbf{t}$, $C_{\mathbf{t}}(\mathbf{r}-\mathbf{r}')$ addresses the question: If a monomer near $\mathbf{r}$ is localized on the scale t^{-1} (or more strongly), how likely is a monomer near $\mathbf{r}'$ to be localized on the same scale (or more strongly)? From Eq. (4a) we see $1/t$ serves as a cutoff to the range of correlation, so that all pairs of monomers that are relatively localized at a length-scale much larger than $1/t$ do not contribute to $C_{\mathbf{t}}(\mathbf{r}-\mathbf{r}')$.

Given $C_{\mathbf{t}}(\mathbf{r}-\mathbf{r}')$, by integrating $C_{\mathbf{t}}(\mathbf{r}-\mathbf{r}')$ over space and passing to the $\mathbf{t}\to\mathbf{0}$ limit we can build a divergent susceptibility: $\lim_{\mathbf{t}\to\mathbf{0}}\int d^d r\, C_{\mathbf{t}}(\mathbf{r})$

$$\sim N \lim_{\mathbf{t}\to\mathbf{0}}\lim_{n\to 0}\langle\Omega(\mathbf{0},\mathbf{t},-\mathbf{t},\mathbf{0},\ldots,\mathbf{0})^{*}\,\Omega(\mathbf{0},\mathbf{t},-\mathbf{t},\mathbf{0},\ldots,\mathbf{0})\rangle^{\mathrm{S}}\sim(-\tau)^{-\gamma}. \tag{5}$$

This quantity is a measure of the spatial extent over which pairs of monomers are relatively localized, no matter how weakly, and thus diverges at the VT.

CONCLUDING REMARKS

As we have seen, the exponents describing the VT obtained via the ε expansion turn out to be numerically equal to those characterizing physically analogous quantities in percolation theory, at least to first order in ε. This equality between exponents seems reasonable, in view of the intimate relationship between percolation theory and the *connectivity* of the system of crosslinked macromolecules. This connection has long been realized, and supports the use of percolative approaches as models of certain aspects of the VT [11,13].

An essential ingredient of the percolative approaches is the Potts model in its one-state limit. It is therefore worth considering similarities and differences between the minimal model of the VT [i.e. Eq. (2b)] and the minimal field theory for the Potts model. Both have a cubic interaction term and both involve $n\to 0$ limits. But, despite the similarities, there exist many differences: (i) The Potts field theory has a multiplet of n real fields on d-dimensional space; the VT field theory has a real singlet field living on $(n+1)d$-dimensional space. (ii) The Potts field theory represents a setting involving a *single* ensemble, whereas the VT field theory describes a physical situation in which *two* distinct ensembles (thermal and disorder) play essential roles. As such, the vulcanization field theory is a natural approach, in the sense that it provides a unified theory of both the transition regime and the structure of the solid state. (iii) The symmetry structures are different. More importantly, the nature of the spontaneous symmetry breaking is different. The percolation transition (in its Potts representation) involves the spontaneous breaking (of the $n\to 0$ limit) of a *discrete* $(n+1)$-fold permutation symmetry. By contrast, the VT involves the spontaneous breaking (of the $n\to 0$ limit) of the *continuous* symmetry of *relative* translations and rotations of the $n+1$ replicas; The fact that the critical exponents near the upper critical dimension are the same for both theories (at least to order ε) despite these apparent distinctions can be understood by delineating between three logically distinct physical properties pertaining to randomly crosslinked macromolecular systems: (i) macroscopic network formation; (ii) the change in nature of thermal motion—random localization; and (iii) the acquisition of rigidity. Within mean-field theory (and hence above six spatial dimensions), these three properties go hand in hand, emerging simultaneously at the VT. At and below six dimensions they appear to continue to go hand in hand (although, we have not yet investigated the issue of rigidity beyond mean-field theory) until one reaches two dimensions where, we believe, this broad picture will change [6,15]. Thus, at least in higher dimensions, it appears that, when both ensembles are considered, disorder fluctuations play a more important role than do thermal fluctuations, as far as the percolative aspects of the critical phenomenon are concerned.

We conclude by simply mentioning that the concepts and strategies featuring in the present discussion of randomly crosslinked macromolecular systems can also be applied to systems such as chemical gels, in which low molecular weight objects (rather than macromolecules) are permanently covalently bonded together at random. These issues are discussed in detail in Refs. [16,17].

REFERENCES

1. For a detailed review, see P. M. Goldbart, H. E. Castillo and A. Zippelius, Adv. Phys. **45**, 393 (1996).
2. W. Peng, H. E. Castillo, P. M. Goldbart, A. Zippelius, Phys. Rev. B **57**, 839 (1998).
3. H. E. Castillo, P. M. Goldbart and A. Zippelius, Phys. Rev. B **60**, 14702 (1999).
4. H. E. Castillo and P. M. Goldbart, Phys. Rev. E **58**, R24 (1998); cond-mat/9909054, Phys. Rev. E (in press, 2000).
5. P. M. Goldbart, J. Phys. Cond. Mat. **12**, 6585 (2000).
6. W. Peng and P. M. Goldbart, Phys. Rev. E **61**, 3339 (2000).
7. R. T. Deam and S. F. Edwards, Phil. Trans. R. Soc. **280A**, 317 (1976).
8. The measure $\overline{\mathcal{D}}^{\dagger}\Omega$ indicates that we integrate $\Omega(\hat{k})$ with $\hat{k} \in$ HRS and with $\hat{k}$ restricted to a half-space via an additional condition $\hat{k} \cdot \hat{m} > 0$ for some suitable unit $(n+1)d$-vector $\hat{m}$. This is due to the fact that $\Omega(-\hat{k}) = \Omega(\hat{k})^*$.
9. S. J. Barsky and M. Plischke, Phys. Rev. E **53**, 871 (1996); S. J. Barsky, Ph.D. Thesis, Simon Fraser University, Canada (unpublished, 1996); S. J. Barsky and M. Plischke (unpublished, 1997).
10. P. G. de Gennes, J. Physique Lett. **38**, L355 (1977); see also Ref. [11].
11. P.-G. de Gennes, *Scaling Concepts in Polymer Physics* (Cornell University Press, Ithaca, 1979).
12. A. B. Harris, T. C. Lubensky, W. K. Holcomb and C. Dasgupta, Phys. Rev. Lett. **35**, 327 (1975); and **35**, 1397 (E).
13. See, e.g., T. C. Lubensky, pp. 405-475, in *Ill-Condensed Matter* (Les Houches XXXI, 1978), edited by R. Balian, R. Maynard and G. Toulouse (North Holland: Amsterdam, 1979); and reference therein.
14. Some initial considerations of these issues were given in A. Zippelius and P. M. Goldbart, pp. 357-389, *Spin Glasses and Random Fields*, edited by A. P. Young (World Scientific, Singapore, 1998).
15. H. E. Castillo, P. M. Goldbart and W. Peng, work in progress (2000).
16. P. M. Goldbart and A. Zippelius, Europhys. Lett. **27**, 599 (1994).
17. K. A. Shakhnovich and P. M. Goldbart, Phys. Rev. B **60**, 3862 (1999).

Optimal Capacity of Ashkin-Teller Neural Networks

P. Kozłowski [1], D. Bollé [2]

Instituut voor Theoretische Fysica, K.U. Leuven, B-3001 Leuven, Belgium

Abstract. The maximal capacity per number of couplings is calculated for Ashkin-Teller type neural networks using a replica-symmetric Gardner approach. The results are compared with numerical simulations. The best value obtained at $\kappa = 0$ is $\alpha_c = 2.26 \pm 0.01$.

INTRODUCTION

The Gardner calculation [1] of the maximal capacity for a perceptron has lead to many new insights in theoretical studies and applications related to learning and generalization [2–5].

Inspired by the recent results obtained for the Ashkin-Teller (AT) neural network with Hebb learning [6,7], suggesting that such a model could be more efficient than a sum of two Hopfield models, we look here at the maximal capacity of AT models using the replica symmetric Gardner approach and numerical simulations with various learning algorithms. In this kind of models there are two types of binary neurons at each site of the network with binary and quartic interactions. For more details and an underlying neurobiological motivation for the introduction of different types of neurons we refer to [7].

THE MODEL

Consider p input patterns $\boldsymbol{\zeta}^\mu = \{\zeta_i^\mu\} = \{\xi_i^\mu, \eta_i^\mu\}$, $i = 1, \ldots, N$ consisting out of two different types of patterns $\boldsymbol{\xi}^\mu = \{\xi_i^\mu\}$ and $\boldsymbol{\eta}^\mu = \{\eta_i^\mu\}$, and a corresponding set of outputs $\zeta_0^\mu = \{\xi_0^\mu, \eta_0^\mu\}$ $\mu = 1, \ldots, p$ which are determined by

$$\xi_0^\mu = \mathrm{sign}(h_1^\mu + \eta_0^\mu h_3^\mu) \tag{1}$$

$$\eta_0^\mu = \mathrm{sign}(h_2^\mu + \xi_0^\mu h_3^\mu) \tag{2}$$

$$\xi_0^\mu \eta_0^\mu = \mathrm{sign}(\eta_0^\mu h_1^\mu + \xi_0^\mu h_2^\mu) \tag{3}$$

---

[1] E-mail: piotr.kozlowski@fys.kuleuven.ac.be
[2] E-mail: desire.bolle@fys.kuleuven.ac.be.

CP553, *Disordered and Complex Systems*, edited by P. Sollich, et al.

where h_r $(r = 1, 2, 3)$ are the local fields acting on the patterns ξ, η and their product $\xi\eta$ respectively

$$h_1^\mu = \frac{1}{n_1}\sum_{i=1}^{N} J_i^{(1)}\xi_i^\mu, \quad h_2^\mu = \frac{1}{n_2}\sum_{i=1}^{N} J_i^{(2)}\eta_i^\mu, \tag{4}$$

$$h_3^\mu = \frac{1}{n_3}\sum_{i=1}^{N} J_i^{(3)}\xi_i^\mu\eta_i^\mu, \quad n_r^2 = \sum_i (J_i^{(r)})^2, \quad r = 1, 2, 3. \tag{5}$$

Both types of input patterns and their corresponding outputs are taken to be independent identically distributed random variables (IIDRV) taking the values +1 or −1 with probability 1/2. The set of three equations (1)-(3) defines a mapping of the inputs ζ_i^μ onto the corresponding outputs ζ_0^μ. We call it the AT perceptron model I. The specific form of the equations (1)-(3) is related to the transition probabilities for a spin-flip in the dynamics [6].

A second model, denoted by AT perceptron model II, is defined by considering only the two equations (1) and (2). When $|h_3| > |h_1|$ and $|h_3| > |h_2|$ then the relations (1)-(2) are satisfied by two (out of the four possible) values of the output ζ_0, otherwise model II gives the same output as model I. In other words, due to the presence of the η_0^μ and ξ_0^μ in the gain functions, model II contains more freedom and, strictly speaking, it is not a mapping.

At this point we remark that when all $J_i^{(3)}$ are equal to zero we find back two independent standard binary perceptron models. In the sequel we take the couplings to satisfy the spherical constraint $n_r = \sqrt{N}$.

REPLICA-SYMMETRIC MAXIMAL CAPACITY

The AT perceptron is trained to store correctly $p = \frac{3}{2}\alpha N$ patterns with α the loading capacity. The factor 3/2 follows naturally from the fact that the capacity is defined as the thermodynamic limit of the ratio of the total number of bits per (input) neuron to be stored and the total number of couplings per (output) neuron. A pattern is stored correctly when the so-called aligning field [9] is bigger than a certain constant $\kappa \geq 0$ whereby the latter indicates the stability. It is a measure for the size of the basin of attraction of that pattern.

In the sequel we only outline explicitly the calculations for the AT perceptron model II. For more details we refer to [10]. Specifically, we require that

$$\lambda_\xi^\mu(\{J\}) = \xi_0^\mu\,(h_1^\mu + \eta_0^\mu h_3^\mu) > \kappa_\xi \geq 0 \tag{6}$$

$$\lambda_\eta^\mu(\{J\}) = \eta_0^\mu\,(h_2^\mu + \xi_0^\mu h_3^\mu) > \kappa_\eta \geq 0\,, \tag{7}$$

with $\{J\} = \{J_i^{(r)}\}$ denoting the configurations in the space of interactions. For $\kappa_\xi = \kappa_\eta = 0$ all patterns that satisfy equations (1)-(2) also satisfy (6) and (7). For the AT perceptron model I we need, of course, an extra inequality [10].

The idea is then to determine the maximal value of the loading α such that couplings satisfying (6)-(7) can still be found. Following refs. [1,8] the problem is formulated as an energy minimization in the space of couplings with the formal energy function given by

$$E(\{J\}) = \sum_{\mu} \left[1 - \Theta(\lambda^{\mu}_{\xi}(\{J\}) - \kappa_{\xi})\Theta(\lambda^{\mu}_{\eta}(\{J\}) - \kappa_{\eta})\right] . \tag{8}$$

This quantity counts the number of weakly embedded patterns, i.e., the patterns with stability less than $\kappa_{\xi}, \kappa_{\eta}$. Therefore, the minimal energy gives the minimal number of patterns that are stored incorrectly. This number is zero below a maximal storage capacity $\alpha_c(\kappa_{\xi}, \kappa_{\eta})$.

The basic quantity to start from is the partition function

$$Z(\beta) = \langle \exp\left[-\beta E(\{J\})\right]\rangle_{\{J\}} \tag{9}$$

where $\langle ... \rangle_{\{J\}} = \int \prod_i dJ_i \prod_r \delta\left(\sum_i (J_i^{(r)})^2 - N\right) ...$ and β is the inverse temperature. As usual it is $\ln Z$ which is assumed to be a self-averaging extensive quantity [1,9]. The related free energy per site is equal, in the limit $\beta \to \infty$, to $\langle E\rangle_{\infty}/N \equiv \lim_{\beta\to\infty}\langle E \exp\left[-\beta E(\{J\})\right]\rangle_{\{J\}}/NZ(\beta)$, which is the minimal fraction of wrong patterns (recall eq. (8)).

The average over the disorder in the input patterns $\boldsymbol{\zeta}^{\mu}$ and the corresponding outputs ζ_0^{μ} is performed by the replica method. The calculations proceed in a standard way although the technical details are more complex. For the order parameters $q^r_{\alpha\beta} = \frac{1}{N}\sum_i J_i^{(r)\alpha} J_i^{(r)\beta}$, $r = 1, 2, 3$ we take the replica-symmetric anzatz $q^r_{\alpha\beta} = q^r$ and, for convenience, we put $q^1 = q^2 = q^3 = q$. Taking the limits $\beta \to \infty$, $N \to \infty$ and $n \to 0$ and following [1] we arrive at

$$\begin{aligned} v = \lim_{N\to\infty} \frac{1}{N} < \ln Z > &= \frac{3}{2}\alpha \int \mathrm{d}s_1 \mathrm{d}s_2 \frac{\exp(-\frac{1}{q}(s_1^2 + \frac{1}{3}s_2^2)}{\pi\sqrt{3q}} \ln \left[\left| \int_{l_1}^{\infty} \int_{l_2}^{l_3} \prod_{\nu} \mathrm{D}u_{\nu} \right|\right] \\ &+ \frac{3}{2}\left(\ln(1-q) + \frac{1}{1-q} + \ln 2\pi\right) \end{aligned} \tag{10}$$

with $\mathrm{D}u_{\nu} = \mathrm{d}u_{\nu} \exp(-\frac{1}{2}u_{\nu}^2)/\sqrt{2\pi}$ a Gaussian measure,

$$\begin{aligned} l_1 = \frac{\sqrt{\frac{2}{3}}(\frac{1}{2}(\kappa_{\xi} + \kappa_{\eta}) - s_2)}{\sqrt{1-q}} \;, \quad l_2 &= \frac{\sqrt{2}(\kappa_{\eta} - s_2 - s_1)}{\sqrt{1-q}} - u_2\sqrt{3} \;, \\ l_3 &= \frac{\sqrt{2}(-\kappa_{\xi} + s_2 - s_1)}{\sqrt{1-q}} + u_2\sqrt{3} \end{aligned} \tag{11}$$

and where q takes values that minimize v, the available volume in the coupling space. Here we do not write down explicitly this fixed-point equation for q.

Taking $\kappa_{\xi} = \kappa_{\eta} = \kappa$ and supposing that the maximal capacity, $\alpha_c = \alpha_{GD}$, is signaled by the Gardner-like criterion $q \to 1$ we obtain the result shown in fig. 1

by the full line on the right. The full line on the left is obtained by an analogous calculation for model I.

One can show that learning antiparallel patterns, i.e., patterns satisfying $(\boldsymbol{\xi}^\mu \xi_0^\mu, \boldsymbol{\eta}^\mu \eta_0^\mu) = -(\boldsymbol{\xi}^\nu \xi_0^\nu, \boldsymbol{\eta}^\nu \eta_0^\nu)$ results in a splitting of the space of couplings into disconnected regions. This suggests that replica symmetry is broken and, hence, the resulting α_{GD} is expected to be an upper bound for the true capacity. For a first step replica symmetry breaking calculation we refer to [10]. For the AT-perceptron model II preliminary results give $\alpha_{RSB1} = 2.28$ (versus $\alpha_{GD} = 2.74$) at $\kappa = 0$.

NUMERICAL SIMULATIONS

The idea is to teach the network as many random patterns as possible by generalizing various algorithms proposed for simple perceptrons [11–14]. The most effective ones appeared to be some particular generalization of the adaptive Gardner algorithm [12] and the Adatron algorithm [13]. We only report on the results obtained with these two algorithms. We remark that we have chosen $\kappa_\xi = \kappa_\eta = \kappa$ in all simulations.

Following the heuristic arguments presented in [12] we have constructed for the AT perceptron model II the following adaptive perceptron learning rule

$$
\begin{aligned}
J_i^{(1)} &\to J_i^{(1)} + \xi_0^\mu \xi_i^\mu \frac{1}{2}\left(\kappa_\xi - \lambda_\xi^\mu\right)\Theta\left(\kappa_\xi - \lambda_\xi^\mu\right) \\
J_i^{(2)} &\to J_i^{(2)} + \eta_0^\mu \eta_i^\mu \frac{1}{2}\left(\kappa_\eta - \lambda_\eta^\mu\right)\Theta\left(\kappa_\eta - \lambda_\eta^\mu\right) \\
J_i^{(3)} &\to J_i^{(3)} + \xi_0^\mu \eta_0^\mu \xi_i^\mu \eta_i^\mu \frac{1}{2}\left[\left(\kappa_\xi - \lambda_\xi^\mu\right)\Theta\left(\kappa_\xi - \lambda_\xi^\mu\right) + \left(\kappa_\eta - \lambda_\eta^\mu\right)\Theta\left(\kappa_\eta - \lambda_\eta^\mu\right)\right] . \qquad (12)
\end{aligned}
$$

This algorithm should be carried out sequentially over the patterns and sequentially or parallel over the couplings as long as one of the arguments of the Θ functions is positive. It appears to have the characteristics of the most efficient, non-linear algorithm discussed in [12]. Up to now, we have failed to prove its convergence analytically.

We have trained networks of sizes $50 \leq N \leq 1000$ neurons (depending on the value of κ) in order to store perfectly as many randomly chosen patterns as possible. For each value of κ we have calculated the maximal capacity for different N and extrapolated the results to $N = \infty$. Results for a given value of κ and N are averages over 1000 samples. As shown in fig. 1 this algorithm performs very well for small values of κ for both the models I and II.

The Adatron algorithm [13] works differently. Instead of searching the maximal capacity for a given stability it tries to find the maximal stability for a given capacity. Below we describe a learning rule that tries to incorporate the ideas of the Adatron approach. The standard analytic convergence proof is based upon the assumption that the problem can be formulated as a quadratic optimization with linear constraints such that it cannot be generalized to the problem at hand.

We assume that the couplings can be written in the form ([13] and references therein)

$$J_i^{(1)} = \frac{1}{N}\sum_\mu^p x_1^\mu \xi_0^\mu \xi_i^\mu \,, \qquad J_i^{(2)} = \frac{1}{N}\sum_\mu^p x_2^\mu \eta_0^\mu \eta_i^\mu \,, \qquad J_i^{(3)} = \frac{1}{N}\sum_\mu^p x_3^\mu \xi_0^\mu \eta_0^\mu \xi_i^\mu \eta_i^\mu \,, \quad (13)$$

where $i = 1, ..., N$ and x_r^μ $(r = 1, 2, 3)$ are the so called embedding strengths of pattern μ. Then, the couplings are updated by modifying x_r^μ with the following increments

$$\delta x_1^\mu = \frac{1}{2}\max\{-x_1^\mu - x_3^\mu, \gamma(1 - n_1\lambda_\xi^\mu)\} \,,$$
$$\delta x_2^\mu = \frac{1}{2}\max\{-x_2^\mu - x_3^\mu, \gamma(1 - n_2\lambda_\eta^\mu)\} \,,$$
$$\delta x_3^\mu = \frac{1}{4}(\max\{-x_1^\mu - x_3^\mu, \gamma(1 - n_3\lambda_\xi^\mu)\} + \max\{-x_2^\mu - x_3^\mu, \gamma(1 - n_3\lambda_\eta^\mu)\})$$

This is done sequentially over the patterns. Again, for each value of the capacity we have considered system sizes $50 \leq N \leq 500$ and extrapolated the results to $N = \infty$. Results for each size are averages over 1000 samples. The best results were obtained for a learning rate $\gamma \in (0, 2)$.

The algorithm works better for small values of the capacity, both in the case of models I and II than the adaptive perceptron algorithm does. This can be seen in fig. 1. For larger values of the capacity, however, it performs worse. We remark that the results for the Adatron algorithm are displayed only in the region where they are better than those for the adaptive Gardner algorithm.

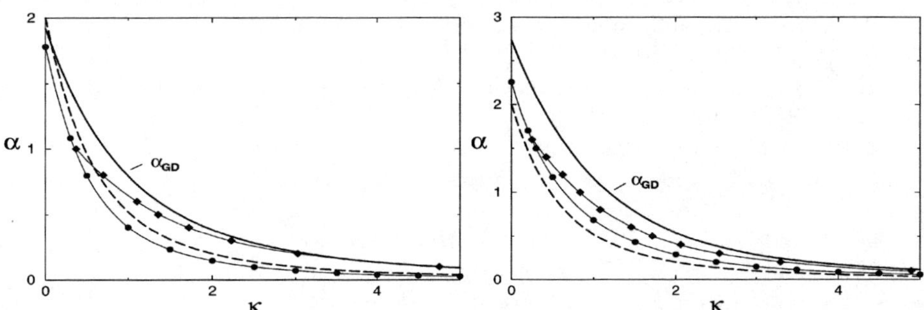

FIGURE 1. The maximal capacity of the AT perceptrons as a function of κ. Left: model I; right: model II. Theoretical result, α_{GD}, versus simulations. The circles are the results for the adaptive Gardner algorithm, the diamonds for the Adatron algorithm. The error bars are smaller than the size of the symbols. The solid thin lines are polynomial fits to these results. The maximal capacity of a simple perceptron is indicated with a broken line.

CONCLUDING REMARKS

We have calculated the maximal capacity per number of couplings for two AT perceptron models using the replica-symmetric Gardner approach. Since the space of couplings for both models is disconnected and, hence, replica symmetry is expected to be broken, the theoretical results obtained are upper bounds for the true capacity.

Numerical simulations have been performed for finite systems and extrapolated to $N = \infty$. The adaptive Gardner algorithm and the Adatron algorithm give the best, but different, results suggesting that they are not really convergent. Hence, the results of the simulations can be considered only as lower bounds for the true maximal capacity.

For $\kappa = 0$ the following can be concluded. The maximal capacity for model I satisfies $1.78 < \alpha < 1.92$ and for model II $2.26 < \alpha < 2.74$. A first step replica symmetry calculation gives a better estimate of the upper bounds. For model II preliminary results indicate that $\alpha_{RSB1} = 2.28$. In any case the value obtained for model II is bigger than the maximal capacity of the standard binary perceptron. However, it is not a strict mapping such that it allows for more freedom in the determination of the couplings.

ACKNOWLEDGMENTS

We would like to thank M. Bouten and J. van Mourik for critical discussions.

REFERENCES

1. Gardner E., *J. Phys. A: Math. Gen.* **21**, 257 (1988).
2. Hertz J., Krogh A., and Palmer R.G., *Introduction to the Theory of Neural Computation*, Addison-Wesley, Redwood City, 1991.
3. Müller B., Reinhardt J. and Strickland M.T., *Neural Networks: An Introduction*, Springer, Berlin, 1995.
4. Opper M., and Kinzel W., in *Models of Neural Networks III*, edited by Domany E., van Hemmen J.L., and Schulten K., Springer: New-York, 1996.
5. Saad D. (ed.), *On-line Learning in Neural Networks*, Cambridge Univ. Press, 1998.
6. Bollé D., and Kozłowski P., *J. Phys. A: Math. Gen.* **31**, 6319 (1998).
7. Bollé D., and Kozłowski P., *J. Phys. A: Math. Gen.* **32**, 8577 (1999).
8. Gardner E., and Derrida B., *J. Phys. A: Math. Gen.* **21**, 271 (1988).
9. Whyte W., and Sherrington D., *J. Phys. A : Math. Gen.* **29**, 3063 (1996).
10. Bollé D., and Kozłowski P., in preparation
11. Krauth W., and Mézard M., *J. Phys. A: Math. Gen.* **20**, L745 (1987).
12. Abbott L.F., and Kepler T.B., *J. Phys. A: Math. Gen.* **22**, L711 (1989).
13. Anlauf J.K., and Biehl M., *Europhys. Lett.* **10**, 687 (1989).
14. Imhoff J., *J. Phys. A: Math. Gen.* **28**, 2173 (1995).

Cluster Derivation of the Parisi Scheme for Disordered Systems

A.C.C. Coolen and J. van Mourik

Department of Mathematics, King's College, University of London
The Strand, London WC2R 2LS, U.K.

Abstract. We propose a general quantitative scheme in which systems are given the freedom to sacrifice energy equi-partitioning on the relevant time-scales of observation, and have phase transitions by separating autonomously into ergodic sub-systems (clusters) with different characteristic time-scales and temperatures. The details of the break-up follow uniquely from the requirement of zero entropy for the slower cluster. Complex systems, such as the Sherrington-Kirkpatrick model, are found to minimise their free energy by spontaneously decomposing into a hierarchy of ergodically equilibrating degrees of freedom at different (effective) temperatures. This leads exactly and uniquely to Parisi's replica symmetry breaking scheme. Our approach, which is somewhat akin to an earlier one by Sompolinsky, gives new insight into the physical interpretation of the Parisi scheme and its relations with other approaches, numerical experiments, and short range models. Furthermore, our approach shows that the Parisi scheme can be *derived* quantitatively and uniquely from plausible physical principles.

INTRODUCTION

The so-called Parisi scheme [1] for replica symmetry breaking (RSB) has been one of the most successful tools in the description of (the statics of) disordered systems in the non-ergodic or glassy phase. Originally proposed as the solution for the Sherrington-Kirkpatrick (SK)-model [2] for mean field spin glasses, it has been successfully applied to a wide range of disordered systems. The physical interpretation of the Parisi scheme has been the subject of many discussions, and has led to the introduction of concepts such as *disparate time-scales* [3], *effective temperatures* [4], *low entropy production* [5] and *non-equilibrium thermodynamics* [6].

In this paper, we will show how a general scheme in which systems are given the freedom to sacrifice energy equi-partitioning by separating autonomously into sub-systems with different characteristic time-scales and temperatures, not only yields the Parisi scheme, but also introduces all the above-mentioned concepts in a very natural way. Our assumptions are simple, and all the quantities that appear in the theory have a clear physical meaning. We briefly discuss systems with disparate

CP553, *Disordered and Complex Systems*, edited by P. Sollich, et al.

time-scales, and then apply this to the benchmark problem of mean field disordered systems, viz. the SK-model [2]. We also present numerical evidence for the existence of multiple disparate time-scales. Finally, we sketch the simple physical picture that naturally emerges from our scheme, and discuss the points that still need further investigation.

SYSTEMS WITH DISPARATE TIME SCALES

In this section we briefly describe how the formalism, as developed and applied for Ising spin systems with slowly evolving bonds [7], can be generalised to arbitrary stochastic systems with two or more disparate time-scales. In the case of systems with two infinitely disparate time-scales, the faster variables $\vec{x}_f$ equilibrate before the slower ones $\vec{x}_s$ can effectively change. Therefore, the $\vec{x}_s$ evolve to an a Boltzmann-type equilibrium distribution with an effective energy which is the free energy of the $\vec{x}_f$ given the $\vec{x}_s$, at an effective inverse temperature β_s, while the $\vec{x}_f$ evolve to the normal Boltzmann equilibrium distribution with $\beta_f = \beta$:

$$\begin{aligned} \mathcal{Z} &= \mathrm{Tr}_{\vec{x}_s} \; (\mathcal{Z}_f(\vec{x}_s))^{\tilde{m}} \; , \\ \mathcal{Z}_f(\vec{x}_s) &\equiv \mathrm{Tr}_{\vec{x}_f} \; \exp\left(-\beta\mathcal{H}(\vec{x})\right) \; . \end{aligned} \tag{1}$$

One thus obtains a theory with a non-negative replica dimension $\tilde{m} \equiv \beta_s/\beta$. This procedure can be generalised to a system with $L+1$ different levels of stochastic variables, time-scales and temperatures $\{(\vec{x}_\ell,\ \tau_\ell,\ \beta_\ell) :\ \ell \in \{0, L\}\ \}$. Assuming that each level is adiabatically slower than the next level ($\tau_\ell/\tau_{\ell-1} = 0$), we obtain the following recursion relations

$$\begin{aligned} \mathcal{Z}_\ell &\equiv \mathrm{Tr}_{\vec{x}_\ell} \; (\mathcal{Z}_{\ell+1})^{\tilde{m}_{\ell+1}} \quad (\ell < L), \\ \mathcal{Z}_L &\equiv \mathrm{Tr}_{\vec{x}_L} \; \exp\left(-\beta_L \; \mathcal{H}(\{\vec{x}\})\right) \; , \end{aligned} \tag{2}$$

where $\tilde{m}_\ell \equiv \beta_{\ell-1}/\beta_\ell$, $\beta_L = \beta$, and the total free energy of the system, defined on the longest time-scale, is given by

$$\mathcal{F} = -\frac{1}{\beta_0} \log(\mathcal{Z}_0) \; . \tag{3}$$

THE SK-MODEL

We will now apply this scheme to the SK-model [2], for which the Parisi scheme was originally developed. Therefore, we briefly recall the definitions:

$$\mathcal{H}(\vec{\sigma}) = -\sum_{i<j}^{N} J_{ij}\sigma_i\sigma_j \; , \tag{4}$$

with Gaussian couplings J_{ij}, $P(J_{ij}) = \mathcal{N}(J_0/N, J/\sqrt{N})$, and Ising spins σ_j.

We assume that there are L+1 levels of spins with corresponding disparate time-scales and temperatures $\{(\vec{\sigma}_\ell = \{\sigma_j \in I_\ell\}, |I_\ell| \equiv N\epsilon_\ell, \tau_\ell, T_\ell), \ell = 0, .., L\}$ in the system, where $\tau_\ell/\tau_{\ell-1} = 0$, such that larger ℓ correspond to faster spins. Although we expect the selection of time-scales for the spins to depend on the specific realisation of the couplings, at present we will make the simplest approximation: the system can only choose the relative sizes ϵ_ℓ of the levels. A more detailed study of the (annealed) selection of levels will be presented elsewere. To deal with the quenched disorder average over the J_{ij} we use the replica trick

$$\overline{\mathcal{F}} = -\frac{1}{\beta_0}\overline{\log \mathcal{Z}_0} = -\lim_{\tilde{n}\to 0} \frac{1}{\tilde{n}\beta_0} \log \overline{\mathcal{Z}_0^{\tilde{n}}} \ . \tag{5}$$

Together with the recursion relations (2) this leads to a nested set of $\tilde{n}\prod_{\ell=1}^{L}\tilde{m}_\ell$ replicas, in which spins at level ℓ carry a set of replica indices $\{a\}_\ell \equiv \{a_0, .., a_\ell\}$. The index $a_0 = 1, .., \tilde{n}$ comes from the disorder average, whereas $a_\ell = 1, .., \tilde{m}_\ell = \beta_{\ell-1}/\beta_\ell$. The asymptotic free energy per spin is then given by

$$f = \lim_{\tilde{n}\to 0} \frac{-1}{\tilde{n}\beta_0} \left[\frac{-J^2\beta^2}{4} \sum_{\{a\}_L, \{b\}_L} \left(q_{\{b\}_L}^{\{a\}_L}\right)^2 + \sum_{\ell=0}^{L} \epsilon_\ell \ \log(\mathcal{K}_\ell) \right] , \tag{6}$$

$$\mathcal{K}_\ell \equiv \mathop{\mathrm{Tr}}_{\sigma^{\{c\}_\ell}} \exp \left[\frac{J^2\beta^2}{2} \sum_{\{a\}_L, \{b\}_L} q_{\{b\}_L}^{\{a\}_L} \ \sigma^{\{a\}_\ell}\sigma^{\{b\}_\ell} \right] , \qquad q_{\{b\}_L}^{\{a\}_L} \equiv \frac{1}{N} \sum_{\ell=0}^{L} \sum_{j\in I_\ell} \sigma_j^{\{a\}_\ell}\sigma_j^{\{b\}_\ell} . \tag{7}$$

With the definitions $m_\ell \equiv \prod_{f=\ell}^{L} \tilde{m}_f = \beta_{\ell-1}/\beta$ we have $\beta_0\tilde{n} = \beta n$, and the connection with the original Parisi scheme becomes clear. Note that $0 \le \tilde{m}_\ell \le 1$, as slower clusters cannot have a lower temperature than faster ones, because otherwise the latter would act as a heat bath. We now assume full ergodicity at each level in the hierarchy of time-scales:

$$q_{\{b\}_L}^{\{a\}_L} = q_\ell \ , \qquad \ell \equiv \sum_{r=0}^{L} r \ \delta_{\{a\}_{r-1}, \{b\}_{r-1}} \ (1 - \delta_{a_r, b_r}) \ , \tag{8}$$

(i.e. ℓ is the slowest level at which $\{a\}_L$ and $\{b\}_L$ differ), to obtain

$$f = \frac{J^2\beta}{2} \sum_{\ell=0}^{L} \left[\frac{m_{\ell+1}}{2} \ (q_{\ell+1}^2 - q_\ell^2) - \epsilon_\ell \ [q]_\ell \right] - \frac{1}{m_1\beta} \sum_{\ell=0}^{L} \epsilon_\ell \int Dz_0 \log(\mathcal{N}_\ell^1) \ , \tag{9}$$

where

$$\mathcal{N}_\ell^r \equiv \begin{cases} \int Dz_r \left(\mathcal{N}_\ell^{r+1}\right)^{\frac{m_r}{m_{r+1}}} & , \qquad r \le \ell \\ 2\cosh(J\beta m_{\ell+1} \sum_{s=0}^{\ell} z_s\sqrt{q_s - q_{s-1}}) & , \qquad r = \ell + 1 \end{cases} \tag{10}$$

$$[q]_\ell \equiv \sum_{r=\ell}^{L} m_{r+1} \ (q_{r+1} - q_r) \ , \tag{11}$$

The physical meaning of the q_ℓ is given by

$$q_\ell = \lim_{N\to\infty} \sum_j \overline{\langle \,..\, \langle \langle \langle \,..\, \langle\sigma_j\rangle_L \,..\, \rangle_{\ell+1} \rangle_\ell^2 \rangle_{\ell-1} \,..\, \rangle_0} \,, \tag{12}$$

where $\langle\cdot\rangle_r$ denotes the average over the level r process, and $\overline{\cdot}$ denotes the disorder average. The minimum of the free energy with respect to the ϵ_ℓ (with $\sum_{\ell=0}^{L} \epsilon_\ell = 1$, for $\tilde{n}, \tilde{m}_\ell$ positive integers), is at $\{\epsilon_L^* = 1,\ \epsilon_\ell^* = 0\ \ \forall \ell < L\}$, and we exactly recover the L-th order Parisi solution. Furthermore, the extremization with respect to the $\tilde{m}_\ell$ is now recognized to express the fact that, for self-consistency, the entropy of the spins slower than level ℓ is zero (they are effectively fixed on the time-scale τ_ℓ).

NUMERICAL EVIDENCE

In numerical simulations of the SK-model we have measured the distribution of the number of flips f at time t per spin: $\rho_{\rm sim}(f,t)$. Assuming a characteristic time-scale τ_j for each spin σ_j and a distribution $W(\tau)$ of these time-scales, we obtain a theoretical prediction of the distribution of the number of flips per spin at time t:

$$\rho_{\rm th}(W,f,t) \simeq \int_0^\infty d\tau\ W(\tau) \begin{pmatrix} t \\ f \end{pmatrix} \left(\frac{1}{\tau}\right)^f (1-\frac{1}{\tau})^{t-f} \,. \tag{13}$$

Minimization of $\sum_{f=0}^{t} \left[\rho_{\rm sim}(f,t) - \rho_{\rm th}(W(\tau),f,t)\right]^2$ with respect to $W(\tau)$ then yields the most probable distribution of time-scales $W^*(\tau)$, see fig. 1.

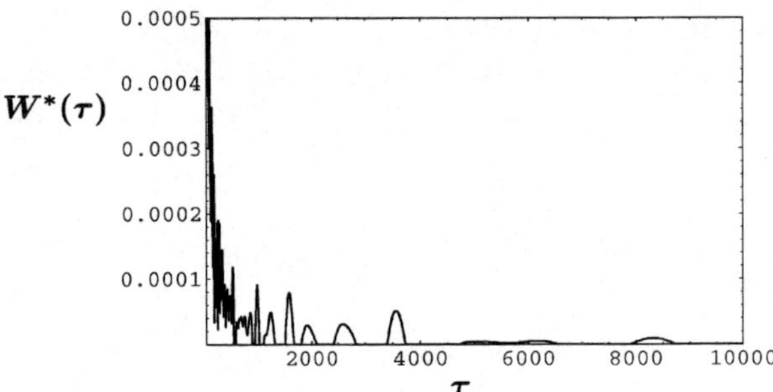

FIGURE 1. The most probable distribution of time-scales for a simulation of the SK-model with $N = 6000$, after $t = 5.10^5$ Monte-Carlo updates per spin, for $T = 0.25$.

We have found that both the number of peaks (in agreement with full RSB), and the separation between peaks (in agreement with infinitely disparate time-scales) seem to grow with increasing system size and/or time. The total fraction of the slow spins, however, seems to remain finite, which implies that a more precise analytical treatment for the choice of clusters or simulations with larger system sizes and/or times are needed.

DISCUSSION

We have shown that the Parisi solution can be derived from simple physical principles, and can be interpreted as describing a system with an infinite hierarchy of time-scales where a vanishingly small fraction of slow spins act as effective disorder for the faster ones. The block-sizes m_ℓ at level ℓ of the Parisi matrix are found to be the ratio of the effective temperature T_ℓ of that level and the ambient temperature T. Extremization with respect to m_ℓ expresses the fact that for self-consistency the entropy of the spins slower than level ℓ is zero (they are frozen on the time-scale τ_ℓ). It follows from physical considerations (i.e. the absence of heat flow in equilibrium) that $m_\ell \leq 1, \ \forall \ell$. The fact that the fraction of slow spins vanishes, indicates that the cumulative entropy of the slow spins is less than extensive, and hence that the so-called *complexity* is zero (at least in this full-RSB model).

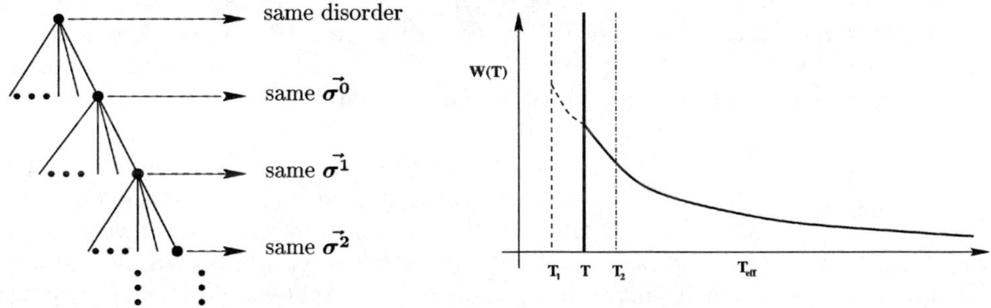

FIGURE 2. Left: the ultra-metric tree, which here is a direct consequence of the hierarchy of time-scales. Right: qualitative picture of the distribution $W(T)$ of (effective) temperatures (time-scales increase with $T_{\rm eff}$) at external temperature T (T_1,T_2, resp.) in the spin glass phase.

Although a more careful treatment of the selection of clusters is obviously required, the main consequences of our interpretation do not crucially depend on it. Firstly, ultra-metricity (see fig. 2a) is a direct consequence of the existence of a hierachy of time-scales. At each level ℓ, the different descendants of a node represent different configurations of the $\vec{\sigma}_{\ell+1}$, which, however, all share the same realisation of disorder and slower spins. Furthermore, a large fraction of the spins (see fig. 2b) evolves at the fastest (microscopic) time-scale at the ambient temperature T, while a small fraction of the spins evolves at (infinitely) slower time-scales at higher effective temperatures. Therefore, cooling to a temperature $T_1 < T$ and heating back to T will leave the spins with $T_{\rm eff} > T$ unchanged, explaining memory effects. On the other hand, after heating to $T_2 > T$ and cooling back to T, the original configuration of the spins with $T \leq T_{\rm eff} \leq T_2$ will be erased, which may explain thermo-cycling experiments [8]. Note that since the spins are discrete variables, finding a small number of flips for a given spin implies long periods of stationarity (persistency) with only short periods of activity (avalanches).

We expect the qualitative features of our picture to survive in short range systems, where the time-scales need not be infinitely disparate due to activated processes. It is as yet unclear whether each level would represent a single cluster or a family of clusters. The origin of the slow time-scales of these clusters can only be understood if they are coupled much stronger internally, than (effectively) to the rest of the system (i.e. a softened version of the completely disconnected clusters which give rise to so-called Griffiths singularities in diluted systems [9]). In short range systems, the clusters would have to be spatially localised, which is in line with the *droplet* picture for short range spin glasses as proposed by Fisher and Huse [10]. The fact that the time-scale of a clusters increases with $T_{\rm eff} - T$, explains why the effective age of a system at a certain T is found to be either older or younger when it spends some time at a $T_1(< T)$, $T_2(> T)$ respectively.

A more careful treatment of the selection of clusters is clearly needed (and currently been carried out), both for full- and 1-RSB models. This may allow us to calculate the complexity in such systems. Furthermore, it needs to be investigated whether slow clusters survive above the thermodynamical (spin) glass temperature $T_{\rm sg}$. Our results suggest further numerical experiments for mean field and short range models, concentrating on quantities such as spin flip frequencies, avalanches, spatial correlations, and cluster persistency.

To conclude, we have shown how a scheme based on an autonomous selection of infinitely disparate time-scales can be used successfully to describe the statics of disordered systems with quenched disorder and discrete and/or continuous stochastic variables. It allows us to *derive* the Parisi scheme from simple physical principles, and interpret its ingredients in such a way that it becomes compatible with the droplet picture for short range models.

It is our pleasure to thank F. Ritort and D. Sherrington for critical comments and stimulating discussions.

REFERENCES

1. G. Parisi, Phys. Lett. **73A**, 203 (1979).
2. D. Sherrington and S. Kirkpatrick, Phys. Rev. Lett. **35**, 1792 (1975).
3. H. Sompolinsky, Phys. Rev. Lett. **47**, 935 (1981).
4. L.L. Bonilla, F.G. Padilla and F. Ritort, Physica A **250**, 315 (1998).
5. L.F. Cugliandolo and J. Kurchan, Phys. Rev. Lett. **71**, 173 (1993); Prog. Theor. Phys. Supp. **126**, 407 (1997).
6. T.M. Nieuwenhuizen, Phys. Rev. E **61**, 26, (2000).
7. R.W. Penney, A.C.C. Coolen and D. Sherrington, J. Phys. A **26**, 3681 (1993).
8. For a recent review see e.g. M. Picco, F. Ricci-Tersenghi and F. Ritort, cond-mat/0005541, and references therein.
9. R. B. Griffiths, Phys. Rev. Lett. **23**, 17 (1969).
10. D.S. Fisher and D.A. Huse, Phys. Rev. Lett. **56**, 1601 (1986).

Statistical Mechanics Models for Jamming in Granular Media

Mario Nicodemi

Imperial College of Science, Technology and Medicine, Department of Mathematics Huxley Building, 180 Queen's Gate, London, SW7 2BZ, U.K.; nicodemi@ic.ac.uk

Abstract. A recently introduced statistical mechanics approach to granular materials is reviewed. Using a lattice model from standard statistical mechanics, within a mean field replica approach and Monte Carlo simulations, we characterise the jamming transition of granular media. In the present unifying framework it is possible to explain a variety of properties of granular materials, ranging from the logarithmic relaxation of granular compaction, typical "memory" phenomena, and anomalous density fluctuations to segregation effects.

INTRODUCTION

Granular packings are media composed of a large number of single grains of typical size above $1\mu m$. They show reproducible macroscopic behaviour whose general properties are not material specific and which are statistically characterised by very few control parameters, such as their density, load or amplitude of external drive [1]. As in the standard systems of statistical mechanics, each of their macroscopic states corresponds to a huge number of microstates. Granular media are "non-thermal" systems since thermal energy is not relevant for these system; however, thermal motion can be replaced by agitation induced by shaking or other driving (an important experimental parameter is $\Gamma_{ex} = a/g$, where a is the shake peak acceleration and g the gravity constant). The introduction of statistical mechanics to describe these systems was, actually, suggested by Edwards more than ten years ago [2].

FRUSTRATED MODELS FOR GRANULAR MEDIA

The full description of the degrees of freedom of grains in a packing is currently a very difficult problem from both a technical and fundamental point of view. We proposed a model [3] which drastically reduces the relevant degrees of freedom of the grains, but retains an essential ingredient of the dynamics of dense granular

CP553, *Disordered and Complex Systems*, edited by P. Sollich, et al.

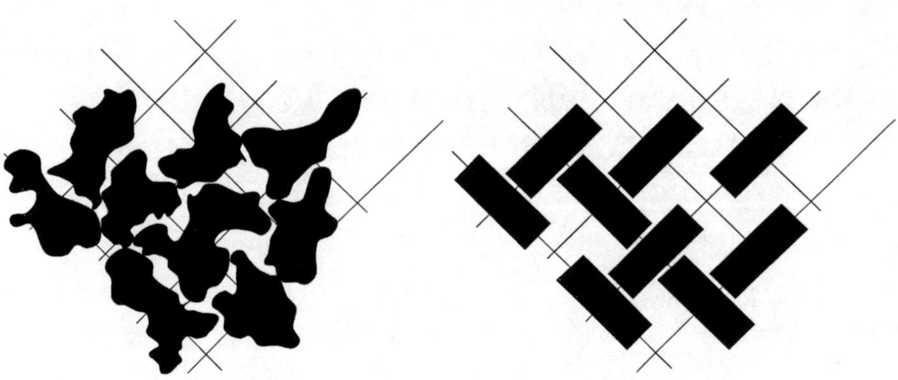

FIGURE 1. A schematic picture of a system of grains and a type of lattice gas like model which we use to describe the grains. The original translational and orientational degrees of freedom of the grains are drastically reduced to those of a lattice system.

media: mechanisms "frustrating" the motion of grains such as steric hindrance, friction, and hard core repulsion.

The centres of mass of the grains are constrained to move on a lattice and their orientational degrees of freedom are assumed to point in the directions of the lattice's main axes (see Fig.1). In the present formulation the hard core repulsion term between grains is easily tractable, and the minimal system Hamiltonian is very simple:

$$H = J \sum_{\langle ij \rangle} f_{ij}(S_i, S_j) n_i n_j - g \sum_i n_i z_i. \tag{1}$$

Here $n_i = 0, 1$ are occupancy variables describing the location of grains on the lattice, $S_i = \pm 1$ are "spin" variables associated with the orientations of the particles, J (with $J \to \infty$) represents the hard core potential, g the gravity constant (masses are set to unity) and z_i the height of site i. The hard core shape factor $f_{ij}(S_i, S_j)$ is 0 or 1 depending whether the configuration S_i, S_j is allowed or not when neighbouring sites i and j are both occupied (see Fig.1). The choice of $f_{ij}(S_i, S_j)$ depends on the level of refinement required in each particular model [3].

We have mainly considered two possible cases for f [3]. One model (called Tetris, and shown in Fig.1) has an Ising like coupling (i.e., $f \propto S_i S_j$), and kinetic constraints on the dynamics to enforce geometric frustration mechanisms. Since a real granular system may contain more disorder than the one schematically depicted in the lattice model of Fig.1, another kind of "frustrated lattice gas" was introduced, made of grains moving in a lattice with quenched geometric disorder (corresponding to the disordered environment generated by other grains). Such a model, the Ising Frustrated Lattice Gas (IFLG), is in the category of Ising Spin Glasses models [4].

The dynamics of the model can be thought as a schematic version of a real

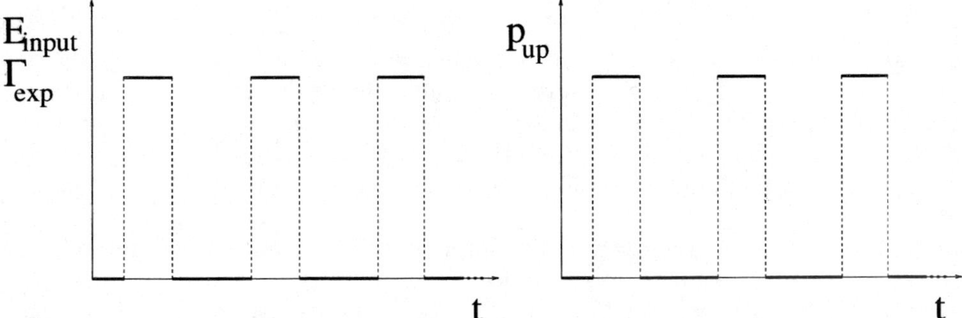

FIGURE 2. Left A schematic diagram of the energy input in a granular pack during a sequence of taps. **Right** Our schematic modelling of the tapping sequence for the lattice model: when the shake is on, particles can also move upwards with a finite probability p_{up}.

"tapping" sequence. Each experimental tap can be divided in two parts: one where the system is supplied with energy and the average kinetic energy of the grains is finite; and a second where, due to dissipation, the system rapidly stops moving. Thus, in our models grains undergo a schematic driven diffusive dynamics: in the absence of vibrations they are subject only to gravity and can only move downwards (always fulfilling the non-overlap condition); the presence of vibration is introduced by also allowing the particles to diffuse upwards with a finite probability p_{up}.

It is easy to show that the above dynamics is equivalent to cyclically putting the Hamiltonian system in contact with a *thermal bath* of temperature zero when "vibrations are off" and of temperature $T/g = 2/[\ln(1 - p_{up}) - \ln(p_{up})]$ when "vibrations are on". $\Gamma \equiv T/g$ is the theoretical quantity related to the experimental tap vibrations intensity, Γ_{ex}.

Notice that T is the temperature of the bath and only in some cases becomes the system equilibrium temperature. Now, an other important question can be asked: is it possible to describe the rest state of the grains (i.e., with a bath at $T = 0$) in terms of a thermodynamic quantity such as an "effective temperature"? This was the point raised by Edwards, who termed such an "effective temperature" *compactivity*, X. The problem of the existence of X and its relations with Γ are dealt with elsewhere [5]. Here I just mention that the statistical mechanics of powders in their rest or "frozen" states (i.e., in absence of driving) is in strict correspondence with the problem of the properties of *inherent structures* in glass formers [6].

A FEW RESULTS

The above models reproduce the logarithmic relaxation of granular compaction, "reversible-irreversible" lines, size segregation phenomena, density fluctuations, and many other properties of granular materials in good agreement with experimental data [3]. In particular, we showed that the slow relaxation, the presence of "memory" effects, and other similar phenomena are related to an underlying "jamming" transition, which can be defined in a similar way to the glassy transition in real glass formers (see Figs. 3 and 4). I recall that it has been shown in a mean field approximation [4] and numerically for finite dimensional systems [3], that the IFLG model, in absence of gravity, exhibits a spin glass (SG) transition at high density (or low temperature) similar to the one found in the p-spin model. At low Γ the system is frozen in a SG-like phase, but at higher Γ it separates in a frozen SG phase at the bottom and a fluid phase on top. Above a critical Γ the system is entirely fluid.

In particular, in the high density region, at ρ_m, the models undergo a structural arrest where the self-diffusivity of the grains vanishes. This suppression of self-diffusion at ρ_m signals that, above such a density, it is impossible to obtain a macroscopic rearrangement of grains without increasing the system volume, a fact interestingly similar to the phenomenon of the Reynolds dilatancy transition [3].

An important question, experimentally still open, is the possibility of formulating a link between the system response and fluctuations, i.e., the analogue of a "fluctuation dissipation relation" (FDR) for the present non-thermal systems. This perspective was explored in Ref. [7] in the context of the above models, where a generalised version of the FDR was found. In typical situations the FDR coincides neither with its usual version for equilibrium thermal systems nor with its extensions valid in off equilibrium thermal systems in the so called "small entropy production" limit [8]. The origin of universalities in off equilibrium dynamics in the above heterogeneous classes of materials is an important issue.

CONCLUSIONS

The schematic description of granular media in terms of ordinary Hamiltonian systems of statistical mechanics, originally proposed by Edwards and sketched above, establishes a general framework within which a variety of seemingly different properties of these systems can be understood. These range from their "jamming" transition with slow compaction, to "memory" effects and many others [3,4,7].

REFERENCES

1. H.M. Jaeger, S.R. Nagel and R.P. Behringer, Rev. Mod. Phys. **68**, 1259 (1996).

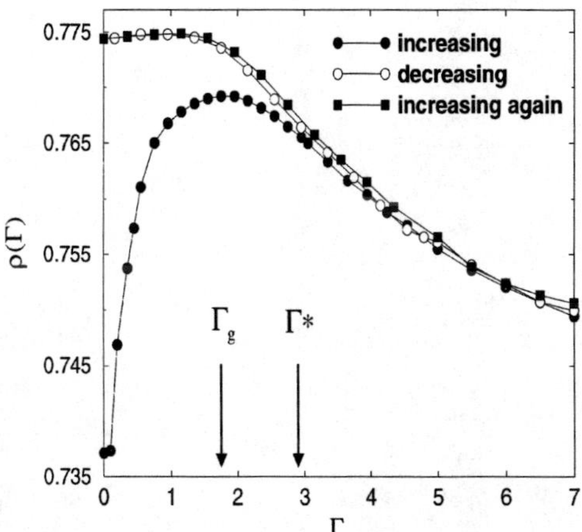

FIGURE 3. The static bulk density, $\rho(\Gamma)$, of the IFLG model as a function of the vibration amplitude, Γ, in cyclic vibration sequences. The system is shaken with an amplitude Γ which at first is increased (filled circles), then is decreased (empty circles) and, finally, increased again (filled squares) with a given "annealing-cooling" velocity $\gamma \equiv \Delta\Gamma/\tau$. Here we fixed $\gamma = 1.25 \cdot 10^{-3}$. The data compare rather well with the experimental data. Γ^* is approximately the point where the "irreversible" and the "reversible" branches meet. Γ_g signals the location of a "jamming transition".

2. S.F. Edwards and R.B.S. Oakeshott, Physica A **157**, 1080 (1989). S.F. Edwards, in *Granular Matter: an interdisciplinary approach*, A. Mehta ed. (Springer-Verlag, New York, 1994).
3. A. Coniglio and H.J. Herrmann, Physica A **225**, 1 (1996). M. Nicodemi, A. Coniglio, H.J. Herrmann, Phys. Rev. E **55**, 3962 (1997); *ibid.* **59**, 6830 (1999). E. Caglioti, H.J. Herrmann, V. Loreto, M. Nicodemi, Phys. Rev. Lett. **79**, 1575 (1997).
4. J.J. Arenzon, M. Nicodemi and M. Sellitto, J. Physique I **6**, 1 (1996). J.J. Arenzon, J. Phys. A, **32**, L107 (1999).
5. A. Coniglio and M. Nicodemi, preprint (1999).
6. F.H. Stillinger and T.A. Weber Phys. Rev. A **25**, 978 (1982). S. Sastry, P.G. Debenedetti and F.H. Stillinger, Nature **393**, 554 (1999). F. Sciortino, W. Kob and P. Tartaglia, Phys. Rev. Lett. **83**, 3214 (1999).
7. M. Nicodemi, Phys. Rev. Lett. **82**, 3734 (1999).
8. L.F. Cugliandolo, D. Dean, J. Kurchan, Phys. Rev. Lett. **79**, 2168 (1997).

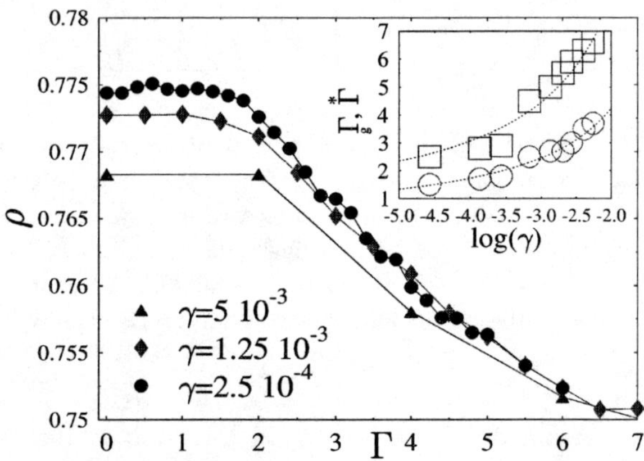

FIGURE 4. *Main frame:* As in Fig.3, the density, $\rho(\Gamma)$, is plotted as a function of the vibration amplitude, Γ for three values of the "cooling" velocity γ. Here only the descending parts of the cycles are shown. As in glasses, a too fast cooling drives the system out of equilibrium. The position of the shoulder, $\Gamma_g(\gamma)$, schematically individuates a "jamming transition". *Inset:* numerical estimate, in the IFLG model of the dependence of $\Gamma_g(\gamma)$ (circles) and $\Gamma^*(\gamma)$ (squares) on the cooling rate γ. Superimposed are logarithmic fits in analogy to those for the glass transition temperature, $T_g(\gamma)$, in glasses [4]. When $\gamma \to 0$ we roughly find $\Gamma_g(0) = \Gamma^*(0)$.

Absence of Replica Symmetry Breaking in a Region of the Phase Diagram of the Ising Spin Glass

Hidetoshi Nishimori* and David Sherrington†

*Department of Physics, Tokyo Institute of Technology
Oh-okayama, Meguro-ku, Tokyo 152-8551, Japan
†Department of Physics, University of Oxford
Theoretical Physics, 1 Keble Road, Oxford OX1 3NP, UK

Abstract. We prove that the distribution functions of magnetization and spin glass order parameter coincide on the Nishimori line in the phase diagram of the $\pm J$ Ising model in any dimension. This implies absence of replica symmetry breaking because the distribution function of magnetization consists only of two delta functions, suggesting the same simple structure for the distribution of the spin glass order parameter. It then follows that the mixed (glassy) phase, where ferromagnetic order coexists with a complex phase space structure, should lie below the Nishimori line if it exists at all. We also argue that the AT line that marks the onset of RSB with a continous distribution of the spin glass order parameter (again, if it exists) would start with an infinite slope from the multicritical point where paramagnetic, ferromagnetic and spin glass phases merge.

INTRODUCTION

The existence and characteristics of the spin glass phase are currently being actively investigated for finite-dimensional random spin systems. Closely related is the problem of the mixed (glassy) ferromagnetic phase. If the mean-field picture applies to finite-dimensional systems, the ferromagnetic phase would split up into two regions, one with a simple structure (the replica-symmetric (RS) phase in the mean-field framework) and the other with a complex phase space (replica-symmetry broken (RSB) state). The boundary between these two phases would be the finite-dimensional counterpart of the AT (de Almeida-Thouless) line found for the infinite-range SK (Sherrington-Kirkpatrick) model.

Investigations of these and related problems are currently being carried out almost exclusively by numerical methods because of a lack of reliable analytical methods. We show in the present contribution that simple symmetry arguments lead to a strong constraint on the possible location and shape of the AT line in the

CP553, *Disordered and Complex Systems*, edited by P. Sollich, et al.

phase diagram of the Ising spin glass on an arbitrary lattice with arbitrary range of interactions. More precisely, it is possible to prove that there is nothing like RSB on the Nishimori line in the phase diagram and that the AT line, if it ever exists, should start as a vertical line from the multicritical point where paramagnetic, ferromagnetic and spin glass phases merge. Our method is an application of the gauge theory which has been used to derive exact energy and many other exact/rigorous results.

GAUGE THEORY OF THE ISING SPIN GLASS

Let us consider the $\pm J$ Ising model ($S_i = \pm 1$) with the Hamiltonian

$$H = -\sum_{\langle ij \rangle} J_{ij} S_i S_j, \tag{1}$$

where the sum is over pairs of sites on an arbitrary lattice with arbitrary range of interactions. The exchange interactions are quenched random variables with the distribution function

$$P(J_{ij}) = p\delta(J_{ij} - J) + (1-p)\delta(J_{ij} + J). \tag{2}$$

The Hamiltonian (1) is invariant under gauge transformation

$$J_{ij} \to J_{ij}\sigma_i\sigma_j, \quad S_i \to S_i\sigma_i, \tag{3}$$

where σ_i is an arbitrarily fixed Ising spin.

The gauge invariance of the Hamiltonian leads to a number of exact and/or rigorous results [1,2]. For example, consider the internal energy

$$E(K, K_p) = [\langle H \rangle_K]_{K_p}, \tag{4}$$

where the inner triangular brackets denote a thermodynamic average and the outer square brackets represent the configurational average at a fixed value of p. The suffixes denote the temperature and p through $K = \beta J = J/k_B T$ and $K_p = \frac{1}{2}\log p/(1-p)$. The exact value of the internal energy can be calculated explicitly under the constraint $K_p = K$ to give

$$E(K, K) = -N_B J \tanh K, \tag{5}$$

where N_B is the number of bonds on the lattice under consideration. The constraint $K = K_p$ defines a curve in the phase diagram shown dashed in Figure 1, which is often called the Nishimori line.

There are many other results including an upper bound on the specific heat

$$k_B T^2 C(K, K) \le \frac{J^2 N_B}{\cosh^2 K}, \tag{6}$$

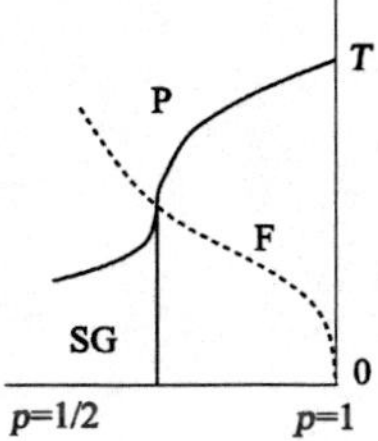

FIGURE 1. Phase diagram of the $\pm J$ model and the Nishimori line (dashed).

a correlation identity

$$[\langle S_i S_j \rangle_K]_K = \left[\langle S_i S_j \rangle_K^2\right]_K, \tag{7}$$

and a correlation inequality

$$[\langle S_i S_j \rangle_K]_{K_p} \leq \left[\left|\langle S_i S_j \rangle_{K_p}\right|\right]_{K_p}. \tag{8}$$

The correlation identity (7) means that the Nishimori line ($K_p = K$) avoids the spin glass phase because the left-hand side reduces to the square of the ferromagnetic long range order m^2 as the distance between i and j tends to infinity whereas the right-hand side approaches the square of the spin glass order parameter q^2 in the same limit. The spin glass phase is characterized by $q > 0, m = 0$ which is inconsistent with the relation $m = \pm q$. The correlation inequality (8) places a constraint on the possible shape of the boundary between the ferromagnetic and spin glass phases: This boundary can either be vertical or re-entrant in the phase diagram. The ferromagnetic phase is not allowed to lie below the spin glass phase [1,2].

DISTRIBUTION OF ORDER PARAMETERS

The gauge theory summarized above can be applied to derive an identity between the distribution functions of the ferromagnetic and spin glass order parameters on the Nishimori line [2,3].

Definitions and the main result

The distribution function of magnetization is defined as

$$P_m(x;K) = \left[\frac{\sum_S \delta(x - N^{-1}\sum_i S_i)e^{-\beta H(S)}}{\sum_S e^{-\beta H(S)}}\right]_{K_p}. \tag{9}$$

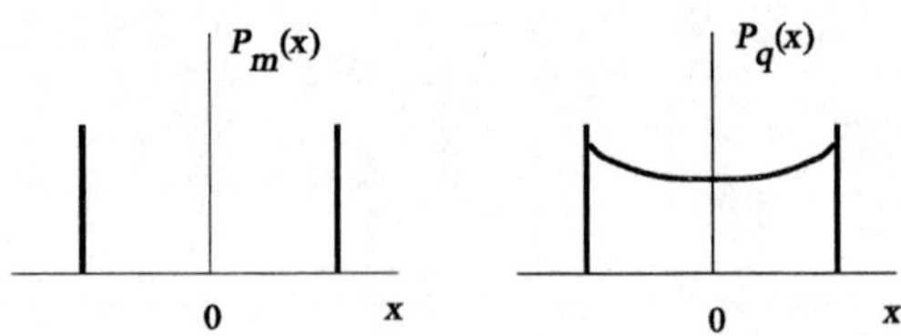

FIGURE 2. Distribution of magnetization (left) and the SG order parameter (right).

To define the distribution function of the spin glass order parameter, it is convenient to introduce two replicas of the same system with spins $\{S_i\}$ and $\{\sigma_i\}$ and the couplings K_1 and K_2, respectively:

$$P_q(x; K_1, K_2) = \left[\frac{\sum_S \sum_\sigma \delta(x - N^{-1} \sum_i S_i \sigma_i) e^{-\beta_1 H(S)} e^{-\beta_2 H(\sigma)}}{\sum_S \sum_\sigma e^{-\beta_1 H(S)} e^{-\beta_2 H(\sigma)}} \right]_{K_p}, \tag{10}$$

where $K_1 = \beta_1 J$ and $K_2 = \beta_2 J$. The distribution of the spin glass order parameter is defined in terms of two replicas with the same temperature, $P_q(x; K) \equiv P_q(x; K, K)$. Note that the distributions $P_m(x; K), P_q(x; K, K_p)$ and $P_q(x; K)$ are all functions of p as well.

It is well established that the distribution of magnetization has a simple structure with two delta functions as depicted in the left part of Figure 2. The spin glass order parameter, on the other hand, has a more complex structure. For example, if there exists full RSB, the distribution has a continuous part as depicted schematically in the right part of Figure 2. Or, it may have delta peaks at a few additional locations in the case of finite-step RSB. In any case, the complexity of the system is closely related to the structure of the distribution function $P_q(x)$ being different from that of $P_m(x)$.

Our main result is the following relation:

$$P_m(x; K) = P_q(x; K, K_p) \tag{11}$$

which holds for any combination of x, K and K_p. In particular, if we set $K_p = K$, Equation (11) reduces to the relation between the usual distribution functions of magnetization and spin glass order parameter:

$$P_m(x; K_p) = P_q(x; K_p). \tag{12}$$

This immediately proves that the distribution function of the spin glass order parameter on the right-hand side has a simple structure represented by the magnetization distribution on the left-hand side if $K = K_p$.

The simplicity of the exact energy (5) and a few other results [2,1] suggest that the phase space of the system would not be complicated. The above result is regarded as definite evidence to support this anticipation.

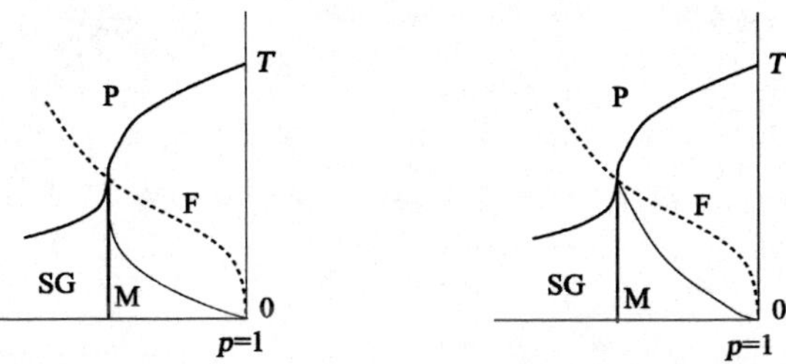

FIGURE 3. Phase diagrams compatible (left) and incompatible (right) with the relations proved in the text when the mixed phase is characterized by a continuous part in $P_q(x)$.

It also follows from differentiation of Equation (11) that the derivatives of the distribution functions $P_m(x; K)$ and $P_q(x; K)$ with respect to K and p vanish when $K = K_p$:

$$\frac{\partial}{\partial K} P_q(x; K) = 0, \quad \frac{\partial}{\partial p} P_q(x; K) = 0 \tag{13}$$

for almost all x when $K = K_p$. These relations have been derived from the fact that the distribution of magnetization $P_m(x; K)$ consists of just two delta functions and therefore its derivatives with respect to the parameters K and p vanish for almost all x. Equation (13) then follows from (11).

The relation (13) shows that the distribution function of the spin glass order parameter remains vanishing at almost all x if the parameters K and p deviate infinitesimally from their values satisfying $K = K_p$. Thus the structure of $P_q(x; K)$ does not develop a continous part under an infinitesimal deviation from the Nishimori line. An interesting consequence is that the mixed phase with a continuous part in $P_q(x)$, if it starts to exist at the multicritical point, does not have a finite extension under a very small deviation from the multicritical point. Stated otherwise, the phase boundaries (the AT line and the boundary between the mixed and spin glass phases) should merge smoothly as they approach the multicritical point as depicted on the left part of Fig. 3. A natural generic situation would be that both of these boundaries are vertical around the multicritical point as in the case of the SK model. It should be noted that this argument does not apply when the RSB in the mixed phase is characterized by several delta functions, in which case a phase boundary with a structure as sketched on the right-hand side of Fig. 3 does not violate (13).

Outline of the proof

The proof of the relation (11) is not very difficult [3]. We start from writing out the configurational average of the distribution function of magnetization

$$P_m(x;K) = \frac{1}{(2\cosh K_p)^{N_B}} \sum_{\{\tau_{ij}=\pm1\}} e^{K_p \sum \tau_{ij}} \frac{\sum_S \delta(x - N^{-1}\sum_i S_i) e^{K\sum \tau_{ij} S_i S_j}}{\sum_S e^{K\sum \tau_{ij} S_i S_j}}. \quad (14)$$

The weight of the configurational average (p or $1-p$ for each bond) is taken care of by the factor $e^{K_p \sum \tau_{ij}}$ [1,2]. The gauge transformation (3) changes the above equation into the form

$$P_m(x;K) = \frac{1}{(2\cosh K_p)^{N_B}} \sum_{\{\tau_{ij}=\pm1\}} e^{K_p \sum \tau_{ij}\sigma_i\sigma_j} \frac{\sum_S \delta(x - N^{-1}\sum_i S_i \sigma_i) e^{K\sum \tau_{ij} S_i S_j}}{\sum_S e^{K\sum \tau_{ij} S_i S_j}}. \quad (15)$$

It is relatively straightforward to see that the right-hand side is equal to $P_q(x; K_p, K)$ by summing it up over all possible $\{\sigma_i\}$, dividing the result by 2^N, and finally inserting the identity $1 = \sum_\sigma e^{K_p \sum \tau_{ij}\sigma_i\sigma_j} / \sum_\sigma e^{K_p \sum \tau_{ij}\sigma_i\sigma_j}$ after the summation symbol over $\{\tau_{ij}\}$.

CONCLUSION

We have proved that the AT line, if it exists, should lie below the Nishimori line in the phase diagram of the $\pm J$ Ising model on an arbitrary lattice. It has also been argued that the mixed phase with a continuous part in the distribution function of the spin glass order parameter, again if it exists, does not have a finite extension around the multicritical point. These results have been derived by a simple application of gauge transformations.

The present work was supported by the Anglo-Japanese Collaboration Programme between the Japan Society for the Promotion of Science and The Royal Society.

REFERENCES

1. Nishimori H., *Prog. Theor. Phys.* **69**, 1169 (1981).
2. Nishimori H., *Spin Glass Theory and Statistical Mechanics of Information*, Iwanami: Tokyo, 1999 (in Japanese).
3. Gillin P., Sherrington D., and Nishimori H., in preparation.
4. Mézard M., Parisi G., and Virasoro M. A., *Spin Glass Theory and Beyond*, Singapore: World Scientific, 1987.

Stochastic Stability

Giorgio Parisi

Dipartimento di Fisica, Sezione INFN and Unità INFM,
Università di Roma "La Sapienza", Piazzale Aldo Moro 2, I-00185 Rome (Italy)

Abstract. In this talk I will introduce the principle of stochastic stability and discuss its consequences both at equilibrium and off-equilibrium.

INTRODUCTION

In this talk I will underline the physical meaning of replica symmetry breaking [1–3]. In this framework I will introduce the principle of Stochastic Stability: I will present its far reaching consequences on the equilibrium properties of the system and on the off-equilibrium behaviour.

THE COEXISTENCE OF MANY PHASES

Usually, if a system has different phases which are separated by a first order transition, just at the phase transition point a very interesting phenomenon is present: phase coexistence. This usually happens if we tune one parameter. This behaviour is summarized by the Gibbs rule which states that, in absence of symmetries, we have to tune n parameters in order to have the coexistence of $n+1$ phases.

In the case of complex systems we have that the opposite situation is valid: the number of phases is very large for a generic choice of parameters. It is usual to assume that all these states are globally very similar: states can be separated only by comparing one state with an other. An example of this phenomenon would be a very long heteropolymer, e.g, a protein or RNA, which folds in many different structures. However quite different foldings may have a very similar density. Of course you will discover that two proteins have folded in two different structures if we compare them.

In order to be precise we should consider a large but (*finite*) system [3]. We want to decompose the phase state in valleys (phases, states) separated by barriers. If the free energy as function of the configuration space has many minima, the number of states will be very large.

CP553, *Disordered and Complex Systems*, edited by P. Sollich, et al.

Let us consider for definitiveness a spin system with N points (spins are labeled by i, which in some cases will be a lattice point). States (labeled by α) are characterized by different local magnetizations: $m_\alpha(i) = \langle\sigma(i)\rangle_\alpha$, where $\langle\cdot\rangle_\alpha$ is the expectation value in the valley labeled by α. The average done with the Boltzmann distribution is denoted as $\langle\cdot\rangle$ and it can be written as linear combinations of the averages inside the valleys. We have the relation:

$$\langle\cdot\rangle \approx \sum_\alpha w_\alpha \langle\cdot\rangle_\alpha \ . \tag{1}$$

We can also write the relation $w_\alpha \propto \exp(-\beta F_\alpha)$, where by definition F_α is the free energy of the valley labeled by α.

In the rest of this talk I will call J the control parameters of the systems. The average over J will be denoted by a bar (e.g. $\overline{F}$). I will consider here the case where a quenched disorder is present: the variables J parametrize the quenched disorder.

THE OVERLAP AND ITS PROBABILITIES

We have already remarked that states may be separated by making a comparison among them. To this end it is convenient to consider their mutual overlap. Given two configurations, we define their overlap:

$$q[\sigma,\tau] = \frac{1}{N} \sum_{i=1\ldots N} \sigma(i)\tau(i) \ . \tag{2}$$

The overlap among the states is defined as

$$q(\alpha,\gamma) = \frac{1}{N} \sum_{i=1\ldots N} m_\alpha(i) m_\gamma(i) \approx q[\sigma,\tau] \ , \tag{3}$$

where σ and τ are two generic configurations that belong to the states α and γ respectively.

We define $P_J(q)$ as probability distribution of the overlap q at given J, i.e. the histogram of $q[\sigma,\tau]$, where σ and τ are two equilibrium configurations. Using eq. (1), one finds that

$$P_J(q) = \sum_{\alpha,\gamma} w_\alpha w_\gamma \delta(q - q_{\alpha,\gamma}) \ , \tag{4}$$

where in a finite volume system the delta functions are smoothed. If there is more than one state, $P_J(q)$ is not a single delta function:

$$P_J(q) \neq \delta(q - q_{EA}) \ . \tag{5}$$

If this happens we say that the replica symmetry is broken: two identical replicas of the same system may be in a quite different state. However at zero magnetic

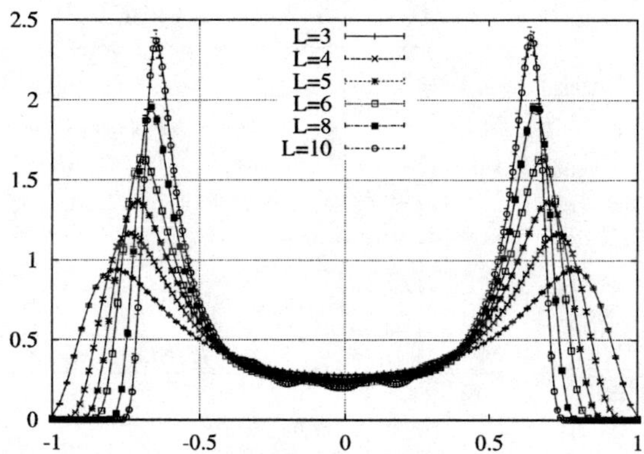

FIGURE 1. The function $P(q)$ after averaging over many samples in four dimensions for system of size L^4, (L=3...10).

field $P_J(q)$ is an even function, so that at low temperature it will contains two delta functions at $|q| \neq 0$.

There are many models where the function $P_J(q)$ is non trivial: a well known example is given by Ising spin glasses [1,4,5]. In this case the Hamiltonian is given by

$$H = -\sum_{i,k} J_{i,k}\sigma_i\sigma_k - \sum_i h_i\sigma_i \ , \tag{6}$$

where $\sigma = \pm 1$ are the spins. The variables J are random couplings (e.g. Gaussian or ± 1) and the variables h_i are the magnetic fields, which may be point dependent.

Let is consider two different models for spin glasses:

a) The Sherrington Kirkpatrick model (infinite range): all N points are connected: $J_{i,k} = O(N^{-1/2})$. Eventually N goes to infinity.

b) Short range models: i belongs to a L^D lattice. The interaction is nearest neighbour (the variables J are or zero or of order 1) and eventually L goes to infinity at fixed D (e.g. $D = 3$).

Analytic studies have been done in the case of the SK model, where one can prove rigorously that the function $P_J(q)$ is non-trivial. In the finite dimensional case an analogous theorem has not been proved and in order to answer to the question if the function $P_J(q)$ is trivial we must resort to some numerical simulations [6] or to experiments.

In this case numerical simulations present ample evidence that in three and four dimensions the function $P_J(q)$ is non-trivial, i.e. at zero magnetic field it is not the sum of two delta functions: this can be seen looking to the average of $P_J(q)$, shown fig. (1) from [7]. We will see later that the the function $P_J(q)$ strongly depends on J.

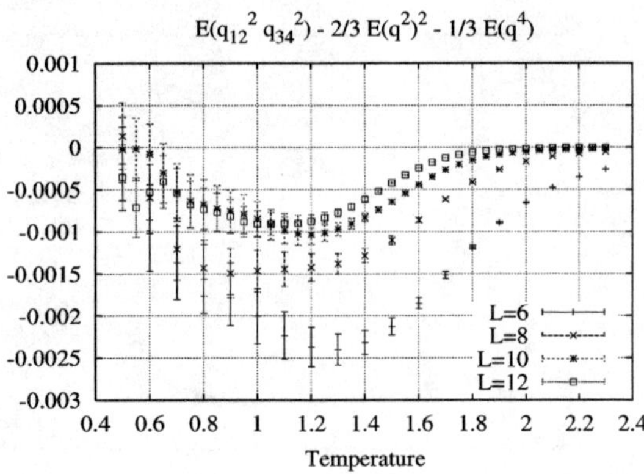

FIGURE 2. The quantity $\overline{\langle q^2\rangle^2} - (\frac{1}{3}\overline{\langle q^4\rangle} + \frac{2}{3}\overline{\langle q^2\rangle}^2)$ as function of the temperature for different values of L in $D = 3$.

STOCHASTIC STABILITY

In the nutshell stochastic stability states that the system we are considering behaves like a generic random system [8–10]. Technically speaking in order to formulate stochastic stability we have to consider the statistical properties of the system with Hamiltonian given by the original Hamiltonian (H) plus a random perturbation (H_R):

$$H(\epsilon) = H + \epsilon H_R \; . \tag{7}$$

Stochastic stability states that all the properties of the system are smooth functions of ϵ around $\epsilon = 0$, after the appropriate averages over the original Hamiltonian and the random Hamiltonian.

Typical examples of random perturbations are (we can chose the value of r in an arbitrary way):

$$H_R^{(r)} = N^{(r-1)/2} \sum_{i_1 \ldots i_r} R(i_1 \ldots i_r)\sigma(i_1) \ldots \sigma(i_r) \; , \tag{8}$$

where for simplicity we can restrict ourselves to the case where the variables R are random uncorrelated Gaussian variables.

It useful to remark that if a symmetry is present, a system cannot be stochastically stable. Indeed spin glasses may be stochastically stable only in the presence of a finite, non zero magnetic field which breaks the $\sigma \leftrightarrow -\sigma$ symmetry. If a symmetry is present, stochastic stability may be valid only for those quantities which are invariant under the action of the symmetry group. It is also remarkable that the union of two non-trivial uncoupled stochastically stable systems is *not* stochastically stable. Therefore a non-trivial stochastically stable system cannot be decomposed as the union of two or more parts whose interaction can be neglected.

In the general case stochastic stability implies that

$$P(q_1,q_2) \equiv \overline{P_J(q_1)P_J(q_2)} = \frac{2}{3}P(q_1)P(q_2) + \frac{1}{3}P(q_1)\delta(q_1-q_2) \ . \tag{9}$$

A particular case of the previous relation is the following one:

$$\overline{\langle q^2\rangle^2} = \frac{1}{3}\overline{\langle q^4\rangle} + \frac{2}{3}\overline{\langle q^2\rangle}^2 \ . \tag{10}$$

We have tested the previous relations in three dimensions as function of the temperature at different values of L [3]. In fig. 2 we plot the quantity $\overline{\langle q^2\rangle^2} - (\frac{1}{3}\overline{\langle q^4\rangle} + \frac{2}{3}\overline{\langle q^2\rangle}^2)$, which should be equal to zero. Indeed it is very small and its values decreases with L. The two quantities in eq. (10) in the low temperature region are a factor of 10^3 bigger of their difference. I believe that there should be few doubts on the fact that stochastic stability is satisfied for three dimensional spin glasses.

OFF-EQUILIBRIUM DYNAMICS

The general problem that we face is to find what happens if the system is carried in a slightly off equilibrium situation. There are two ways in which this can be done.

a) We rapidly cool the system starting from a random (high temperature) configuration at time zero and we wait a time much longer than the microscopical one. The system orders at distances smaller that a coherence distance $\xi(t)$ (which eventually diverges when t goes to infinity) but remains always disordered at distances larger than $\xi(t)$.

b) A second possibility consists in forcing the system in an off-equilibrium state by gently *shaking* it. This can be done for example by adding a small time dependent magnetic field, which should however strong enough force a large scale rearrangement of the system [11].

In the first case we have the phenomenon of ageing. This effects may be evidenziated if we define a two time correlation function and two time relaxation function (we cool the system at time 0) [12,13]. The correlation function is defined to be

$$C(t,t_w) \equiv \frac{1}{N}\sum_{i=1}^{N}\langle\sigma_i(t_w)\sigma_i(t_w+t)\rangle \ , \tag{11}$$

which is equal to the overlap $q(t_w, t_w+t)$ among a configuration at time t_w and one at time t_w+t. The relaxation function $S(t,t_w)$ is a just given by

$$S(t,t_w) = \beta^{-1}\lim_{\delta h\to 0}\frac{\delta m(t+t_w)}{\delta h} \ , \tag{12}$$

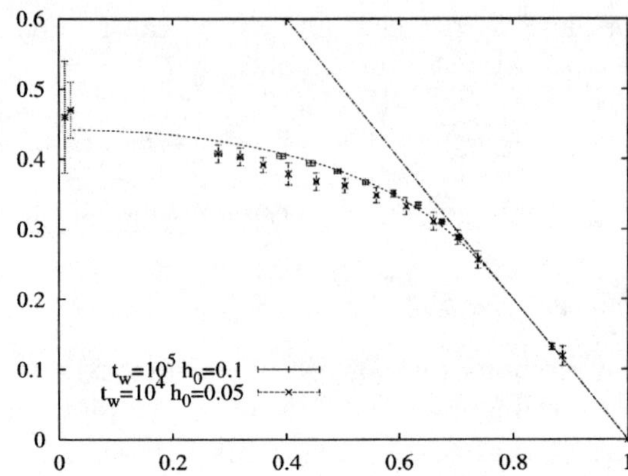

FIGURE 3. Relaxation function versus correlation in the Edwards-Anderson (EA) model in $D = 3$ $T = 0.7 \simeq \frac{3}{4}T_c$ and theoretical predictions from eq. (16) .

where δm is the variation of the magnetization when we add a magnetic field δh starting from time t_w. More generally we can introduce the time dependent Hamiltonian:

$$H = H_0 + \theta(t - t_w) \sum_i h_i \sigma_i \ . \tag{13}$$

The relaxation function is thus defined as:

$$\beta S(t, t_w) \equiv \frac{1}{N} \sum_{i=1}^{N} \langle \frac{\partial \sigma_i(t_w + t)}{\partial h_i} \rangle \ . \tag{14}$$

We can distinguish two situations:

- For $t \ll t_w$ we stay in the *quasi-equilibrium* regime [14], $C(t, t_w) \simeq C_{\text{eq}}(t)$, where $C_{\text{eq}}(t)$ is the equilibrium correlation function; in this case $q_{EA} \equiv \lim_{t\to\infty} \lim_{t_w\to\infty} C(t, t_w)$.
- For $t = O(t_w)$ or larger we stay in the aging regime. If simple aging holds $C(t, t_w) \propto \mathcal{C}(t/t_w)$.

In the equilibrium regime, if we plot parametrically the relaxation function as function of the correlation, we find that

$$\frac{dS}{dC} = -1 \ , \tag{15}$$

which is a compact way of writing the fluctuation-dissipation theorem.

Generally speaking the fluctuation-dissipation theorem is not valid in the off-equilibrium regime. In this case one can use stochastic stability to derive a relation among statics properties and the form of the function $S(C)$ measured in off-equilibrium experiments [12,13,15]:

$$-\frac{dS}{dC} = \int_0^C dqP(q) \equiv X(C) \ . \tag{16}$$

The validity of these relation has been intensively checked in numerical experiments (see for example fig. (3) from [16]).

In spin glasses the relaxation function has been experimentally measured many times in the aging regime, while the correlation function has not been measured: it is a much more difficult experiment in which one has to measure thermal fluctuations. Fortunately enough measurements of both quantities for spin glasses are in now progress. It would be extremely interesting to see if they agree with the theoretical predictions.

REFERENCES

1. M. Mézard, G. Parisi and M. A. Virasoro, *Spin Glass Theory and Beyond* (World Scientific, Singapore, 1987).
2. G. Parisi, *Field Theory, Disorder and Simulations* (World Scientific, Singapore, 1992).
3. E. Marinari, G. Parisi, F. Ricci-Tersenghi, J. Ruiz-Lorenzo, F. Zuliani, J. Stat. Phys. **98** 973 (2000).
4. K. Binder and A. P. Young, Rev. Mod. Phys. **58** 801 (1986).
5. K. H. Fischer and J. A. Hertz, *Spin Glasses* (Cambridge U. P., Cambridge 1991).
6. For a recent review see E. Marinari, G. Parisi and J. J. Ruiz-Lorenzo, in A. P. Young (ed.), *Spin Glasses and Random Fields* (World Scientific, Singapore, 1998).
7. E. Marinari, F Zuliani, J. Phys. A **32** 7447 (1999).
8. F. Guerra, Int. J. Mod. Phys. B **10** 1675 (1997).
9. M. Aizenman and P. Contucci, J. Stat. Phys. **92** 765 (1998).
10. G. Parisi, *On the probabilistic formulation of the replica approach to spin glasses*, cond-mat/9801081, submitted to Europhys. J..
11. L. F. Cugliandolo, J. Kurchan and L. Peliti, Phys. Rev. **E55** 3898 (1997).
12. L. F. Cugliandolo and J. Kurchan, Phys. Rev. Lett. **71** 173 (1993), J. Phys. A **27** 5749 (1994).
13. S. Franz and M. Mézard, Europhys. Lett. **26** 209 (1994); Physica A **210** 48 (1994).
14. S. Franz and M. Virasoro, J. Phys. A **33** 891 (2000).
15. S. Franz, M. Mézard, G. Parisi and L. Peliti, Phys. Rev. Lett. **81** 1758 (1998); J. Phys. A (in press).
16. E. Marinari, G. Parisi, F. Ricci-Tersenghi and J. Ruiz-Lorenzo, J. Phys. A **31** 2611 (1998).

Configurational Entropy and the One-Step RSB Scenario in Glasses

A. Crisanti* and F. Ritort†

* *Dipartimento di Fisica, Università di Roma "La Sapienza", and Istituto Nazionale Fisica della Materia, Unità di Roma I P.le Aldo Moro 2, I-00185 Roma, Italy.*
† *Physics Department, Faculty of Physics University of Barcelona, Diagonal 647, 08028 Barcelona, Spain*

Abstract. In this talk we discuss the possibility of constructing a fluctuation theory for structural glasses in the non-equilibrium aging state. After reviewing well known results in a toy model we discuss some of the key assumptions which support the validity of this theory, in particular the role of the configurational entropy and its relation to the effective temperature. Recent numerical results for mean-field finite-size glasses agree with this scenario.

THE GLASS STATE

A theoretical understanding of the physical mechanisms behind the glass transition remains an open problem [1]. What is commonly understood as a glass (for instance, window glass) is a metastable phase, with free energy higher than that of the crystal, obtained by the continuation of the liquid line below the melting transition temperature T_M. Such a glassy state may be experimentally achieved by cooling the liquid fast enough. Glasses share physical properties with both liquids and solids, making it difficult to decide in which phase (if any) they are. On the one hand, atom positions are randomly located in the glass much as in a liquid, and apparently there is no long-range order. On the other hand, under compression glasses behave like solids, showing a very low mobility. Glasses thus constitute a state of matter in between liquids and solids, not completely classifiable as either of them. They show the following generic features:

- *The viscosity anomaly.* Glasses show a very rapid increase of the viscosity η in a relatively narrow range of temperatures. For instance, in window glass the viscosity increases over nearly 20 orders of magnitude by changing the temperature in a range within 10% of the value of the melting transition temperature. This increase is often well fitted by the Vogel-Tamman-Fulcher law, $\eta \sim \exp(\Delta/(T-T_0))$ where T_0 and Δ are free parameters. Fragile glasses

CP553, *Disordered and Complex Systems*, edited by P. Sollich, et al.

are those where T_0 is finite while for strong glasses T_0 is small (compared to T_M) and the viscosity displays Arrhenius behavior. By convention the glass transition temperature T_g is defined such that $\eta(T_g) = 10^{13}$ Poise which for many liquids corresponds to a relaxation time of several minutes .

- *Two-step relaxation.* In the supercooled regime glasses show relaxation functions with a characteristic two-step form. Intermediate scattering functions show a first decay to a *plateau* (β-process) followed by a secondary and slower relaxation (α-process) which defines the longest activated time. Although both process are activated their characteristic times turn out to be well separated below the mode-coupling transition temperature.
- *Aging.* Suppose a glass is quenched to a temperature T_f below the glass transition temperature T_g. If one were able to measure the viscosity as a function of time one would observe that it grows with time t according to the approximate law: $\eta \sim \eta_0 + at$ where η_0 is the initial value reached soon after the quenching and a is a temperature dependent parameter $a \sim \exp(-B/T_f)$ where B is an activation barrier. This growth can be extremely slow since a can be very small depending on the value of the ratio B/T_f. The growth of the viscosity manifests itself as a dependence of intermediate scattering functions on the time of the measurement (usually called *waiting time* t_w), with the longest decorrelation time depending on the value of the viscosity η at t_w. Such a dependence on absolute time of the response of the system to an external perturbation is commonly referred as *aging.*

A SIMPLE SOLVABLE MODEL OF A GLASS

Before describing the key assumptions for understanding glasses it is convenient to consider an instructive example. One of the simplest solvable models which shows key features of structural glasses is the oscillator model with parallel dynamics introduced by one of us in [2]. This model contains some essential features of non-equilibrium thermodynamics applied to glassy systems. The model is defined by a set of N non-interacting harmonic oscillators with energy

$$E = \frac{K}{2} \sum_{i=1}^{N} x_i^2, \qquad -\infty < x_i < \infty \tag{1}$$

where K is a coupling constant. An effective interaction between oscillators is introduced through a parallel Monte Carlo dynamics characterized by small jumps $x_i \to x_i' = x_i + \frac{r_i}{\sqrt{N}}; 1 \le i \le N$ where the variables r_i are extracted from a Gaussian distribution $P(r) = \frac{1}{\sqrt{2\pi\Delta^2}} \exp(-r^2/(2\Delta^2))$. At each Monte Carlo step all oscillators are updated following the previous rule and the move is accepted according to the Metropolis algorithm with probability $W(\Delta E) = \min[1, \exp(-\beta\Delta E)]$ where $\Delta E = E(\{x'\}) - E(\{x\})$. The smallness of the jumps in the variables x_i is required in order to give a finite change in the energy so the acceptance does not vanish in the $N \to \infty$ limit.

This model (as well as some modifications proposed afterwards [3]) has been extensively studied and shows the following behavior. Because of the finiteness of the moves the ground state cannot be reached in a finite amount of time. Despite the absence of interactions in the Hamiltonian, the Monte Carlo dynamics induces entropy barriers corresponding to flat directions in energy space that the system hardly ever finds when the acceptance rate is low. A simple calculation at finite temperature shows activated behavior for the relaxation time, $\tau_{relax} \sim \exp(\frac{K\Delta^2}{8T})$, despite the absence of energy barriers in the model (the potential is a single well in N dimensions). This makes the dynamics of this model quite reminiscent of (but simpler than) that of the Backgammon model [4].

The interesting dynamics in this model is found when studying the relaxation after quenching to zero temperature. In what follows we summarize the main findings for this case [2]:

- *Slow decay of the energy:* The evolution equation for the energy is Markovian. This simplicity allows for an asymptotic large-time expansion. The energy asymptotically decays logarithmically, $E(t) \sim 1/\log(t)$, and the acceptance ratio decays faster, $A(t) \sim 1/(t \log(t))$.

- *The effective temperature.* The fluctuation-dissipation (FDT) ratio [5] can be exactly computed and depends only on the smaller of its two time arguments, s. At zero temperature one gets in the large s limit,

$$T_{\text{eff}}(s) = \frac{\partial C(t,s)/\partial s}{G(t,s)} \to \frac{2E(s)}{N} \tag{2}$$

  i.e., equipartitioning is obeyed off-equilibrium thus allowing an effective temperature to be defined in terms of the FDT ratio.

- *The role of the configurational entropy.* The effective temperature previously obtained allows for the definition of a time dependent configurational entropy. At time t the number of configurations explored by the system is given by the surface of an N-dimensional sphere of radius $R = E^{\frac{1}{2}}$ where E is the energy at time t. The number of configurations is then given by $\Omega \sim R^{N-1} = E^{\frac{N-1}{2}}$ leading to an extensive configurational entropy $S_c = \log(\Omega) = \frac{N}{2}\log(E)$ which satisfies, using (2), the canonical thermodynamic relations,

$$\frac{\partial S_c}{\partial E} = \frac{N}{2E} = \frac{1}{T_{\text{eff}}} \tag{3}$$

- *Heat-flow driven by the effective temperature.* Following [6] we can obtain the resulting heat-current by coupling two identical harmonic oscillator systems with a term $\epsilon x_i y_i$ and measuring the energy flow to order ϵ^2. One gets [7],

$$J(t) = \lim_{t \to s} \left(\frac{\partial Q(t,s)}{\partial t} - \frac{\partial Q(t,s)}{\partial s} \right) \tag{4}$$

with $Q(t,s) = \frac{1}{N}\sum_{i=1}^{N} x_i(t) y_i(s)$. If one of the two systems acts as a thermometer with characteristic frequency $\omega \gg 1/t$ then the measured temperature coincides with the effective temperature and the Onsager relation is satisfied,

$$J = L_{QQ} \nabla \left(\frac{1}{T}\right) \tag{5}$$

with $L_{QQ} \simeq 1/(t \log(t))$, and $\lambda \simeq \log(t)/t$ for the thermal conductivity [7] in the Fourier law. This example generalizes the Fourier law (and by extension, all linear relations between fluxes and forces initially derived in the vicinity of equilibrium [8]) to the case where fluctuations are present around the off-equilibrium aging state.

THE CANONICAL THERMODYNAMIC SCENARIO FOR GLASSES

In the previous example we have seen how equipartitioning in the off-equilibrium state is satisfied when the acceptance rate is small. This can be easily understood: at zero temperature only configurations with the same or lower energy are accepted. Because the size of the moves Δ is finite the system samples all the configurations of equal or lower energy with the same probability. Given that their number is a monotonically increasing function of the energy and because the energy cannot increase at $T = 0$ we conclude that the relevant fluctuations are those which explore the constant energy surface. After quenching to $T = 0$ and for long times, the probability to explore a different state with energy $E < E^* = E(t)$ is uniform for all states with a given energy and given by,

$$P \sim \Omega(E) = \exp(N\, S_c(E)) \sim \exp\left(\frac{N}{2}\log(E)\right) =$$
$$\exp\left(\frac{N}{2}\log(E^*) + \frac{N}{2E^*}(E - E^*)\right) \propto \exp(\beta_{\text{eff}}(E - E^*))\, \Theta(E^* - E) \quad . \tag{6}$$

Fluctuations are then described in terms of an effective temperature given by the configurational entropy in eq.(3). In the last years Th. M. Nieuwenhuizen, inspired by results for this oscillator model as well as for the p-spin spherical spin glass, has proposed that a similar scenario could apply for generic glassy systems [9]. Although it is not clear in what conditions these results are valid for realistic systems the idea is suggestive enough to merit a more detailed investigation. A more elaborate picture along the same line has been presented by Franz and Virasoro who, using well known results from spin glasses and assuming a separation of timescales, have proposed that a modified version of the Onsager hypothesis should bring the configurational entropy into play [10]. Although their analysis is based on some mean-field models it can probably be extended to realistic systems under some fairly mild and general assumptions.

The idea goes as follows. Imagine a glass that is obtained by quenching the liquid to a temperature T_f below T_g. As already discussed, one of the main experimental facts regarding relaxational processes (for which mode-coupling theory gives a fairly good description in a large range of temperatures) is the emergence of well separated timescales (the α and β processes). Dynamics can then be viewed as if the system were jumping among different metastable states or *basins*, each basin defined as the set of configurations explored by the system during a time t^* of the order of the relaxation time itself. The definition of the precise value of t^* does not matter as soon as the timescales separation becomes very strong, a condition which emphasizes the robustness of the present scenario. Fluctuations around a given configuration have two contributions: 1) Thermal fluctuations *within* a basin, i.e., typical fluctuations occurring on a timescale less than t^* and driven by the temperature T_f and 2) Activated jumps between different basins typically occurring for times larger than t^*. If basins with a given free energy are more or less similar (they are neither too big nor too small) then their typical lifetime is of order t^* implying that the probability to visit them at subsequent times $t > t^*$ is simply proportional to their number $\exp(S_c(F))$ where $S_c(F)$ defines the configurational entropy,

$$P(F) \propto \Omega(F) = \exp\Big(S_c(F)\Big) = \exp\Big(S_c(F^*) + (\frac{\partial S_c(F)}{\partial F})_{F=F^*}(F-F^*) \\ +O(F-F^*)^2\Big) \propto \exp\Big((\frac{\partial S_c(F)}{\partial F})_{F=F^*}(F-F^*)\Big) \tag{7}$$

where F^* is the *threshold* value of the free energy, i.e., the free energy of the basins at time t. Note that the present threshold we define here is conceptually different from that defined in mean-field p-spin models since now F^* is a time dependent quantity. From (7) we can infer that the probability distribution for basin-to-basin fluctuations is given by the configurational entropy $S_c(F)$ which satisfies the relation,

$$\frac{1}{T_{\text{eff}}} = \Big(\frac{\partial S_c(F)}{\partial F}\Big)_{F=F^*} \tag{8}$$

and defines an effective temperature for interbasin fluctuations. The essentials of the physics contained in eq.(7) are the same as that contained in eq.(6) for the oscillator model with the difference that in the oscillator model fluctuations above the threshold E^* are forbidden because $T_f = 0$ while now fluctuations are dominated by jumps to basins with free energies around the threshold F^* and differing by a finite amount. We must stress that eq.(7) gives a good description of the fluctuations when T_f is much smaller than T_{eff}, otherwise thermal activation driven by T_f will modify the simple entropic contribution entering in (7). Biroli and Kurchan [11] have recently addressed the notion of configurational entropy in the framework of the Gaveau-Schulmann formalism showing that S_c is a time-dependent quantity; let us call it $S_c^*(F)$. This is in complete agreement with the present scenario: basins with free energies F above the time-dependent *threshold* F^* are completely washed out as time goes leading to the appearance of a time-dependent cutoff: $S_c^*(F) = S_c(F)$ for $F < F^*$ and $S_c^*(F) = -\infty$ for $F > F^*$, with $S_c(F)$ a monotonically increasing function of F. This keeps the full meaning of (8).

The conclusion of this discussion is that, as soon as the formula (7) gives a faithful description of basin-to-basin fluctuations occurring beyond a timescale t^* (which changes itself as a function of time), then it is possible to write down a dynamical free energy which controls both fast and slow processes:

$$F_{dyn} = F_{basin} - T_{\rm eff}\, S_c(F) \quad . \tag{9}$$

Keeping in mind that both terms contribute to physically different processes occurring in different (well separated) timescales this corresponds to the formula obtained in the framework of p-spin models. We must stress that in models where free energy basins are not *uniformly* sampled the simple results (8,9) cannot hold. Examples of such a class of systems are generic coarsening models with a finite number of absorbing states [12].

This formula has recently been checked in finite-size mean-field models for glasses where activated processes occur with finite probability [13,14]. One can show the presence of an activated regime different from that observed in the $N \to \infty$ limit (where ergodicity breaks below the threshold). To verify the above scenario it is necessary to estimate the configurational entropy. A good method for this is to partition the phase space into basins by counting the number of energy minima. These are also called inherent structures and were proposed by Stillinger and Weber as a powerful tool to investigate the energy 'landscape' of liquids [15]. The above scenario has been verified in the *ROM* model [16] which is a faithful microscopic realization of the random energy model introduced by Derrida [17]. In this model, $F_{basin} \simeq E_{basin}$ because the intra-basin entropy is very small ($q_{EA} \sim 0.96$ for the metastable states) and the effective temperature can then be obtained using eq. (8) with $S_c(F) \simeq S_c(E)$. The results for the configurational entropy and the effective temperature (obtained through FDT plots) are shown in figure (1), panels (a) and (b) respectively. The agreement between equation (8) and the numerics is excellent. Recent results [18] in the case of Lennard-Jones glasses, where the contribution F_{basin} turns out to be important, also confirm these results.

The simplicity of the hypothesis presented here suggests that some generic and simple principles are behind the statistics of fluctuations around the aging state for structural glasses. Surely we will see fast developments in the forthcoming years which will show to what extent these hypotheses are correct.

Acknowledgements We are grateful to A. Garriga for a careful reading of the manuscript.

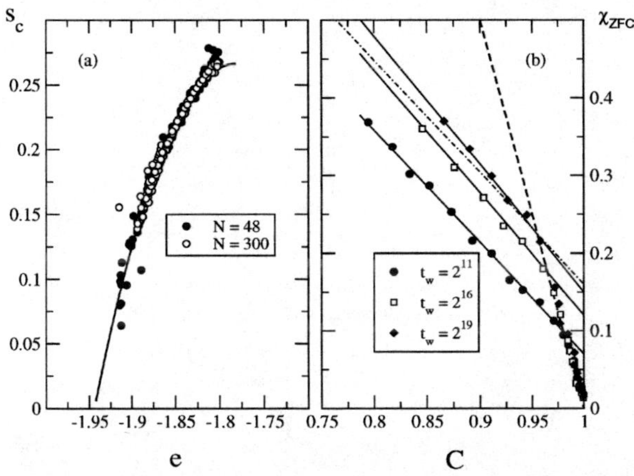

FIGURE 1. (a) Configurational entropy as a function of energy. The data are for temperatures $T = 0.4, 0.5, 0.6, 0.7, 0.8, 0.9$ and 1.0. The line is the quadratic best fit. From Ref. [13]. (b) FDT plot for the ROM model (integrated response function as a function of IS correlation function). The dashed line has slope $\beta_f = \frac{1}{T_f} = 5.0$, while the full lines are the predictions from eq.(8) with $S_c(E)$ taken from Ref. [14]: $T_{\rm eff}(2^{11}) \simeq 0.694$, $T_{\rm eff}(2^{16}) \simeq 0.634$ and $T_{\rm eff}(2^{19}) \simeq 0.608$. The dot-dashed line is the same line for $t_w = 2^{11}$ but shifted.

REFERENCES

1. See the special issue of J. Phys. Condensed Matter **12**, issue number 29 (2000), ed. by S. Franz, S. C. Glotzer and S. Sastry.
2. L.L. Bonilla, F.G. Padilla and F. Ritort, Physica A **250**, 315 (1998).
3. T. M. Nieuwenhuizen, Phys. Rev. Lett. **80**, 5580 (1998); Phys. Rev. E **61**, 267 (2000).
4. F. Ritort, Phys. Rev. Lett. **75**, 1190 (1995).
5. L. F. Cugliandolo and J. Kurchan, J. Phys. A (Math. Gen.) **27**,5749 (1994); S. Franz and M. Mezard, Physica A **210**, 48 (1994).
6. L. F. Cugliandolo, J. Kurchan and L. Peliti, Phys. Rev. E **55**, 3898 (1997).
7. A. Garriga and F. Ritort (unpublished)
8. S. R. de Groot and P. Mazur, *Non-equilibrium thermodynamics*, Dover, 1984.
9. T. M. Nieuwenhuizen, Phys. Rev. Lett. **79**, 1317 (1997); J. Phys. A (Math. Gen.) **31**, L201 (1998).
10. S. Franz and M. A. Virasoro, J. Phys. A (Math. Gen.) **33**, 891 (2000).
11. G. Biroli and J. Kurchan, Preprint cond-mat/**0005499**.
12. A. Crisanti, F. Ritort, A. Rocco and M. Sellitto, Preprint cond-mat/**0006045**
13. A. Crisanti and F. Ritort, Europhys. Lett. **51**, 147 (2000);
14. A. Crisanti and F. Ritort, Preprint cond-mat/**9911226**.
15. F. H. Stillinger and T. A. Weber, Phys. Rev. A **25**, 978 (1982).
16. E. Marinari, G. Parisi and F. Ritort, J. Phys. A (Math. Gen.) **27**, 7647 (1998)
17. B. Derrida, Phys. Rev. Lett. **45**, 79 (1980)
18. F. Sciortino and P. Tartaglia, Preprint cond-mat/**0007208**.

Public Key Cryptography and Error Correcting Codes as Ising Models

David Saad*, Yoshiyuki Kabashima† and Tatsuto Murayama†

*The Neural Computing Research Group, Aston University, Birmingham B4 7ET, UK.
†Dept. of Comp. Intel. & Syst. Sci., Tokyo Institute of Technology, Yokohama 2268502, Japan.

Abstract. We employ the methods of statistical physics to study the performance of Gallager type error-correcting codes. In this approach, the transmitted codeword comprises Boolean sums of the original message bits selected by two randomly-constructed sparse matrices. We show that a broad range of these codes potentially saturate Shannon's bound but are limited due to the decoding dynamics used. Other codes show sub-optimal performance but are not restricted by the decoding dynamics. We show how these codes may also be employed as a practical public-key cryptosystem and are of competitive performance to modern cyptographical methods.

Error-correcting codes are of significant practical importance as they provide mechanisms for retrieving the original message after possible corruption due to noise during transmission. They are being used extensively in most means of information transmission from satellite communication to the storage of information on hardware devices. The coding efficiency, measured in the fraction of informative transmitted bits, plays a crucial role in determining the speed of communication channels and the effective storage space on hard-disks. Rigorous bounds [1] have been derived for the maximal channel capacity for which codes, capable of achieving arbitrarily small error probability, can be found. However, most existing practical error-correcting codes are significantly far from saturating this bound and the quest for more efficient error-correcting codes has been going on ever since.

One family of codes, introduced originally by Gallager [2], and abandoned in favour of other codes due to the limited computing facilities of the time, has recently been re-introduced [3], showing excellent performance with respect to most existing codes. In fact, it has recently been discovered that irregular constructions of Gallager's code result in better performance than any other method [4,5] and nearly saturate Shannon's bound for infinite message size. Gallager-type methods are generally based on the introduction of random sparse matrices for generating the transmitted codeword as well as for decoding the received corrupted codeword. Various decoding methods have been successfully employed; we will mainly focus here on the leading technique of Belief Propagation (BP) [6].

Most studies of Gallager-type codes conducted so far have been carried out via numerical simulations. Some analytical results have been obtained via methods of information theory [3], setting bounds on the performance of certain code types, and by combinatorial/statistical methods [4]. Here we analyze their typical performance for several parameter choices via the methods of statistical physics, and validate the analytical solutions against results obtained by the Thouless-Anderson-Palmer (TAP) approach [7] to diluted systems and via numerical methods.

In a general scenario, the N dimensional Boolean message $\boldsymbol{\xi}$ is encoded to the M dimensional vector $\boldsymbol{J}^0$ which is then transmitted through a noisy channel with flipping probability p per bit (other noise types may also be considered). The received message $\boldsymbol{J}$ is decoded to retrieve the original message.

One can identify several slightly different versions of Gallager-type codes. The one used here, termed the MN code [3] is based on choosing two randomly-selected sparse matrices A and B of dimensionality $M \times N$ and $M \times M$ respectively; these are characterized by K and L non-zero unit elements per row and C and L per column respectively. The finite, usually small, numbers K, C and L define a particular code; both matrices are known to both sender and receiver. Encoding is carried out by constructing the modulo 2 inverse of B and the matrix $B^{-1}A$ (modulo 2); the vector $\boldsymbol{J}^0 = B^{-1}A\,\boldsymbol{\xi}$ (modulo 2, $\boldsymbol{\xi}$ in a Boolean representation) constitutes the codeword. Decoding is carried out by taking the product of the matrix B and the received message $\boldsymbol{J} = \boldsymbol{J}^0 + \boldsymbol{\zeta}$ (modulo 2). The equation

$$A\boldsymbol{\xi} + B\boldsymbol{\zeta} = A\boldsymbol{S} + B\boldsymbol{\tau} \pmod 2, \tag{1}$$

is solved via the iterative methods of BP [3] to obtain the most probable Boolean vectors $\boldsymbol{S}$ and $\boldsymbol{\tau}$; BP methods in this context have recently been shown to be identical to a TAP based solution of a similar physical system [8].

The similarity between error-correcting codes of this type and Ising spin systems was first pointed out by Sourlas [9], who formulated the mapping of a simpler code onto an Ising spin system Hamiltonian. To facilitate the current investigation we first map the problem onto that of an Ising model with finite connectivity. We employ the binary representation (± 1) of the dynamical variables $\boldsymbol{S}$ and $\boldsymbol{\tau}$ and of the vectors $\boldsymbol{J}$ and $\boldsymbol{J}^0$ rather than the Boolean $(0,1)$ one; the vector $\boldsymbol{J}^0$ is generated by taking products of the relevant binary message bits $J^0_\mu = \prod_{i\in\mu} \xi_i$, which correspond to the non-zero elements of $B^{-1}A$, producing a binary version of $\boldsymbol{J}^0$. As we use statistical mechanics techniques, we consider the message and codeword dimensionality (N and M respectively) to be infinite, keeping the ratio $R = N/M$, which constitutes the code rate, finite. Using the thermodynamic limit is natural as Gallager-type codes are used to transmit long ($10^4 - 10^5$) messages, where finite size corrections are likely to be negligible. We examine the Hamiltonian

$$\mathcal{H} = \sum_{\mu,\sigma} \mathcal{D}_{\mu\sigma}\, \delta\Big[-1\;;\; \mathcal{J}_{\mu\sigma} \prod_{i\in\mu} S_i \prod_{j\in\sigma} \tau_j\Big] - \frac{F_s}{\beta}\sum_{i=1}^{N} S_i - \frac{F_\tau}{\beta}\sum_{j=1}^{M} \tau_j \,. \tag{2}$$

The tensor product $\mathcal{D}_{\mu\sigma}\mathcal{J}_{\mu\sigma}$, where $\mathcal{J}_{\mu\sigma} = \prod_{i\in\mu}\xi_i \prod_{j\in\sigma}\zeta_j$ and $\sigma = \langle j_1, \ldots j_L\rangle$, is the binary equivalent of $A\boldsymbol{\xi} + B\boldsymbol{\zeta}$, treating both signal ($\boldsymbol{S}$ and index i) and noise

($\boldsymbol{\tau}$ and index j) simultaneously. Elements of the sparse connectivity tensor $\mathcal{D}_{\mu\sigma}$ take the value 1 if the corresponding indices of both signal and noise are chosen (i.e., all corresponding indices of A and B are 1) and 0 otherwise; it has C unit elements per i-index and L per j-index, representing the system's degree of connectivity. The δ function provides 1 if the selected sites' product $\prod_{i\in\mu} S_i \prod_{j\in\sigma} \tau_j$ is in disagreement with $\mathcal{J}_{\mu\sigma}$, recording an error, and 0 otherwise. Note that this term is not frustrated, as there are $M+N$ degrees of freedom and M constraints (1). The two terms on the right represent our prior knowledge in the case of biased messages F_s and of the noise level F_τ, and require assigning certain values to these additive fields. The choice of $\beta \to \infty$ imposes the restriction of Eq.(1), while the last two terms remain finite. Note that the noise dynamical variables $\boldsymbol{\tau}$ are irrelevant to measuring the retrieval success $m = \frac{1}{N} \left\langle \sum_{i=1}^{N} \xi_i \,\mathrm{sign}\, \langle S_i \rangle_\beta \right\rangle_\xi$. The latter monitors the normalized mean overlap between the Bayes-optimal retrieved message, corresponding to the alignment of $\langle S_i \rangle_\beta$ to the nearest binary value [9], and the original message; the subscript β denotes thermal averaging. The selection of elements in $\mathcal{D}$ introduces disorder to the system; we calculate the partition function $\mathcal{Z}(\mathcal{D}, \boldsymbol{J}) = \mathrm{Tr}_{\{\boldsymbol{S},\boldsymbol{\tau}\}} \exp[-\beta \mathcal{H}]$ averaged over $\mathcal{D}$, $\boldsymbol{\xi}$ and $\boldsymbol{\zeta}$ using the replica method [8]. We employ the replica symmetry ansatz to obtain a set of saddle point equations with respect to the emerging continuous order parameters, representing local field probability distributions and the respective conjugate distributions [10].

For unbiased messages and either $K \geq 3$ ($L \geq 2$) or $L \geq 3$ ($K \geq 2$) we obtain both the ferromagnetic and paramagnetic solutions either by applying the TAP approach or by solving the saddle point equations numerically. The former was carried out at the values of F_τ and $F_s = 0$ which correspond to the true noise and input bias levels (for unbiased messages $F_s = 0$) and thus to Nishimori's condition [11]. This is equivalent to having the correct prior within the Bayesian framework [9].

The most interesting quantity to examine is the maximal code rate, for a given corruption process, for which messages can be perfectly retrieved. This is defined in the case of $K, L \geq 3$ by the value of $R = K/C = N/M$ for which the free energy of the ferromagnetic solution becomes smaller than that of the paramagnetic solution, constituting a first order phase transition. The critical code rate obtained, $R_c = 1 - H_2(p) = 1 + (p \log_2 p + (1-p) \log_2 (1-p))$, coincides with *Shannon's capacity.*

The MN code for $K, L \geq 3$ seems to offer optimal performance. However, the main drawback is rooted in the co-existence of the stable $m = 1, 0$ solutions, which implies that from most initial conditions the system will converge to the undesired paramagnetic solution. Studying the ferromagnetic solution numerically shows a highly limited basin of attraction, which becomes smaller as K and L increase, while the paramagnetic solution at $m = 0$ *always* enjoys a wide basin of attraction.

Studying the case of $K = L = 2$ indicates the existence of paramagnetic and ferromagnetic solutions depicted in the inset of Fig.1. For corruption probabilities $p > p_s$ one obtains either a dominant paramagnetic solution or a mixture of ferromagnetic ($m = \pm 1$) and paramagnetic ($m = 0$) solutions. Reliable decoding may only be obtained for $p < p_s$, which corresponds to a spinodal point, where a unique

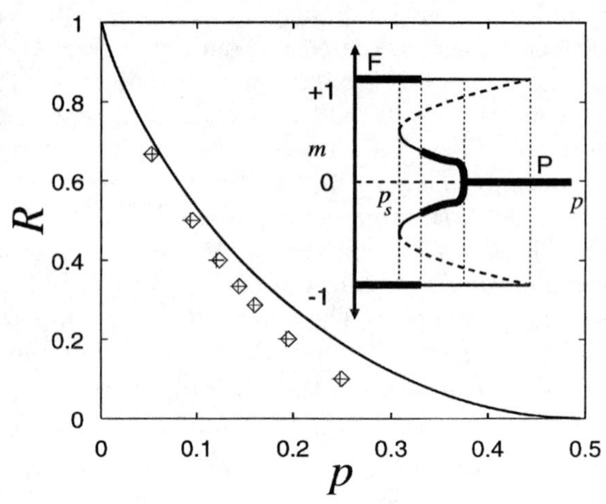

FIGURE 1. Critical transmission rate as a function of p, obtained analytically ($\diamond$) and via BP ($+$) iterative solutions ($N = 10^4$) for unbiased messages (averaged over 10 different initial conditions); error bars are smaller than the symbol size. Shannon's bound (solid line) is shown for reference. Inset: The ferromagnetic (F) and paramagnetic (P) solutions as functions of p; thick and thin lines denote stable solutions of lower and higher free energies respectively, dashed lines correspond to unstable solutions. Lines between the $m = \pm 1$ and $m = 0$ axes correspond to sub-optimal ferromagnetic solutions.

ferromagnetic solution emerges at $m = 1$ (plus a mirror solution at $m = -1$). Initial conditions for both simulations (TAP/BP) and the numerical solutions were chosen almost randomly, with a slight bias of $\mathcal{O}(10^{-12})$, in the initial magnetization. The results obtained point to the existence of a unique pair of global solutions to which the system converges (below p_s) from *all initial conditions.*

The main question that emerges is the possibility of producing more complicated constructions for which the spinodal point is closer to Shannon's critical flip rate. This has been based mainly on the introduction of irregular constructions [4,5], where the number of unit elements per row/column in the matrices A and B is not fixed. Analytical investigation [12] aimed at optimising the construction are yet to provide a principled method for carrying out the optimisation.

The study of parity check codes and the insight gained from the analysis led us to suggest the potential use of a similar system as a public-key cryptosystem [13].

Public-key cryptography is based on a distribution of a public key which may be used to encrypt messages in a manner that can only be decrypted, on practical time scales, by the service provider. Several quite safe and efficient cryptosystems are currently in use such as RSA, elliptic-curves and the McEliece cryptosystem [15], most of which are based on number theory methods. Public-key cryptography plays an important role in many aspects of modern information transmission, for instance, in the areas of electronic commerce and internet-based communication. It enables the service provider to distribute a public key which may be used to encrypt messages in a manner that can only be decrypted by the service provider.

In the suggested cryptosystem, a plaintext represented by an N dimensional Boolean vector $\boldsymbol{\xi} \in (0,1)^N$ is encrypted to the M dimensional Boolean ciphertext $\boldsymbol{J}$ using a predetermined Boolean matrix G, of dimensionality $M \times N$, and a corrupting M dimensional vector $\boldsymbol{\zeta}$, whose elements are 1 with probability p and 0 otherwise, in the following manner: $\boldsymbol{J} = G\boldsymbol{\xi} + \boldsymbol{\zeta}$ (where all operations are modulo 2). The matrix G and the probability p constitute the public key; the corrupting vector $\boldsymbol{\zeta}$ is chosen at the transmitting end. The matrix G, which is at the heart of the encryption/decryption process is constructed by choosing two randomly-selected sparse matrices A $(M\times N)$ and B $(M\times M)$, and a dense matrix D $(N\times N)$, defining $G=B^{-1}AD \pmod 2$. The matrices A and B are generally characterised by K and L non-zero unit elements per row and C and L per column respectively; all other elements are set to zero. The finite, usually small, numbers K, C and L define a particular cryptosystem. The dense invertible Boolean matrix D is arbitrary and is added for improving the system's security. It may be constructed as $D = TP$, where T and P are $N \times N$ triangular and random permutation matrices, respectively, for minimising the computational costs. All matrices are known only to the authorised receiver. Suitable choices of probability p will depend on the maximal achievable rate for the particular cryptosystem as discussed below.

The authorised user may decrypt the ciphertext $\boldsymbol{J}$ by taking the (mod 2) product $B\boldsymbol{J} = A(D\boldsymbol{\xi})+B\boldsymbol{\zeta}$, and finding the most probable solution to Eq.(1) using the methods of BP; the estimate of $\boldsymbol{\xi}$ is obtained by taking the product of the $(D\boldsymbol{\xi})$ estimate and D^{-1}. Studying the case of $K=L=2$ and $p < p_s$ we found that iterative BP decoding converges to the ferromagnetic solution from *any* initial conditions. Cryptosystems with other K, L values generally suffer from progressively smaller basins of attraction as K and L increase, although specific matrices with higher K and L values (such as in [5]) may still be used successfully.

The cryptosystem offers a guaranteed convergence to the plaintext solution, in the thermodynamic limit $N \to \infty$, as long as $p < p_s$. The main consequence of finite plaintexts would be a decrease in the allowed corruption rate. Experimental results with system sizes as small as $N=1000$ still show good performance.

The unauthorised receiver, on the other hand, faces the task of decrypting the ciphertext $\boldsymbol{J}$ knowing only G and p. The straightforward attempt to try all possible $\boldsymbol{\zeta}$ constructions is clearly doomed, provided that p is not vanishingly small, giving rise to only a few corrupted bits. We study the problem by exploiting the similarity between the task at hand and the error-correcting model suggested by Sourlas [9]. In this case, the matrix G generated in the case of $K=L=2$ is dense and has a certain distribution of unit elements per row. The fraction of rows with a low (finite, not of $\mathcal{O}(N)$) number of unit elements vanishes as N increases, allowing one to approximate this scenario by the diluted random energy model studied in [8].

To investigate the typical properties of this (frustrated) model, we calculate again the partition function and the free energy by averaging over the randomness in choosing the plaintext, the corrupting vector and the choice of the random matrix G (being generated by a product of two sparse random matrices). To assess the likelihood of obtaining spin-glass/ferromagnetic solutions, we calculated the free energy landscape (per plaintext bit - f) as a function of overlap m. This can be carried out straightforwardly using the analysis of [10], and provides the golf-course-like energy landscape with a relatively flat area around the one-step replica symmetry breaking (frozen) spin-glass solution and a very deep but extremely narrow area, of $\mathcal{O}(1/N)$, around the ferromagnetic solution [13].

It is worthwhile mentioning that this free-energy landscape may be related directly to the marginal posterior $P(S_i = 1|\boldsymbol{J})$ $1 \leq i \leq N$ and is therefore indicative of the difficulties in obtaining ferromagnetic solutions when the starting point for the search is not infinitesimally close to the original plaintext (which is clearly highly unlikely). Numerical studies of similar energy landscapes show that the time required increases exponentially with the system size [14].

Most attacks on this cryptosystems, by an unauthorised user, will face the same difficulty: without explicit knowledge of the current plaintext and/or the decomposition of G to the matrices A, B and D it will require an exponentially long time to decipher a specific ciphertext. We investigated attacks of several types, some of which appear in [13], concluding that the suggested system is secure.

A brief comparison of our method and the leading technique of RSA [15] shows that: 1) RSA decryption takes $\mathcal{O}(N^3)$ operations while our method only requires $\mathcal{O}(N \log N)$ operations. 2) Encryption costs are of $\mathcal{O}(N^2)$ (as in RSA); inverting the matrices B and D is carried out only once and is of $\mathcal{O}(N^3)$. Two drawbacks of our method: 1) The public key is a dense matrix of dimensionality $M \times N$. However, as public key transmission is carried out only once we do not expect it to be of great significance. 2) The ciphertext/plaintext bit ratio is greater than one (as is the case in RSA). Choosing the N/M ratio is in the hands of the user and is related to the security level required. In addition, the increased transmission time is compensated by a very fast decryption and the added robustness against noise.

We discussed the relation between Ising models, certain error-correcting codes and public-key cryptosystems. Important aspects that are yet to be investigated include the relation between our results and the bounds obtained in the information theory literature for error-correcting codes, finite size effects and methods for alleviating the drawbacks of the new cryptosystem. Support by JSPS-RFTF (YK), The Royal Society and EPSRC-GR/N00562 (DS) is acknowledged.

REFERENCES

1. C.E. Shannon, *Bell Sys. Tech. J.*, **27**, 379 (1948); **27**, 623 (1948).
2. R.G. Gallager, *IRE Trans. Info. Theory*, **IT-8**, 21 (1962).
3. D.J.C. MacKay, *IEEE Trans. IT*, **45**, 399 (1999).
4. T. Richardson, A. Shokrollahi and R. Urbanke, unpublished (1999).
5. I. Kanter and D. Saad, *Phys. Rev. Lett.* **83**, 2660 (1999); *J. Phys. A*, **33**, 1675 (2000).
6. J. Pearl, *Probabilistic Reasoning in Intelligent Systems* (Morgan Kaufmann) 1988.
7. D.J. Thouless, P.W. Anderson and R.G. Palmer, *Philos. Mag.*, **35**, 593 (1977).
8. Y. Kabashima and D. Saad, *Europhys.Lett.*, **44**, 668 (1998); **45**, 97 (1999).
9. N. Sourlas, *Nature*, **339**, 693 (1989); *Europhys.Lett.*, **25**, 159 (1994).
10. Y. Kabashima, T. Murayama and D. Saad, *Phys. Rev. Lett.*, **84**, 1355 (2000).
11. H. Nishimori, *Progr. Theor. Phys.*, **66**, 1169 (1981).
12. R. Vicente, D. Saad and Y. Kabashima, *J. Phys. A* (2000) in press.
13. Y. Kabashima, T. Murayama and D. Saad, *Phys. Rev. Lett.*, **84**, 2030 (2000)
14. E. Marinari, G. Parisi and F. Ritort, *J. Phys. A*, **27**, 7615 and 7647 (1994).
15. D.R. Stinson, *Cryptography: Theory and Practice* (CRC press, New York, 1995).

Statistical Physics of Adaptive Correlation of Agents in a Market

David Sherrington, Juan P. Garrahan and Esteban Moro

Theoretical Physics, University of Oxford, 1 Keble Road, Oxford, OX1 3NP, UK

Abstract. Recent results and interpretations are presented for the thermal minority game, concentrating on deriving and justifying the fundamental stochastic differential equation for the microdynamics.

Market economics poses several problems of potential interest and challenge to statistical physics, involving the co-operative behaviour of many agents whose actions involve mutual frustration and disorder, both quenched and stochastic. In a nutshell, speculators in an idealized stock market are made up of buyers and sellers, each having personal gain as their objectives, trying to buy low and sell high, making their decisions based on commonly available information using individual strategies, with their collective actions determining the (time-varying) 'right choices' and learning from experience. From the point of view of the market regulator, however, preference is for low volatility and market efficiency.

The *minority game* (MG) is a simple encapsulation of some of the ingredients and issues of a market. It consists of N agents each of whom at each step of a parallel dynamical process makes either of two choices, with the objective of being in the minority overall. The agents have no direct knowledge of one another and make their decisions based on purely global information $\vec{I}(t)$, available equally to all. Their decisions are determined through the application to $\vec{I}(t)$ of individual strategy functions, each agent having a small number of such strategies, drawn randomly and independently from a large distribution at the outset and fixed throughout the game. At each time-step each agent employs (just) one of his or her strategies. Adaptation occurs through the development of functions which determine their choices of strategy.

In the original formulation [1] the information $\vec{I}(t)$ was the minority choice over the last m time-steps and the adaptation was achieved through the cumulative award of points at each time-step to the strategies which would have yielded the actual minority choice at that step. The strategy played by any agent at any time-step was that of his/her strategies which currently had the largest point-score.

A remarkable observation in simulations [2] was that the variance in the minority

CP553, *Disordered and Complex Systems*, edited by P. Sollich, et al.

choice became smaller than that of random choice for large enough m, indicating correlation of the agents' actions. A critical memory length m_c was observed for minimum variance, with agents appearing to be frozen in their choices for $m > m_c$, non-frozen for $m < m_c$. Moreover, it was shown that the dependence on m was through the scaling variable D/N, where $D = 2^m$ was the dimension of the space of strategies [2]. Further simulations showed (i) these results are unaffected by replacing the true history by a random $\vec{I}(t)$ [3], indicating that as far as macroscopic observables are concerned the 'information' merely effectuates the correlation; (ii) replacing the deterministic strategy-choices by stochastic ones can significantly reduce the volatility for information vectors of less than the critical length [4].

Here we consider the determination of a fundamental analytic theory and report the derivation of the underlying stochastic differential equation for the microdynamics [5]. We concentrate on a continuous formulation in which $\vec{I}(t)$ is a stochastically randomly chosen unit-length vector on a D-dimensional hypersphere, the strategies are quenched random vectors of length $\sqrt{D}$ in the same space, $\vec{R}_i^\alpha, i = 1, \ldots, N$ labeling the agents and the $\alpha = 1, \ldots, s$ their strategies. The analogues of the binary choices above are bids $b_i^\alpha(t) = \vec{R}_i^\alpha \cdot \vec{I}(t)$. The strategies which are actually used are indicated by $\vec{R}_i^*(t)$. The total bid at time t is $A(t) = \sum_i \vec{R}_i^*(t) \cdot \vec{I}(t)$. The point update rule is

$$P_i^\alpha(t+1) = P_i^\alpha(t) - b_i^\alpha(t)A(t)/N. \tag{1}$$

For simplicity we specialize to $s = 2$ and define

$$\vec{\xi}_i \equiv (\vec{R}_i^1 - \vec{R}_i^2)/2, \quad \vec{\omega}_i \equiv (\vec{R}_i^1 + \vec{R}_i^2)/2; \quad p_i(t) = P_i^1(t) - P_i^2(t) \tag{2}$$

In a generalized *thermal minority game* (TMG) the probability of strategy use is

$$\pi_i^{1,2}(t) \equiv [1 + \exp(\mp\beta f(p_i(t)))]^{-1} \tag{3}$$

and it is useful to define a 'spin'

$$s_i(t) \equiv \pi_i^1(t) - \pi_i^2(t) = \tanh(\beta f(p_i(t))). \tag{4}$$

In [4] the choice $f(p) = p$ was employed, but here we consider $f(p) = \mathrm{sgn}(p) \equiv z$ [5].

We are interested in coarse-grained average behaviour on a time-scale greater than the step-length in order to pass to a continuum-time theory. Equivalently, we take a time-scale Δt with $\vec{I}(t)$ a differential random noise $\vec{I}(t) = \Delta\vec{W}(t)$ with zero mean and variance Δt. In the limit $\Delta t \to \infty$ a Kramers-Moyal expansion yields [5]

$$dp_i(t) = -(ND)^{-1}\vec{R}_i^*(t) \cdot \vec{\xi}_i dt + \mathcal{O}(dt^2) \tag{5}$$

so that to $\mathcal{O}(dt)$ the information noise has been eliminated in favour of an effective interaction between the agents and the averaged variance becomes

$$\sigma^2 \equiv N^{-1}\overline{\langle A(t)^2 \rangle} = (ND)^{-1} \sum_{ij} \langle R_i^*(t) \cdot R_j^*(t) \rangle, \tag{6}$$

where the $\langle \cdot \rangle$ refer to a temporal average and the bar to an average over the quenched disorder of the strategies.

At $T = 0$ Eqs. (2) are deterministic and to leading order in dt reduce to

$$d\boldsymbol{p}/dt = -\boldsymbol{\nabla}_{\boldsymbol{s}} \mathcal{H}|_{\boldsymbol{s}=\boldsymbol{z}}; \qquad \boldsymbol{p} \equiv (p_1, ...p_N) \tag{7}$$

$$\mathcal{H} = \sum_i h_i s_i + \frac{1}{2} \sum_{i \neq j} J_{ij} s_i s_j, \tag{8}$$

$$h_i = (ND)^{-1} \sum_j \vec{\omega}_j \cdot \vec{\xi}_i, \quad J_{ij} = (ND)^{-1} \vec{\xi}_i \cdot \vec{\xi}_j \tag{9}$$

At finite temperature correlations between the fluctuations of the right hand sides of Eqs. (5) are of the same order as the mean and Eq. (7) must be replaced by a set of stochastic differential equations [5]

$$d\boldsymbol{p} = -\boldsymbol{\nabla}_{\boldsymbol{s}} \mathcal{H} dt + \mathcal{M} \cdot d\boldsymbol{W} \tag{10}$$

where $\mathcal{M} \equiv \{M_{ij}\}$ is the covariance matrix

$$M_{ij}[\boldsymbol{p}(t)] = \sum_{\kappa} J_{ik} J_{jk} [1 - s_k^2(t)] \tag{11}$$

and $\boldsymbol{W}(t)$ is an N-dimensional Wiener process of unit scale; for $T = 0, \quad s_k(t)^2 = 1$ so the Wiener term has no weight. Correspondingly, the Fokker-Planck equation for the probability distribution of the p_i is

$$\frac{\partial P}{\partial t} = -\sum_i \frac{\partial}{\partial p_i} \left(\frac{\partial \mathcal{H}}{\partial s_i} P \right) + \frac{1}{2} \sum_{ij} \frac{\partial^2}{\partial p_i \partial p_j} (M_{ij} P). \tag{12}$$

The average volatility is given by

$$\sigma^2 \equiv N^{-1}\overline{\langle A(t)^2 \rangle} = 1 + 2\overline{\langle \mathcal{H} \rangle} \tag{13}$$

Eq. (10) is thus the fundamental microscopic equation from which the macrodynamics should be calculable. To check this we have compared numerical evaluations of the volatility and the density of frozen agents (those for whom $p_i(t)$ does not change sign after an initial transient) from Eqs. (7) and (10) with corresponding direct simulations from Eqs. (1) and (3). They are in perfect accord. This is shown explicitly in Fig. 1 for $T = 0$. Figs. 2a and 2b show the effect of temperature as given by Eq. (10); direct simulation gave results identical within statistical error.

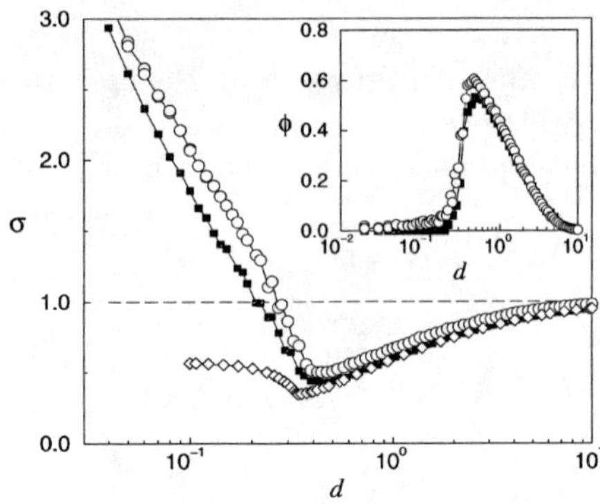

FIGURE 1. Volatility σ as a function of the reduced dimension $d = D/N$. Squares correspond to the original dynamics of Eq. (1), circles to Eq. (7); $T = 0$ and $p_i(0) \sim 0$. Diamonds correspond to minimization of $\overline{\mathcal{H}}$. Inset: fraction of frozen agents.

In [6] $\overline{\langle \mathcal{H} \rangle}$ was evaluated for $T = 0$ on the assumption that the system equilibrated and hence was equivalent to minimizing $\overline{\mathcal{H}}$. The result is also exhibited in Fig. 1 and can be seen to be good (and probably correct) for $d = D/N > d_c$ but in error for $d < d_c$. In fact, however, Eq. (7) does not describe a simple descent dynamics since the variables on the right and left hand sides are different and a metric is needed to relate p and s. Substitution shows that the dynamics is non-Markovian in s. An explicit demonstration of non-equilibration for $d < d_c$ follows from a simulation starting with $|p_i(0)| \sim \mathcal{O}(1) \gg dt$, where dt is the time-step. This is illustrated in Fig. 3a [5].

It is also tempting to compare with a Hopfield neural network which is characterizable by an effective Hamiltonian

$$H = -\frac{1}{2} \sum_{i \neq j} J_{ij}^{H} \sigma_i \sigma_j; \quad J_{ij}^{H} = N^{-1} \sum_{\mu=1}^{p=\alpha N} \vec{\xi}_i^{\mu} \cdot \vec{\xi}_j^{\mu} \tag{14}$$

where the $\{\xi_i^{\mu}\}$ are quenched random patterns; indeed it was by the application of techniques devised for (14) that [6] minimized $\overline{\mathcal{H}}$. Clearly there is a difference of sign between Eqs. (8) and (14) but one might be tempted to anticipate that this will merely suppress the retrieval attractors while maintaining the spin glass state, in analogy with the SK model, and then attribute the reduction in energy compared with the random state to spin-glass binding. However, this is false; the Hopfield spin-glass solution is not symmetric under change of sign of β. Rather, the random-field term of Eq. (8) is crucial in reducing the ground state energy of $\mathcal{H}$ and the volatility below their random-state values. This is demonstrated explicitly in Fig. 3b which shows the effect of choosing $\vec{R}_i^1$ randomly but $\vec{R}_i^2$ as

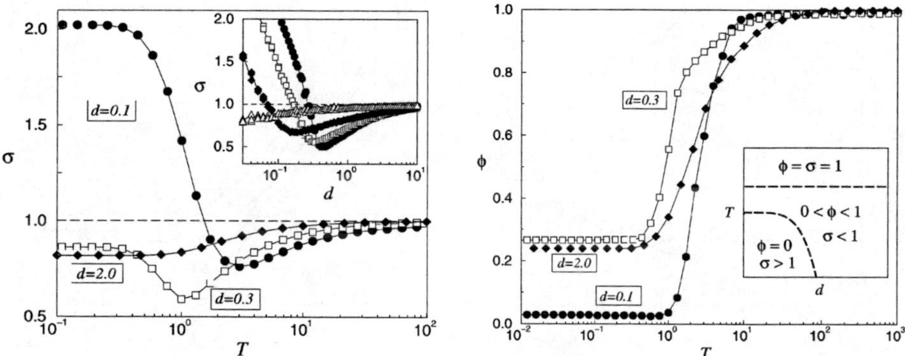

FIGURE 2. Left: Volatility as a function of the temperature from the continuous dynamics Eq. (10); $f(p) = \text{sgn}(p)$, $p_i(0) \sim 0$. Inset: volatility as a function of d for $T = 10^{-3}$, 1, 2, and 10. Right: Fraction of frozen agents as a function of T. Inset: schematic cross-over phase diagram.

$$\vec{R}_i^2 = -(1-\lambda)\vec{R}_i^1 + \lambda\vec{\tilde{R}}_i \tag{15}$$

where $\vec{\tilde{R}}_i$ is also chosen randomly [10]. For $\lambda = 0, \vec{\omega}_i = 0, h_i = 0$ and the volatility never falls below random, although again there appears a (different) critical d_c separating regimes (worse-than-random and random).

As noted, above we have used $f(p) = \text{sgn}(p)$ in the simulations. If instead $f(p) = p$ is employed, then for $d > d_c$ the system iterates over a long time to its zero-temperature behaviour [7] since the mean $|p_i(t)|$ grows quasi-continuously and $s_i(t)$ saturates to its zero-temperature value, which being ± 1 eliminates the effects of the second term of Eq. (10). However, for $d < d_c$ there continues to be an improvement with temperature, to an optimal value which is better than random and is reached at a temperature of $\mathcal{O}(1)$ [8], but without any further rise to the random value; in this case $p_i(t)$ fluctuates around $p_i(0)$.

Finally we remark on the relationship with the crowd-anticrowd concept of [9], where a crowd is a group of agents playing the same strategy and the corresponding anticrowd play the opposite strategy. From Eq. (6)

$$\sigma^2 = D^{-1}\sum_{\mu}\langle n_\mu(t)\rangle^2; \quad n_\mu = N^{-1/2}\sum_i \vec{R}_i^*(t)\cdot\vec{e}_\mu \tag{16}$$

where $\vec{e}_\mu$ is a unit vector in the μth Cartesian direction of the D-dimensional space. n_μ then formalizes the notion of the number of agents in crowd μ minus the number in the corresponding anti-crowd. The qualitative difference of $d < d_c$ and $d > d_c$ then follows from the recognition that for $d \ll d_c$ the vectors $\vec{R}_i^\alpha$ are densely distributed on the D-sphere permitting $n_\nu \sim \mathcal{O}(N)$, while for $d \gg d_c$ they are sparsely distributed so all $n_\nu \sim \mathcal{O}(1)$.

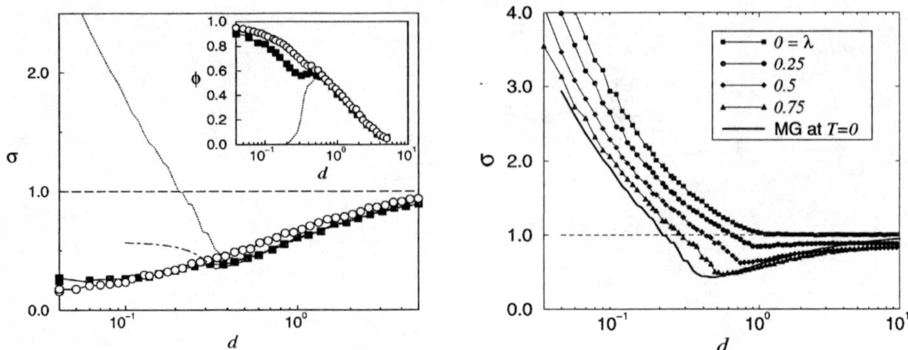

FIGURE 3. Left: Volatility as a function of d for random initial conditions $|p_i(0)| \sim \mathcal{O}(1) \gg dt$ for original dynamics and from Eq. (10). Right: Volatility for partially anticorrelated $\vec{R}_i^{1,2}$; $\vec{R}_i^2 = -(1-\lambda)\vec{R}_i^1 + \lambda\vec{\tilde{R}}_i$, with $\vec{R}_i^1$ and $\vec{\tilde{R}}_i$ random.

ACKNOWLEDGMENTS

We are grateful to Andrea Cavagna, Irene Giardina and Matteo Marsili for useful comments and discussions. We acknowledge financial support from EPSRC Grant No. GR/M04426 and EC Grant No. ARG/B7-3011/94/27.

REFERENCES

1. Challet, D. and Zhang, Y-C., *Physica A* **246**, 407 (1997).
2. Savit, R., Manuca, R. and Riolo, R., *Phys. Rev. Lett.* **82**, 2203 (1999).
3. Cavagna, A., *Phys. Rev. E* **59**, R3783 (1999).
4. Cavagna, A., Garrahan, J.P., Giardina, I. and Sherrington D., *Phys. Rev. Lett.* **83**, 4429 (1999).
5. Garrahan, J.P., Moro, E. and Sherrington, D., *Phys. Rev. E* **85**, R9 (2000).
6. Challet, D., Marsili, M. and Zecchina, R., *Phys. Rev. Lett.* **84**, 1824 (2000).
7. Challet, D., Marsili, M. and Zecchina, R., cond-mat/0004308.
8. Cavagna, A., Garrahan, J.P., Giardina, I. and Sherrington, D., cond-mat/0005134.
9. Hart, M., Jeffries, P., Johnson, N.F. and Hui, P.M., cond-mat/0003486.
10. Garrahan, J.P., Moro, E. and Sherrington, D., in preparation.

Random Field Ising Chains and Neural Networks with Synchronous Dynamics

N.S. Skantzos and A.C.C. Coolen

Department of Mathematics, King's College, University of London
The Strand, London WC2R 2LS, U.K.

Abstract. We first present an exact solution of the one-dimensional random-field Ising model in which spin-updates are made fully synchronously, i.e. in parallel (in contrast to the more conventional Glauber-type sequential rules). We find transitions where the support of local observables turns from a continuous interval into a Cantor set and we show that synchronous and sequential random-field models lead asymptotically to the same physical states. We then proceed to an application of these techniques to recurrent neural networks where 1D short-range interactions are combined with infinite-range ones. Due to the competing interactions these models exhibit phase diagrams with first-order transitions and regions with multiple locally stable solutions for the macroscopic order parameters.

INTRODUCTION

In this paper we study the 1D random-field Ising model (RFIM) in which spin dynamics (the familiar stochastic alignment to local fields) is executed fully synchronously. Models with synchronous spin dynamics have been studied already a long time ago [1] as an alternative to the more familiar sequential Glauber-type rule and have found many applications e.g. in the context of neural networks and probabilistic cellular automata. Interestingly, what is known about the relation between the two types of dynamics concerns mostly the equilibrium physics of infinite-range models. For instance, it is known that the phase diagram of the SK model appears unaffected by the choice of dynamics whereas the phase diagram of the Hopfield model does change. Far fewer results are known concerning disordered systems in finite dimensions. In this paper we first show how one can solve the synchronous dynamics RFIM by adapting sequential random-field techniques and then compare the equilibrium physical states of the model to those found for Glauber-type dynamics. As an application of these techniques we also show how one can derive equilibrium properties of recurrent neural networks models in which 1D short-range interactions compete with infinite-range ones. It turns out that frustration effects induce phenomena quite novel in associative memories, such as first-order transitions separating regions with multiple locally stable states.

CP553, *Disordered and Complex Systems*, edited by P. Sollich, et al.

SYNCHRONOUS DYNAMICS RFIM TECHNIQUES

Our first model is defined as a collection of N Ising spin variables $\boldsymbol{\sigma} = (\sigma_1, \ldots, \sigma_N)$ arranged in a 1D chain and coupled with nearest-neighbour interactions $J_{ij} = J(\delta_{i,j+1} + \delta_{i,j-1})$. At each site i an external field θ_i is drawn at random from $w(\theta_i) = \frac{1}{2}[\delta[\theta_i + \tilde{\theta}] + \delta[\theta_i - \tilde{\theta}]]$ where $\tilde{\theta} > 0$ denotes the strength of the field. The spin dynamics is the familiar stochastic alignment to local fields $h_i(\boldsymbol{\sigma}) = \sum_j J_{ij}\sigma_j + \theta_i$, $\text{Prob}[\sigma_i(t+1) = \pm 1] = \frac{1}{2}[1 \pm \tanh[\beta h_i(\boldsymbol{\sigma}(t))]]$, but, in contrast to the more conventional Glauber-type dynamics, spin updates are here made fully synchronously. The parameter β controls the amount of stochasticity in the dynamics. The above process can be shown to lead to a non-Gibbsian equilibrium state [2] with corresponding β-dependent Hamiltonian $H_\beta(\boldsymbol{\sigma}) = -\frac{1}{\beta}\sum_i \log\cosh[\beta h_i(\boldsymbol{\sigma})] - \sum_i \sigma_i\theta_i$.

The three key ingredients which in combination render the model non-trivial are: the short-range connectivity, the random external fields and the synchronous dynamics leading to a non-Boltzmann equilibrium. If one tries to evaluate the partition sum via the transfer matrix method one will immediately find that the matrix multiplications involve (due to the random fields) an N-fold product of random matrices. To evaluate this product and subsequently the asymptotic free energy we adapt the technique of [3] and define the following 'conditioned' partition functions:

$$Z_{\pm\pm}^{(N)} = \sum_{\sigma_1 \ldots \sigma_N} e^{-\beta H_\beta(\boldsymbol{\sigma})}\, \delta_{\sigma_{N-1},\pm 1}\, \delta_{\sigma_N,\pm 1}$$

The key observation enabling the evaluation of the partition function is that the four quantities $\{Z_{**}^{(N)}\}$ can be mapped onto $\{Z_{**}^{(N+1)}\}$ by the recurrence relation

$$\begin{pmatrix} Z_{++}^{(N+1)} \\ Z_{+-}^{(N+1)} \end{pmatrix} = \boldsymbol{M}_1[\theta_{N+1}, \theta_N] \begin{pmatrix} Z_{++}^{(N)} \\ Z_{-+}^{(N)} \end{pmatrix} \qquad \begin{pmatrix} Z_{-+}^{(N+1)} \\ Z_{--}^{(N+1)} \end{pmatrix} = \boldsymbol{M}_2[\theta_{N+1}, \theta_N] \begin{pmatrix} Z_{+-}^{(N)} \\ Z_{--}^{(N)} \end{pmatrix} \tag{1}$$

where $\boldsymbol{M}_{1,2}[\theta, \theta']$ are 2×2 matrices involving the random fields. In order to evaluate the partition sum $Z_N = \sum_{\lambda\tau=\pm} Z_{\lambda\tau}^{(N)}$ it will turn out helpful to define the ratios

$$k_j^{(1)} = e^{2\beta\theta_j}\frac{Z_{+-}^{(j)}}{Z_{++}^{(j)}} \qquad k_j^{(2)} = \frac{\cosh[\beta(\theta_j + J)]}{\cosh[\beta(\theta_j - J)]}\frac{Z_{--}^{(j)}}{Z_{+-}^{(j)}} \qquad k_j^{(3)} = e^{2\beta\theta_j}\frac{Z_{--}^{(j)}}{Z_{-+}^{(j)}} \tag{2}$$

which from the recurrence relation (1) can be found to be generated by three coupled stochastic processes: $k_{j+1}^{(\ell)} = \psi_\ell(\{k_j^{(a)}\}_{a=1,2,3}; \theta_j)$ for all $\ell = 1, 2, 3$. Using symmetry properties of the functions ψ_ℓ the three ratios are found to reduce further to a single stochastic variable obtained by

$$k_{j+2} = \psi[k_j; \theta_{j+1}, \theta_j] \qquad \psi[k'; \theta, \theta'] = \frac{\cosh[\beta\tilde{\theta}] + e^{-2\beta\theta'}k'\cosh[\beta(\theta - 2J)]}{\cosh[\beta(\theta + 2J)] + e^{-2\beta\theta'}k'\cosh[\beta\tilde{\theta}]} \tag{3}$$

which allows one to express the asymptotic free energy per site and observable quantities such as the local magnetization $\langle\sigma_\ell\rangle$ as integrals over the distribution $P(k) = \lim_{N\to\infty}\frac{1}{N}\sum_i \langle\delta[k - \psi(k_{i-1}; \theta_i, \theta_{i-1}]\rangle_{\text{eq}}$.

Cantor Spectra & Link with Glauber-type systems

In figure 1 we show the support of the stochastic variable $\tilde{k} = \frac{k-1}{k+1} \in [-1,1]$ (left picture) and the support of local magnetizations $\langle\sigma_\ell\rangle \in [-1,1]$ (right picture) as functions of the ratio $\tilde{\theta}/J$. We see that for small values of $\tilde{\theta}/J$, i.e. when couplings dominate over random fields, the supports of $\tilde{k}$ and $\langle\sigma_\ell\rangle$ are continuous intervals. Upon increasing $\tilde{\theta}$, however, these intervals appear to break into more and more disconnected pieces (white regions) the union of which consititutes the relevant supports of $\tilde{k}$ and $\langle\sigma_\ell\rangle$. An interesting transition occurs at $\tilde{\theta}/J > 1$ (for $\tilde{k}$, left) and $\tilde{\theta}/J > 2$ (for $\langle\sigma_\ell\rangle$, right) where a self-similar structure appears: the supports of $\tilde{k}$ and $\langle\sigma_\ell\rangle$ become fully discretized and can be identified as Cantor sets.

If one tries to simplify the present model, either by considering the less comlicated ∞-range version where $J_{ij} = J/N$ (with random fields present) or by replacing the random fields θ_i by a uniform field where $\theta_i = \theta$ for all i (with short-range interactions present), one finds simple exact identities relating synchronous and sequential models, e.g. for the free energies: $f_{syn}(m) = 2f_{seq}(m)$ with $m = \frac{1}{N}\sum_i \langle\sigma_i\rangle_{\text{eq}}$, or for the transfer matrices: $\boldsymbol{T}_{syn} = \boldsymbol{T}^2_{seq}$. These relations immediately raise the question of whether synchronous and sequential models are still closely related in the (complex) RFIM regime. To answer this question we inspect the solution as it follows for the stochastic processes which for sequential dynamics takes the form [3]

$$k_{j+1} = \psi_{\text{seq}}[k_j; \theta_j] \qquad \psi_{\text{seq}}[k'; \theta'] = \frac{e^{-\beta J} + k' e^{\beta J} e^{-2\beta\theta'}}{e^{\beta J} + k' e^{-\beta J} e^{-2\beta\theta'}} \tag{4}$$

while for the synchronous case it is given by (3). Although synchronous and sequential dynamics lead to quite different stochastic processes, cf. equations (3)-(4) we find that the two are related by the following identity:

$$\psi_{\text{seq}}[\psi_{\text{seq}}[k'; \theta']; \theta] = \psi_{\text{syn}}[k'; \theta, \theta'] \tag{5}$$

from which it follows that equilibrium observables become for $N \to \infty$ identical [4].

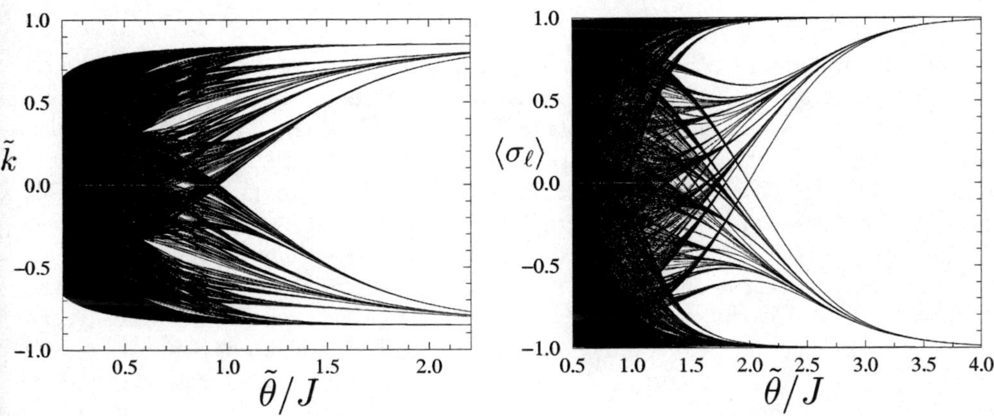

FIGURE 1. The supports of $\tilde{k} = \frac{k-1}{k+1}$ and $\langle\sigma_\ell\rangle$ as functions of the ratio $\tilde{\theta}/J$.

$1+\infty$ DIM ATTRACTOR NEURAL NETWORKS

We will now apply our techniques to models of recurrent neural networks in which Ising neurons $\boldsymbol{\sigma} = (\sigma_1, \ldots, \sigma_N)$ are coupled by interactions of the form:

$$J_{ij} = \left[\frac{J_\ell}{N} + J_s(\delta_{j,i+1} + \delta_{j,i-1})\right] w_{ij} \qquad \text{with} \qquad w_{ij} = \sum_{\mu=1}^{p} \xi_i^\mu \xi_j^\mu$$

This corresponds to the result of having stored p binary patterns $\{\xi_i^1, \ldots, \xi_i^p\}$ with $\xi_i^\mu \in \{-1, 1\}$ in a Hebbian-type fashion with nearest-neighbour interactions of the form $J_s w_{ij}$ in combination with infinite-range ones of strength $(J_\ell w_{ij})/N$. We will consider the so-called low-storage regime where $\lim_{N\to\infty} p/N = 0$. Our main order parameter $m_\mu(\boldsymbol{\sigma}) = \lim_{N\to\infty} \frac{1}{N}\sum_i \langle\sigma_i\rangle_{\text{eq}}\, \xi_i^\mu \in [-1, 1]$ measures the overlap between the equilibrium neuron configuration and a nominated pattern $\{\xi_i^\mu\}$ with $m_\mu(\boldsymbol{\sigma}) = 0$ corresponding to null recall and $m_\mu(\boldsymbol{\sigma}) = 1$ corresponding to full recall. To isolate our order parameters we insert $1 = \int dm_\mu\ \delta[m_\mu - \frac{1}{N}\sum_i \sigma_i \xi_i^\mu]$ in the expression for the free energy per site $f = -\lim_{N\to\infty} \frac{1}{\beta N} \log \sum_{\boldsymbol{\sigma}} \exp[-\beta H(\boldsymbol{\sigma})]$ where we will consider both synchronous and sequential dynamics with

$$H_{\text{seq}}(\boldsymbol{\sigma}) = -\frac{1}{2}\sum_{i\neq j} \sigma_i J_{ij} \sigma_j \qquad H_{\text{syn}}(\boldsymbol{\sigma}) = -\frac{1}{\beta}\sum_i \log\cosh[\beta h_i(\boldsymbol{\sigma})]$$

Upon replacing the δ-functions by their integral representations we then arrive at

$$f = \text{extr}_{\{\boldsymbol{m},\hat{\boldsymbol{m}}\}}\ \phi(\boldsymbol{m}, \hat{\boldsymbol{m}}) \qquad \begin{cases} \phi_{\text{seq}}(\boldsymbol{m}, \hat{\boldsymbol{m}}) = -i\,\boldsymbol{m}\cdot\hat{\boldsymbol{m}} - \frac{1}{2}J_\ell \boldsymbol{m}^2 - \frac{1}{\beta N} R_{\text{seq}} \\ \phi_{\text{syn}}(\boldsymbol{m}, \hat{\boldsymbol{m}}) = -i\,\boldsymbol{m}\cdot\hat{\boldsymbol{m}} - \frac{1}{\beta N} R_{\text{syn}} \end{cases} \tag{6}$$

where $\boldsymbol{m} = (m_1, \ldots, m_p)$ and $\hat{\boldsymbol{m}} = (\hat{m}_1, \ldots, \hat{m}_p)$. The partition sum lies in

$$R_{\text{seq}} = \sum_{\boldsymbol{\sigma}} e^{-i\beta\sum_i \sigma_i \hat{\boldsymbol{m}}\cdot\boldsymbol{\xi}_i}\ e^{\beta\sum_i \sigma_i J^s_{i,i+1}\sigma_{i+1}}$$

$$R_{\text{syn}} = \sum_{\boldsymbol{\sigma}} e^{-i\beta\sum_i \sigma_i \hat{\boldsymbol{m}}\cdot\boldsymbol{\xi}_i}\ e^{\sum_i \log[2\cosh[\beta(J_i^\ell + \sigma_{i-1}J^s_{i-1,i} + \sigma_{i+1}J^s_{i,i+1})]]}$$

with the short-hand notation $J^s_{i,i+1} = J_s\,\boldsymbol{\xi}_i\cdot\boldsymbol{\xi}_{i+1}$ and $J_i^\ell = J_\ell\,\boldsymbol{\xi}_i\cdot\boldsymbol{m}$. In the simplest scenario where $p = 1$ the Mattis transformation $\sigma_i\xi_i \to \sigma_i$ allows one to eliminate the disorder and evaluate the asymptotic free energy exactly. The resulting phase diagrams for $p = 1$ are shown in figure 2 (upper row). For $p > 1$ however, the random variables $\{\xi_i^\mu\}$ cannot be transformed away and the evaluation of the partition function involves a product over N random matrices. This problem is similar to the one of the 1D RFIM, with the pattern variables $\{\xi_i^\mu\}$ playing here the role of random-bonds. It turns out that one can evaluate the non-trivial part of the free energy R_{seq} and R_{syn} by deriving recurrence relations between (appropriately defined) 'conditioned' partition sums. As in the 1D RFIM, this allows one to express observable quantities as integrals over the distribution of a characteristic quantity which represents the ratio of 'conditioned' partition functions.

Phase Diagrams

Our main order parameter $\boldsymbol{m}(\boldsymbol{\sigma})$ is to be determined from the extremisation condition (6). We will assume that solutions of the type $\boldsymbol{m} = (m, 0, \ldots, 0)$ give the dominant states of the system (an assumption which is supported by simulations). In figure 2 we plot the corresponding phase diagrams for $p=1$ and $p=2$ in the plane $(\beta J_s, \beta J_\ell)$ for sequential (left pictures) and synchronous dynamics (right pictures). For $p=1$ (upper row) we obtain: (i) a region where $m=0$ only, (ii) a region where $m \neq 0$ and (iii) a region where $m=0$ and $m \neq 0$ can be both locally stable states (depending on initial conditions). The creation of a new ergodic component in region (iii) is the result of competing interactions: negative short-range ones vs. positive infinite-range ones while regions (i) and (iii) are separated by a first-order transition. Solid lines denote second-order transitions. For synchronous dynamics (right) we observe that a second pair of transition lines has emerged which can be shown to be exact reflections in the origin of the sequential pair. For $p = 2$ (lower row) we observe two 'pairs' of first-order transition lines in the competition quadrant $(\beta J_s < 0, \beta J_\ell > 0)$ separating regions with different number of locally stable states: region B with 1 $m > 0$, region C with 2 $m > 0$ and region F with 3 $m > 0$ states. The transition A (where $m=0$ only) to B is a second-order one.

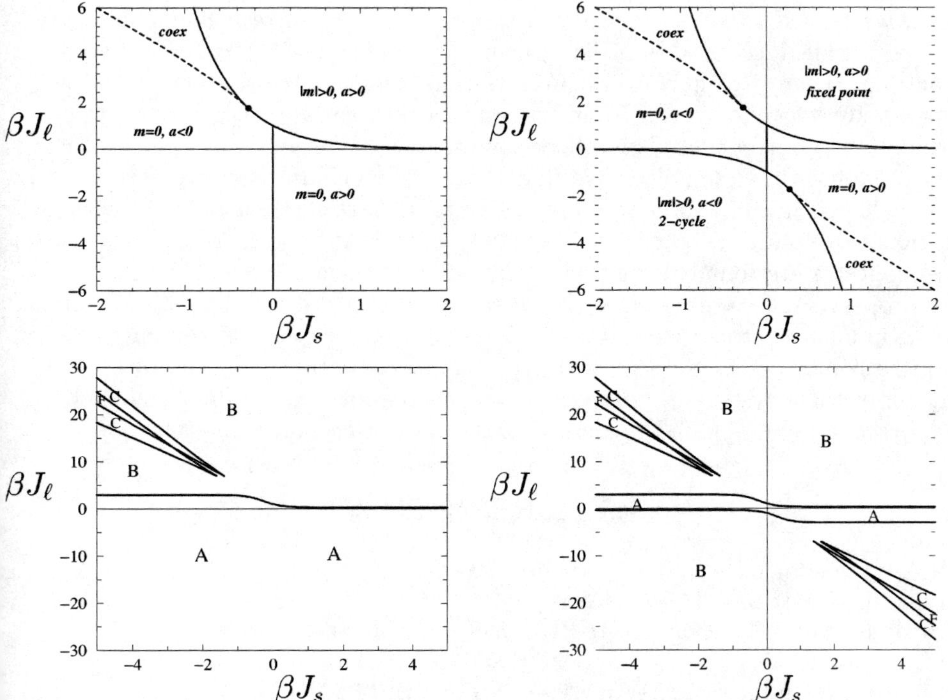

FIGURE 2. Phase diagrams for $p = 1$ (upper row) and $p = 2$ (lower row) for sequential (left) and synchronous dynamics (right) for $1+\infty$ dimensional neural network models.

DISCUSSION

In this paper we have performed an equilibrium analysis of the 1D random field Ising model with synchronous (rather than the more familiar sequential Glauber-type) spin dynamics. Although this process leads to a non-Boltzmann equilibrium state we show that one can derive analytic expressions for the asymptotic free energy by adapting 1D random-field techniques. These are based on the derivation of a recurrence process relating 'conditioned' partition sums, which allow one to perform the (random) matrix multiplications in the partition function.

For sufficiently small values of the random field strength relative to spin-couplings, we find that the support of macroscopic observables is a continuous interval; local fields almost fully determine spin states. However, for increasingly large random field strengths the support of local observables becomes a union of disjoint intervals until a transition occurs into a regime where the support is a zero-measure Cantor set. The equilibrium picture of the synchronous RFIM closely resembles that found for the sequential RFIM [3] which raises the question of whether the physics of the two models (synchronous vs. sequential RFIM) are related. It turns out that one can rigorously prove equivalance of the physics of the two models [4]; in the present paper we derive the key identity based on a functional relation of the two RFIM stochastic processes.

Next, we turned to an application of these techniques to recurrent neural networks: we studied 1D models of Ising neurons coupled via two different types of interactions: nearest-neighbour vs. infinite-range ones. Our equilibrium observables measure 'recall' of a stored pattern, and pattern variables can be seen to play the role of random bonds. This allows one to derive an analytic solution based on 1D random-field techniques. We find that, due to competing interactions and entropic effects, this type of model induces first-order transitions between parameter regions with multiple locally stable states, whereas synchronous and sequential phase diagrams are found to be related by reflection symmetries. Another interesting property of these systems is the increasing complexity of the phase diagram with the number of stored patters p which is due to the explicit appearance of p in the macroscopic laws even in 'pure states', due to short-range interactions [5]. An extension of these systems to include distant neighbour interactions has also been studied recently and is found to lead to particularly rich phase diagrams [6].

REFERENCES

1. W A Little (1974) *Math. Biosci.* **19** 101-120
2. P Peretto (1984) *Biol. Cybern.* **50** 51-62
3. R Bruinsma and G Aeppli (1983) *Phys. Rev. Lett.* **50** 1494-1497
4. N S Skantzos and A C C Coolen (2000) *J. Phys. A* **33** 1841-1855
5. N S Skantzos and A C C Coolen (2000) *J. Phys. A* **33** 5785-5807
6. N S Skantzos and A C C Coolen (2000) *J. Phys. A.* submitted, preprint KCL-MTH-00-49, cond-mat/0008049

PART II:
INFORMATION GEOMETRY

Geometrical Methods in Learning Theory

G. Burdet*, Ph. Combe*, H. Nencka†

*Centre de Physique Théorique, CNRS-Luminy, Case 907,
F 13288 Marseille Cedex 9, FRANCE [1]
†Center of Mathematical Sciences, University of Madeira,
9000 Funchal, Madeira - Portugal [2]
E-mail: burdet@cpt.univ-mrs.fr,combe@cpt.univ-mrs.fr,nencka@cpt.univ-mrs.fr

Abstract. The methods of information theory provide natural approaches to learning algorithms in the case of stochastic formal neural networks. Most of the classical techniques are based on some extremization principle. A geometrical interpretation of the associated algorithms provides a powerful tool for understanding the learning process and its stability and offers a framework for discussing possible new learning rules. An illustration is given using sequential and parallel learning in the Boltzmann machine.

INTRODUCTION

After Fisher [13] and Cramér [8] had given some rigorous presentation of parametric statistics, C.R. Rao [18] was able to link statistics to differential geometry and provide new mathematical tools for solving complex statistical problems. The introduction of affine connections [11,12] allows one to interpret the important notion of statistical curvature and leads to a new branch of mathematics: the *theory of info-manifolds* [3,4,10,17,19].

Let $(X, \mathcal{F}, \mu)$ be a measure space with μ a σ-finite measure. Consider a parametric family of probabilities, $P_\mu(\Theta) = \{P_\theta,\, \theta \in \Theta,\, P_\theta \ll \mu\}$, absolutely continuous with respect to the measure μ, where $\Theta \in \mathbb{R}^d$ is an open subset (or a family of open overlapping subsets) defining uniquely the elements of the family, with coordinates $\theta = \{\theta^1, \theta^2, \ldots, \theta^d\}$ in the canonical basis of $\mathbb{R}^d$. The Radon-Nikodym derivative $\rho_\theta = dP_\theta/d\mu$ exists and $P_\mu(\Theta)$ can be identified with the family of random variables $\mathcal{P}_\mu(\Theta) = \{\rho_\theta = dP_\theta/d\mu\,,\, \theta \in \Theta\}$. The mapping $\varphi : P_\theta \mapsto \theta$ defines a chart $\varphi : \mathcal{P}_\mu(\Theta) \to \Theta$. In the following, let us assume that P_θ and $P_{\theta'}$ are equivalent, i.e. for all θ and θ', $P_\theta \equiv P_{\theta'}$ if $P_\theta \ll P_{\theta'}$ and $P_{\theta'} \ll P_\theta$.

---

1) Ph. C. thanks Prof. R. Streater for the kind invitation to King's College London.
2) Ph.C. and H. N. are partially supported by the projects plurianual and Praxis XXI.

CP553, *Disordered and Complex Systems*, edited by P. Sollich, et al.

Under smoothness conditions $\mathcal{P}_\mu(\Theta)$ can be equipped with a fully symmetric 2-covariant tensor $g_{ij} = \mathbb{E}_{P_\theta}[\partial_i \ell_\theta \, \partial_j \ell_\theta] = -\mathbb{E}_{P_\theta}[\partial_i \partial_j \ell_\theta]$, where $\ell_\theta = \ln \rho_\theta$: the Fisher metric [13,18]; and a fully symmetric 3-covariant tensor $t_{ijk} = \mathbb{E}_{P_\theta}[\partial_i \ell_\theta \, \partial_j \ell_\theta \, \partial_k \ell_\theta]$: the skewness tensor [3,6]. The triplet $\mathcal{S} = \{\mathcal{P}_\mu(\Theta), g, t\}$ is called a (parametric) info-manifold on $(X, \mathcal{F}, \mu)$.

The tangent plane T_{ρ_θ} at $\rho_\theta \in \mathcal{P}_\mu(\Theta)$ is generated by the tangent vectors at ρ_θ, $A = \sum_k A^k(\rho_\theta) \, \partial_k$, $\partial_k = \frac{\partial}{\partial \theta^k}$. The score vectors $\partial_k \rho_\theta$ being the Radon-Nikodym derivatives of a signed-measure absolutely continuous w.r.t. P_θ, verifying $\mathbb{E}_{P_\theta}[\partial_k \ell_\theta] = 0$, generate a d-dimensional Euclidean space $T_\theta = \left\{A_\theta \in \mathbb{R}^X : A_\theta = \sum_{k=1}^d A^k(\rho_\theta) \partial_k \ell_\theta \right\}$, which can be identified with T_{ρ_θ}. The natural scalar product on T_θ is $\langle A_\theta, B_\theta \rangle = \mathbb{E}_{P_\theta}[A_\theta B_\theta]$. The cotangent plane [5] $T^\star_{\rho_\theta}$ generated by the differential forms $\alpha = \sum_{i=1}^d \alpha_i(\theta) d\theta^i$, can be realized on T_θ through $\lambda(\alpha) = \sum_{i=1}^d \alpha_i(\theta) a^i$, with $\mathbb{E}_{P_\theta}[a^i \partial_j \ell] = \delta_{ij}$. A family of affine metric geometries is provided by α-connections of Amari-Chensov [3,6]: $\overset{\alpha}{\nabla} g = \alpha t$, $\alpha \in \mathbb{R}$. The case $\alpha = 0$ corresponds to a Riemannian geometry, with $\overset{0}{\nabla}$ being the Levi-Cività connection. The α-connections are torsion free and verify the duality relation

$$A g_{|\rho_\theta}(B, C) = g_{|\rho_\theta}(\overset{\alpha}{\nabla}_A B, C) + g_{|\rho_\theta}(B, \overset{-\alpha}{\nabla}_A C), \qquad A, B, C \in T_{\rho_\theta}\,.$$

In the natural coordinate system the α-connections are defined through the α-coefficients by

$$\overset{\alpha}{\Gamma}{}_{ij}{}^k = \overset{0}{\Gamma}{}_{ij}{}^k - \frac{\alpha}{2} t_{ij}{}^k, \quad \overset{\alpha}{\Gamma}{}_{ij}{}^k = \sum_\ell g^{k\ell} \overset{\alpha}{\Gamma}{}_{ij\ell}, \quad \overset{\alpha}{\Gamma}{}_{ij\ell} = \overset{\alpha}{\Gamma}{}_{ji\ell}$$

with the Christoffel symbols $\overset{0}{\Gamma}{}_{ij\ell} = \frac{1}{2}[\partial_i g_{j\ell} + \partial_j g_{i\ell} - \partial_\ell g_{ij}]$. The associated α-Riemannian curvature is given by

$$\overset{\alpha}{R}{}_{ijk\ell} = g_{\ell m}(\partial_i \overset{\alpha}{\Gamma}{}_{jk}{}^m - \partial_j \overset{\alpha}{\Gamma}{}_{ik}{}^m) + \overset{\alpha}{\Gamma}{}_{im\ell} \overset{\alpha}{\Gamma}{}_{jk}{}^m - \overset{\alpha}{\Gamma}{}_{jm\ell} \overset{\alpha}{\Gamma}{}_{ik}{}^m,$$

and the α-scalar curvature $\overset{\alpha}{R} = \sum_{i=1}^d \overset{\alpha}{R}{}_i{}^i$ where $\overset{\alpha}{R}_{ij} = \sum_{k=1}^d \overset{\alpha}{R}{}_{kij}{}^k$ is the Ricci α-curvature tensor. If $\overset{\alpha}{R} = 0$ the family is called α-flat. It can be proven that an α-flat family is also $-\alpha$-flat.

The α-self-parallel curves $\gamma : s \to \gamma(s)$ are such that $\overset{\alpha}{\nabla}_{\dot\gamma} \dot\gamma = 0$ with $\dot\gamma := d\gamma/ds$. In the θ-coordinate system the equation for α-self-parallel curves takes the form

$$\ddot\theta^k + \overset{\alpha}{\Gamma}{}_{ij}{}^k \, \dot x^i \dot x^j = 0\,.$$

The α-self-parallel curves verify $\frac{d}{ds} g(\dot\gamma, \dot\gamma) = \alpha t(\dot\gamma, \dot\gamma, \dot\gamma)$; they do not minimize the Riemannian distance if $\alpha \neq 0$. Only for $\alpha = 0$ is $g(\dot\gamma, \dot\gamma) = \text{const}$ a first integral.

The next section treats the geometry associated with the Kullback-Leibler relative entropy [15,16]. We apply these results to learning in the Boltzmann machine [1,2] in the last section.

EXPONENTIAL FAMILIES

Let $\mathcal{P}(\Theta) = \{\rho_\theta\}_{\theta\in\Theta}$ be an exponential family (of equivalent probabilities) on $(X, \mathcal{F}, \mu)$, i.e. if $\ln \rho_{\theta'}/\rho_\theta \in T_\theta$, $\forall \theta, \theta' \in \Theta$. Such a family is (± 1)-flat [5].

Let us denote by $\Upsilon_{\rho_\theta} = \{u \in \mathbb{R}^X \mid u - \mathbb{E}_{\rho_\theta}[u] \in T_\theta\}$, the set of (centered) random variables over $(X, \mathcal{F}, P_\theta)$ with $\rho_\theta = dP_\theta/d\mu$, which admits an expansion in terms of the scores. If $\rho_\theta, \rho_{\theta'} \in \mathcal{P}(\Theta)$ then $\ell_\theta, \ell_\theta - \ell_{\theta'} \in \Upsilon_{\rho_\theta}$, $\ell_\theta = \ln \rho_\theta$. Some algebraic computations lead to the following propositions.

Proposition 1 *Let P and P^* be two probability distributions of an exponential family, then*

$$\overset{-1}{\nabla} \mathrm{d}_{|P}\ K(P;P^*) = g_{|P}, \quad \overset{1}{\nabla} \mathrm{d}_{|P^*} K(P;P^*) = g_{|P^*}, \tag{1}$$

where $K(P;P^*) = \mathbb{E}_\rho[\ell - \ell^*]$ is the Kullback-Leibler relative entropy.

Let $\gamma(s)$ be a (-1)-self-parallel, then

$$\frac{d^2}{ds^2} K(\gamma(s);P^*)|_{s=s_0} = g_{|\rho}(\gamma_s, \gamma_s). \tag{2}$$

The length of the velocity field appears on the right hand side and does not depend on P^*, so the differential in P^* is null

$$\mathrm{d}_{|P^*}\ (\frac{d^2}{ds^2} K(\gamma(s);P^*) = \frac{d^2}{ds^2}(\mathrm{d}_{|P^*} K(\gamma(s);P^*)) = 0. \tag{3}$$

Therefore, $\mathrm{grad}_{|P^*} K(\gamma(s);P^*)$ is a vector field linear in s. Now, to express that the differential in P^* of the relative entropy vanishes if $P^* = P$, the origin of γ is fixed at $\gamma(0) = P^*$. Hence, the set of points $\gamma(s)$, satisfying the relation

$$(\mathrm{grad}_{|P^*} K)(\gamma(s);\gamma(0)) = -\gamma_0 s\ ,$$

belongs to the (-1)-self-parallel curve γ, with γ_0 the initial velocity. Interchanging the roles of P and P^* one gets:

Proposition 2 *The parametrized (± 1)-self-parallel curves of a 1-flat statistical manifold are explicitly described by the gradients of the relative entropy,*

$$\begin{array}{lll} (-1) - \mathrm{self} - \mathrm{parallel\,curves} & : & (\mathrm{grad}_{|P^*} K)(\gamma(s);\gamma(0)) = -\gamma_0 s \\ (+1) - \mathrm{self} - \mathrm{parallel\,curves} & : & (\mathrm{grad}_{|P} K)(\gamma^*(0);\gamma^*(t)) = -\gamma_0^* t \end{array}$$

Let P be a current probability on a curve γ in the info-manifold $\mathcal{S}$. Given $P^* \in \mathcal{S}$, with $P^* \notin \gamma$, the position of P which locally minimizes $K(P;P^*)$, if it exists, is given by the $s = s_0$ solutions of

$$g(\gamma_s, (\mathrm{grad}_{|P} K)(P;P^*)) = 0. \tag{4}$$

On the other hand, $\frac{d^2}{ds^2} K(P(s);P^*) = g_{|P}(\gamma_s,\gamma_s) + g(\gamma_{ss}, \mathrm{grad}_{|P} K(P;P^*))$, and the first term on the left hand side is positive. From the minimization condition the second term vanishes at s_0. Hence the extrema with $\gamma_s \neq 0$ are minima. Thus we have:

Proposition 3 *Let γ^* (γ) be a 1-self-parallel (-1-self-parallel) curve parametrized by affine parameter t (s), crossing for vanishing affine parameter a curve γ (γ^*) in $P = \gamma(s_0)$ ($= \gamma^*(t_0)$).*

1. *For any t, $K(\gamma(s_0); \gamma^*(t))$ locally minimizes $K(\gamma(s); \gamma^*(t))$ if the curves γ and γ^* are orthogonal, i.e. $g(\gamma_{s_0}, \gamma_0^*) = 0$.*
2. *For any s, $K(\gamma(s); \gamma^*(t_0))$ locally minimizes $K(\gamma(s); \gamma^*(t))$ if the curves γ and γ^* are orthogonal, i.e. $g(\gamma_0, \gamma_{t_0}^*) = 0$.*

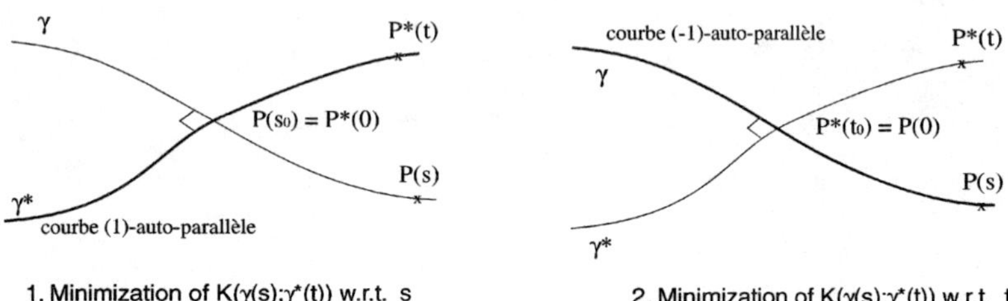

FIGURE 1. Minimization of Kullback-Leibler relative entropy

These properties can be extended to congruences of curves, i.e. to submanifolds of $\mathcal{S}$, allowing one for instance to define the orthogonal (± 1)-projections minimizing the relative entropy locally. This is the geometrical version of the Pythagorean theorem of Csiszár [9,10], see also Amari [3].

LEARNING IN THE BOLTZMANN MACHINE

A symmetric Boltzmann machine is a probabilistic formal neural network, its architecture being given by a non-directed simplicial labelled graph $\mathcal{G} = (\mathcal{V}, \mathcal{L})$, with a set of vertices $\mathcal{V} = \{v_1, v_2, \ldots, v_n\}$ and a set of edges $\mathcal{E} = \{e_1, e_2, \ldots, e_m\}$. The graph is not necessarily complete but we ask that each vertex has a high degree of connectivity. Because the graph is simplicial and non-directed, there is at most one edge (x_i, x_j) between two vertices x_i, x_j, and the weight $W_{ij} = W(x_i, x_j)$ of the edge is symmetric $W_{ij} = W_{ji}$. On each vertex there is a probabilistic binary automaton with state space $\{0, 1\}$, so the configuration space of the network is $X = \{0, 1\}^n$. For the sake of simplicity we assume in the following that each neuron is an input and output neuron and that there are no hidden neurons. We denote, as previously, by $\mathcal{S}$ the family of probabilities on X and we remark that the log-likelihood $\ell(x) = \ln p(x)$, $x = \{x_1, x_2, \ldots, x_n\} \in X$ depends only on $2^n - 1$ parameters and can be written as follows

$$\ell(x) = \sum_{i_1=1}^{n} \theta_{i_1}^1 x_{i_1} + \sum_{i_1<i_2=1}^{n} \theta_{i_1 i_2}^2 x_{i_1} x_{i_2} + \ldots + \theta_{i_1 i_2 \ldots i_n}^n x_{i_1} x_{i_2} \ldots x_{i_n} - \Psi(\theta), \tag{5}$$

with $i_1 < i_2 < \ldots < i_n$. The normalisation constant Ψ is related to the entropy by

$$\Psi(\theta) = \sum \theta^r_{i_1 i_2 \ldots i_r} \eta_r^{i_1 i_2 \ldots i_r} + H(P), \quad \eta_r^{i_1 i_2 \ldots i_r} = \frac{\partial \Psi}{\partial \theta^r_{i_1 i_2 \ldots i_r}} = \mathbb{E}_P[x_{i_1} x_{i_2} \ldots x_{i_r}]. \tag{6}$$

This family of probabilities is (± 1)-flat and has the θ's as a natural parametrization; the η's define another parametrization, the "momentum parametrization".

An individual neuron at v_i is updated at time $t+1$ according to the state of the connected neurons at time t, such that $x_i = 1$ (0, respectively) with probability p^i_β ($1 - p^i_\beta$, respectively), with $p^i_\beta = [1 + \exp(-\beta h_i(t))]^{-1}$ where β is a characteristic parameter and $h_i(t)$ is the effective field $h_i(t) = \sum_{j=1, j \neq i}^n W_{ij} x_j(t)$. The threshold of the neurons is chosen as zero.

In the sequential Boltzmann machine [1,2,7] one neuron, chosen with the uniform law, is updated at each instant. This procedure determines an irreducible Markov chain $x(t)$ on $X = \{0,1\}^n$ corresponding to the Glauber dynamics [7,14] and having the Gibbs probability measure $G_\beta(u)$ as the unique stationary probability distribution:

$$\ln G_\beta(u) = \beta \sum_{i<j=1}^{n} W_{ij} u_i u_j - \ln Z(\beta), \quad Z(\beta) = \sum_{u \in \{0,1\}^n} e^{\beta \sum_{i<j=1}^n W_{ij} u_i u_j}, \tag{7}$$

The family of all possible states of the Boltzmann machine is defined by the $2^n - 1$ linear equations:

$$\theta^2_{ij} = \beta W_{ij}; \qquad \theta^m_{i_1 i_2 \ldots i_m} = 0, \; m \neq 2. \tag{8}$$

Then the family $\mathcal{G}$ of the Gibbs probabilities inherits the 1-flat structure of $\mathcal{S}$. By the proposition of the previous section there exists a unique (-1)-self parallel ortogonal projection G_p on $\mathcal{G}$ of the probability P. The projection G_p is the best approximation of P by a probability in $\mathcal{G}$. By the Pythagorean theorem if G is a Gibbs measure we have $K(P;G) = K(P;G_P) + K(G_P;G)$. Then the minimal value for the Kullback-Leibler relative entropy is fixed by the projection property and is given by $\min_{G \in \mathcal{G}} K(P;G) = K(P;G_P)$.

The learning procedure is implemented by the steepest descent gradient method which consists in the modification of the synaptic weight W_{ij} to $W_{ij} + \delta W_{ij}$ with

$$\delta W_{ij} = -\epsilon \sum_{\{kl\}} g^{ij,kl} \frac{\partial K(P;G)}{\partial W_{kl}} = \epsilon \beta \sum_{\{k\ell\}} g^{ij,k\ell} \{\mathbb{E}_P[u_k u_\ell] - \mathbb{E}_G[u_k u_\ell]\}, \tag{9}$$

where ϵ is a sufficiently small parameter. Let us remark that under this infinitesimal transformation the momentum coordinates $\eta_2^{ij} = \beta \mathbb{E}_G[u_i u_j]$ of W_{ij} in $\mathcal{G}$ are changed to $\eta_2^{ij} + \delta \eta_2^{ij}$, $\delta \eta_2^{ij} = \sum_{\{k\ell\}} g_{ij,k\ell} \delta W_{k\ell}$. One has

Proposition 4 *The gradient learning method for a sequential symmetric Boltzmann machine provides a minimization algorithm of the Kullback-Leibler relative entropy $K(P;G)$ in $\mathcal{G}$ and consist in following the (1)-self parallel curve in $\mathcal{G}$ from the current point G to the (-1)-self parallel orthogonal projection G_P of $P \in \mathcal{S}$ on $\mathcal{G}$.*

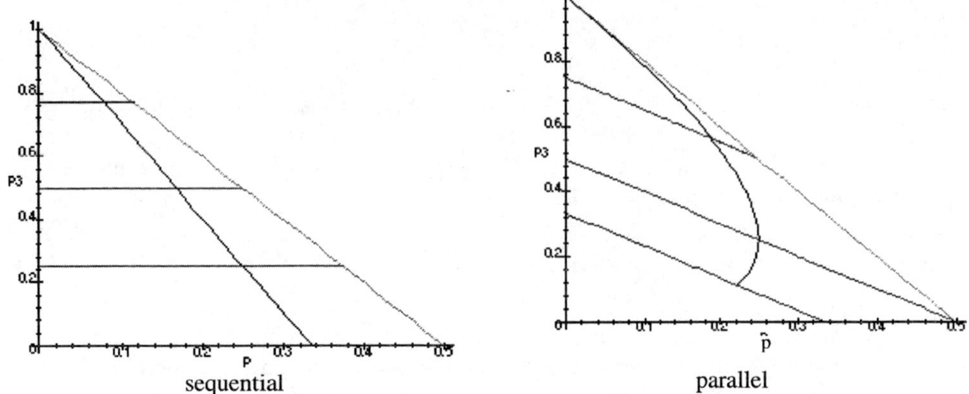

FIGURE 2. Toy model:-1-self-parrallel projection.

In a parallel Boltzmann machine [1,5,7] all the neurons are updated simultaneously. For thresholds taken as zero, the transition probability then takes the form:

$$\Pi[\omega(t+1) = v \mid \omega(t) = u] = \prod_{i=1}^{n} \frac{1}{1 + e^{\beta(1-2v_i)h_i(u)}}, \quad h_i(u) = \sum_{\substack{j=1 \\ j \neq i}}^{n} W_{ij} u_j. \tag{10}$$

In this case the Markov chain defined by this transition probability also has a unique stationary probability measure $\Pi_\beta(u)$,

$$\ln \Pi_\beta(u) = \frac{\beta}{2} \sum_{i=1}^{n} \sum_{\substack{j=1 \\ i \neq j}}^{n} W_{ij} u_j + \sum_{i=1}^{n} \ln\cosh\left(\frac{\beta}{2} \sum_{\substack{j=1 \\ i \neq j}}^{n} W_{ij} u_j\right) - \ln N(\beta) + n \ln 2. \tag{11}$$

with $N(\beta)$ the normalization factor $N(\beta) = \sum_{u \in \{0,1\}^n} \sum_{v \in \{0,1\}^n} e^{\beta \sum_{i=1^n} v_i h_i(u)}$.

Then the submanifold of the output stationary measures of a given parallel Boltzmann machine is neither a (1)-linear submanifold of $\mathcal{S}$ nor a $(+1)$-convex one. Then the uniqueness of the (-1)-projection is not automatically insured and needs a study of the (± 1)-self parallel curves.

However, a learning algorithm based on the minimization by the gradient method of the Kullback-Leibler relative entropy $K(P; \Pi_\beta)$ leads to

$$\begin{aligned} \delta W_{ij} = \epsilon\beta \textstyle\sum_{\{k\ell\}} g^{ij,k\ell} \quad & \{\mathbb{E}_P[u_k(t)u_\ell(t-1)] - \mathbb{E}_\Pi[u_k(t)u_\ell(t-1)] \\ & -\mathbb{E}_P[u_k(t-1)u_\ell(t)] + \mathbb{E}_\Pi[u_k(t-1)u_\ell(t)]\}, \end{aligned} \tag{12}$$

which does not depend on the present configuration but rather on the trajectory; the geometrical interpretation is then no longer simple and needs further study.

The simplest Boltzmann machine with two neurons, taken here as a toy model, has as its state space the 3-simplex, $0 < p_i < 1$, $i = 0, \ldots, 3$, $\sum_{i=0}^{3} p_i = 1$. In Fig. 2 the left plot corresponds to the -1-projections on the Gibbs submanifold given by the equation $p_1 = p_2 = p$, $p_3 = 1 - p$. The right plot of Fig. 2 presents the -1-projections on the equilibrium manifold for parallel dynamics, given by $p_1 = p_2 = p$, $\sqrt{p_3} - p_3 = p$, $p_3 > 1/9$. This plot shows that in the case of parallel dynamics the learning algorithm based on the minimization of the Kullback-Leibler relative entropy may not be convergent on the equilibrium manifold.

REFERENCES

1. E. Aarts, J. Korst, *Simulated Annealing and Boltzmann Machines. A Stochastic Approach to Combinatorial Optimization and Neural Computing*, Wiley and Sons, 1989.
2. D. Ackley, G. Hilton, T. Sejnowski, *Cognitive Science* **9**, 147–169 (1985).
3. S.-I. Amari, *Differential Geometrical Methods in Statistics*, Lecture Notes in Statistics, vol. 28, Springer, 1985; "Differential Geometry Statistical Inference", in *IMS Lecture Notes Monographs series*, vol. 10, edited by S. S. Gupta, 1987, pp. 19–94; *Contemporary Mathematics* **203**, 81–95 (1997).
4. O. Barndorff-Nilsen, *Parametric Statistical Models and Likelihood*, Lecture Notes in Statistics, vol. 50, Springer, 1988.
5. G. Burdet, Ph. Combe, H. Nencka, *Progress in Probability* **45**, 87–99 (Birkhäuser, 1998), "Neural Information Processing", *Mathematical Biology and Medicine* **7**, 172–198 (World Scientific, 1999).
6. N.N. Chensov, *Statistical Decision and Optimal Inference*, in Russian (Nauka, Moscow 1972), English translation *AMS* **53** (Providence, R.I. 1982).
7. Ph. Combe, H. Nencka, *Contemporary Mathematics* **203**, 105–116 (1997).
8. H. Cramér, *Mathematical Methods of Statistics*, Princeton University Press, 1946.
9. I. Csiszár, *Ann. of Probab.* **3**, 146–158 (1975).
10. I. Csiszár, J. Körner, *Information Theory*, Academic Press, 1981.
11. A.P. Dawid, *Ann. Stat.* **3**, 1231–1234 (1975), *Ann. Stat.* **5**, 1242 (1977).
12. B. Efron, *Ann. Stat.* **3**, 1189–1942 (1975), *Ann. Stat.* **6**, 362–376 (1975).
13. R.A. Fisher, *Proc. Camb. Phil. Soc.* **22**, 700–725 (1925).
14. R.J. Glauber, *J. Math. Phys.* **4**, 294–307 (1963).
15. S. Kullback, *Information Theory and Statistics*, Wiley, New York, 1959.
16. S. Kullback, R. Leibler, *Ann. Math. Stat.* **22**, 79–86 (1951).
17. M.K. Murray, J.W. Rice, *Differential Geometry and Statistics*, Chapman & Hall, 1990.
18. C.R. Rao, *Bull. Calcutta Math. Soc.* **37**, 81–91 (1945).
19. I. Vajda, *Theory of Statistical Inference and Information*, Kluwer, 1989.

Geometry of Neural Networks and Models with Singularities

Kenji Fukumizu

The Institute of Statistical Mathematics
4-6-7 Minami-Azabu, Minato-ku, Tokyo 106-8569 Japan

Abstract. This paper discusses maximum likelihood estimation with unidentifiability of parameters. Unidentifiability is formulated as a conic singularity of the model. It is known that the likelihood ratio may have unusually large order in unidentifiable cases. A sufficient condition for such large order is given and applied to neural networks.

INTRODUCTION

This paper discusses the asymptotic behavior of the maximum likelihood estimator (MLE) under the condition that the true parameter is unidentifiable. The asymptotics of MLE is important in statistical theory, and the asymptotic normality under some regularity conditions is well known. However, in the case that the true parameter is not a point, the asymptotics has not been clarified completely.

We formulate the unidentifiability as a conic singularity [1] in the space of probability density functions. The likelihood ratio of the MLE, with the true probability at the singularity, can be described by a Gaussian process over the unit vectors in the tangent cone. This Gaussian process shows various behaviors depending on the functional property of the tangent cone. One interesting feature is the order of the likelihood ratio as the number of samples n goes to infinity. A model satisfying the usual asymptotic theory has likelihood ratio of $O_p(1/n)$. However, in neural networks, a larger order $O_p(\log n/n)$ has been reported in unidentifiable cases [2]. We will give a useful sufficient condition of larger order than $O_p(1/n)$ in terms of the tangent cone, and derive the order for neural network models.

UNIDENTIFIABILITY AND CONIC SINGULARITY

Parametric estimation and neural networks

A *statistical model* $S = \{f(z;\theta) \mid \theta \in \Theta\}$ is a family of probability density functions (p.d.f.) on a measure space $(\mathcal{Z}, \mathcal{B}, \mu)$, where the parameter space Θ is

CP553, *Disordered and Complex Systems*, edited by P. Sollich, et al.

a domain in $\mathbb{R}^d$. We assume that $f(z;\theta) > 0$ for all z and θ, and that $f(z;\theta)$ is C^∞ on θ for each $z \in \mathcal{Z}$. Suppose the probability of i.i.d. random variables $Z_1, \dots, Z_n$ is $P = f(z)\mu$ with $f(z) > 0$. The *likelihood ratio* and *Kullback-Leibler (KL) divergence* are defined by

$$L_n(\theta) = \tfrac{1}{n}\textstyle\sum_{i=1}^n \log \frac{f(Z_i;\theta)}{f(Z_i)} \qquad \text{and} \qquad D(\theta) = \int f(z) \log \frac{f(z)}{f(z;\theta)} d\mu(z) \tag{1}$$

respectively. We consider the *maximum likelihood estimator* (MLE) $\hat{\theta}$ that attains the maximum of the likelihood ratio; $L_n(\hat{\theta}) = \sup_{\theta\in\Theta} L_n(\theta)$.

Feed-forward neural networks like multilayer perceptrons can be described within this framework. Let $x \in \mathbb{R}$, $s(t) = \tanh(t)$, and $\theta = (a_j, c_j, b_j, d)^T$. The *multilayer perceptron* model with H hidden units is a family of functions defined by

$$\varphi(x;\theta) = \textstyle\sum_{j=1}^H b_j\, s(a_j x + c_j) + d, \tag{2}$$

We discuss only one-dimensional input and output for simplicity. For the statistical model, assume a probability $Q = q(x)dx$ and a conditional p.d.f. $r(y|u)$ of $y \in \mathbb{R}$ given $u \in \mathbb{R}$. We assume the existence of their second moment. Define the statistical model by $f(z;\theta) = r(y \mid \varphi(x;\theta))q(x)$, where $z = (x,y) \in \mathcal{Z} = \mathbb{R}^2$. Examples of $r(y \mid u)$ are the Gaussian noise model $\frac{1}{\sqrt{2\pi}\sigma}\exp\{-\frac{1}{2\sigma^2}(y-u)^2\}$, and the binomial distribution $u^y(1-u)^{1-y}$.

Throughout this paper, the true p.d.f. $f(z)$ is assumed to be included in the model $\{f(z;\theta) \mid \theta \in \Theta\}$. Then, there exists $\theta_0 \in \Theta$ such that $f(z;\theta_0) = f(z)$. The true parameter θ_0 may not be unique. Such unidentifiability of the true parameter exists in the multilayer perceptron model. Suppose we have the model with 2 hidden units, and the true function is given by $\varphi_0(x) = b_0\, s(a_0 x)$. In the model, any parameter in $\{\theta \in \Theta \mid a_1 = a_0, b_1 = b_0, c_1 = b_2 = d = 0\}$ gives $\varphi_0(x)$. It is known that if a function can be realized by a network with a smaller number of hidden units than the model, then the set of parameters to give the function is always high dimensional [3–5]. Asymptotic normality does not hold in such cases.

If the true function is identifiable, under some regularity conditions, the asymptotic theory says that $L_n(\hat{\theta})$ and $D(\hat{\theta})$ are the same to leading order, $L_n(\hat{\theta}) = D(\hat{\theta}) + o_p(1/n)$, and their limiting distribution is given by $nL_n(\hat{\theta}) \to \chi^2_d$ in law.

Locally conic model

Following Dacunha-Castelle & Gassiat [1], with some modifications, we employ a conic singularity to formulate unidentifiability. Let $S = \{f(z;\theta) \mid \theta \in \Theta\}$ be a statistical model with $\Theta \subset \mathbb{R}^d$, and $f_0(z)$ be in S. The parameter θ is decomposed as $\theta = (\boldsymbol{\alpha}, \beta)$ with $\boldsymbol{\alpha} \in \mathbb{R}^{d-1}$ and $\beta \in \mathbb{R}$. Then, S is called *locally conic at* f_0 if
(1) $f(z;\theta)$ is C^∞ function of θ for almost every z.
(2) Let $\Theta_0 = \Theta \cap (\mathbb{R}^{d-1} \times \{0\})$, and $\Theta(\boldsymbol{\alpha}) = \Theta \cap (\{\boldsymbol{\alpha}\} \times \mathbb{R})$ for each $\boldsymbol{\alpha}$ such that $(\boldsymbol{\alpha}, 0) \in \Theta_0$. Then, $\Theta = \cup_{(\boldsymbol{\alpha},0)\in\Theta_0} \Theta(\boldsymbol{\alpha})$.

(3) The parameter set to give f_0 is Θ_0, that is, $f(z;(\boldsymbol{\alpha},\beta))\mu = f_0(z)\mu \Leftrightarrow \beta = 0$.
(4) $\frac{\partial \log f(z;\boldsymbol{\alpha},\beta)}{\partial\beta}$ is in $L^2(f_{(\boldsymbol{\alpha},\beta)}\mu)$, and $\left\|\frac{\partial \log f(z;\boldsymbol{\alpha},0)}{\partial\beta}\right\|_{L^2(f_0\mu)} = 1$ $\forall\boldsymbol{\alpha}$ with $(\boldsymbol{\alpha},0)\in\Theta_0$.

Geometrically, S is a d-dimensional set in the space of p.d.f. with a singularity at f_0. The score of the submodel $S_{\boldsymbol{\alpha}} = \{f(z;(\boldsymbol{\alpha},\beta)) \mid \beta\in\Theta(\boldsymbol{\alpha})\}$ at $\beta = 0$,

$$v_{\boldsymbol{\alpha}}(z) = \tfrac{\partial f}{\partial\beta}(z;(\boldsymbol{\alpha},0)), \tag{3}$$

can be seen as a unit tangent vector along $S_{\boldsymbol{\alpha}}$. The set of the score functions $C = \{v_{\boldsymbol{\alpha}}\}$ generates the tangent cone at f_0. We call C the *basis of the tangent cone.* This view of tangent vectors can be rigorously formulated if S is in the maximal exponential model [6], which is an infinite dimensional Banach manifold.

The multilayer perceptron model is an example of a locally conic model. Let $0\le K < H$, and $\varphi(x)$ be a function realizable with K hidden units. In the parameter space of the model with H hidden units, the subset to give $\varphi(x)$ is unidentifiable. We can rewrite this by a conic singularity. Let $\Theta_H^* = \{\theta = (a_j,b_j)\in\mathbb{R}^{2H} \mid a_j\neq 0\ b_j\neq 0\ (1\le j\le H),\ |a_j|\neq|a_h|\ (1\le j<h\le H)\}$ be the parameter space of the multilayer perceptrons with H hidden units. Given a function $\varphi(x) = \sum_{k=1}^K b_k^0\, s(a_k^0 x)$, $((a_k^0,b_k^0)\in\Theta_K^*)$, we slightly restrict the parameter space as $\Theta_H^{**} = \{\theta\in\Theta_H^* \mid |a_j|\neq|a_k^0|\ (K+1\le j\le H, 1\le k\le K)\}$. Define new parameters by $\beta = \operatorname{sgn}(b_{K+1})\sqrt{b_{K+1}^2+\cdots+b_H^2}$, $\xi_k = \frac{a_k-a_0}{\beta}$, $\xi_j = a_j$, $\eta_k = \frac{b_k-b_0}{\beta}$, $\eta_j = \frac{b_j}{\beta}$ $(1\le k\le K,\ K+1\le j\le H)$ for $\theta\in\Theta_H^{**}$, a parameter space Π_H by

$$\begin{aligned}\Pi_H = \{\omega = (\xi_i,\eta_i,\beta) \mid\ & a_k^0+\beta\xi_k\neq 0,\ b_k^0+\beta\eta_k\neq 0\ (1\le k\le K),\ \eta_{K+1}>0,\\ & \textstyle\sum_{j=K+1}^H\eta_j^2 = 1,\ \xi_j\neq 0,\ \eta_j\neq 0\ (K+1\le j\le H),\ \ |\xi_j|\neq|\xi_i|\\ & (K+1\le j<i\le H),\ \ |a_k^0+\beta\xi_k|\neq|a_h^0+\beta\xi_h|\ (1\le k<h\le H),\\ & |a_k^0+\beta\xi_k|\neq|\xi_j|,\ |\xi_j|\neq|a_k^0|\ (1\le k\le K, K+1\le j\le H),\ \beta\in\mathbb{R}\},\end{aligned}$$

and $\Pi_H^{**} = \{\omega\in\Pi_H \mid \beta\neq 0\}$. We can rewrite the multilayer perceptron as

$$\psi(x;\omega) = \textstyle\sum_{k=1}^K(b_k^0+\beta\eta_k)s((a_k^0+\beta\xi_k)x) + \sum_{j=K+1}^H\beta\eta_j s(\xi_j x). \tag{4}$$

It is easy to see that $\varphi(x;\theta) = \psi(x;\omega)$ for the corresponding $\theta\in\Theta_H^{**}$ and $\omega\in\Pi_H^{**}$.

We write $\omega = (\boldsymbol{\alpha},\beta)$, summarizing $(\xi_1,\ldots,\eta_H)$ by $\boldsymbol{\alpha}$. We have $\psi(x;\omega) = \varphi(x)$ iff $\beta = 0$. Assume $r(y|u)$ satisfies $r(y|u_1)dy\neq r(y|u_2)dy$ for $u_1\neq u_2$ and the Fisher information $I(u)$ is positive and finite. Let S_H be $\{f(x,y;\omega) = r(y|\psi(x;\omega))q(x) \mid \omega\in\Pi_H\}$, and $f_{0,K}(x,y)\in S_H$ be given by $\varphi(x)$. Then, we obtain

Proposition 1. *The model S_H is locally conic at $f_{0,K}$ $(0\le K<H)$.*

We omit the proof. We need to normalize β using the L^2-norm of the tangent vectors $\tilde{v}(\boldsymbol{\alpha})$ at $(\boldsymbol{\alpha},0)$, which are given by $\tilde{v}(\boldsymbol{\alpha}) = \frac{\partial r(y|\varphi(x))}{\partial u}\frac{\partial\psi(x;(\boldsymbol{\alpha},0))}{\partial\beta}$, where

$$\tfrac{\partial}{\partial\beta}\psi(x;(\boldsymbol{\alpha},0)) = \textstyle\sum_{j=K+1}^H\eta_j s(\xi_j x) + \sum_{k=1}^K\eta_k s(a_k^0 x) + \sum_{k=1}^K b_k^0\xi_k s'(a_k^0 x)x. \tag{5}$$

If the true function is zero, i.e. $\varphi(x) = 0$, and $r(y|u) = \exp\{y\kappa(u)+\tau(y)-\zeta(u)\}$ is an exponential family such that $\kappa(u)$ is invertible and $\int yr(y|u)dy = u$, the tangent vectors are given by $\tilde{v}(\boldsymbol{\alpha}) = \kappa'(0)\ y\ (\sum_{j=1}^H\eta_j s(\xi_j x))$. We will use this later.

MLE IN LOCALLY CONIC MODELS

MLE as a supremum of a random process

Let $S = \{f(z;(\boldsymbol{\alpha},\beta)) \mid (\boldsymbol{\alpha},\beta) \in \Theta\}$ be a statistical model locally conic at $f_0 \in S$. Suppose $Z_1, \ldots, Z_n$ are i.i.d. with the law $f_0\mu$. For each $\boldsymbol{\alpha}$, the submodel $S_{\boldsymbol{\alpha}}$ is a smooth, one-dimensional statistical model. Consider the MLE $\hat{\beta}_{\boldsymbol{\alpha}}$ in $S_{\boldsymbol{\alpha}}$, then, the MLE in S is given by $\sup_{\theta\in\Theta} L_n(\theta) = \sup_{\boldsymbol{\alpha}} L_n(\boldsymbol{\alpha}, \hat{\beta}_{\boldsymbol{\alpha}})$. Assume each $S_{\boldsymbol{\alpha}}$ satisfies the regularity condition of the asymptotic efficiency, then Taylor expansion leads to

$$L_n(\boldsymbol{\alpha}, \hat{\beta}_{\boldsymbol{\alpha}}) = \tfrac{1}{2n} U_n(\boldsymbol{\alpha})^2 + o_p(1/n), \quad \text{where} \quad U_n(\boldsymbol{\alpha}) = \frac{\frac{1}{\sqrt{n}}\sum_{i=1}^n v_{\boldsymbol{\alpha}}(Z_i)}{\sqrt{\frac{1}{n}\sum_{i=1}^n v_{\boldsymbol{\alpha}}(Z_i)^2}}. \tag{6}$$

The variable $U_n(\boldsymbol{\alpha})$ converges in law to the standard normal distribution for each $\boldsymbol{\alpha}$. If we consider $U_n(\boldsymbol{\alpha})$ over all $\boldsymbol{\alpha}$, it can be seen as a stochastic process over $\boldsymbol{\alpha}$ or C. If the higher order term of $o_p(1/n)$ is bounded uniformly over $\boldsymbol{\alpha}$, and the stochastic process converges uniformly to a Gaussian process, the limit of $\sup_{\boldsymbol{\alpha}} 2nL_n(\boldsymbol{\alpha}, \hat{\beta}_{\boldsymbol{\alpha}})$ is the square of the supremum of the Gaussian process. Dacunha-Castelle and Gassiat [1] discussed this case, assuming that the class $\{v_{\boldsymbol{\alpha}}(z)\}$ is Donsker [9].

Let $(\Omega, \mathcal{A}, P)$ be a probability space, $(\mathcal{Z}, \mathcal{B})$ be a measurable space, and $Z_1, Z_2, \ldots$ be i.i.d. with their value in $\mathcal{Z}$. A family of Borel measurable functions $\mathcal{F} \subset \{v : \mathcal{Z} \to \mathbb{R}\}$ is called *Donsker* if $E_P[v(Z)]$ and $E_P[v(Z)^2]$ exist for all $v \in \mathcal{F}$, the map $z \mapsto \sup_{v\in\mathcal{F}} |v(z)|$ is finite for every $z \in \mathcal{Z}$, and the $\mathcal{F}$-indexed empirical processes $\frac{1}{\sqrt{n}}\sum_{i=1}^n (v(Z_i) - E_P[v(Z)])$, as random elements with their values in the Banach space $\ell^\infty(\mathcal{F})$ of all bounded functions on $\mathcal{F}$ with sup norm, converge in law to a tight Borel measurable random element with its value in $\ell^\infty(\mathcal{F})$.

For Donsker cases, Dacunha-Castelle and Gassiat [1] clarify the limiting distribution of $\sup L_n(\theta)$. Non-Donsker cases are complex, and even the order of $\sup L_n(\theta)$ may be different from the usual $O_p(1/n)$. Hartigan [8] and Hagiwara et al. [2] reported a larger order than $O_p(1/n)$ in specific models. We will derive a useful sufficient condition for such larger order in terms of the tangent cone.

Donsker cases

To apply the theory of Gaussian processes, we need the uniformity over $\boldsymbol{\alpha}$ of the small order in eq.(6), and a uniform version of the consistency and efficiency condition. However, we omit it here, and simply refer it as the *uniformity condition*. The next theorem is due to Dacunha-Castelle and Gassiat [1].

Theorem 1. *Let a statistical model $S = \{f(z;(\boldsymbol{\alpha},\beta))\}$ be locally conic at $f_0(z)$. Assume that a uniformity condition is satisfied, and the basis of the tangent cone $C = \{v_{\boldsymbol{\alpha}}(z) = \frac{\partial f(z;(\boldsymbol{\alpha},0))}{\partial\beta}\}$ is Donsker. Then $n\sup_{(\boldsymbol{\alpha},\beta)} L_n(\boldsymbol{\alpha},\beta)$ converges in law to $\frac{1}{2}\sup_{v\in C} W^2$, where W is a tight Gaussian process over C, to which U_n converges.*

A sufficient condition of Donsker is known [9]. A class $\mathcal{F}$ is Donsker if (i) $F(z) = \sup_{v\in\mathcal{F}} |v(z)|$ is in $L^2(P)$, (ii) the square root of the uniform entropy number is integrable, and (iii) some measurability conditions are satisfied. In these three conditions, the measurability is automatic if $\mathcal{Z}$ is a separable metric space and any function in $\mathcal{F}$ is continuous. The integrability is satisfied if the VC-dimension of $\mathcal{F}$ is finite. These are often satisfied by many models, including multilayer perceptrons.

In Donsker cases, similarly to regular models, the following fact holds.

Theorem 2. *Under the same assumptions as in Theorem 1, we have $D(\hat{\boldsymbol{\alpha}}, \hat{\beta}) = L_n(\hat{\boldsymbol{\alpha}}, \hat{\beta}) + o_p(1/n)$, and both $D(\hat{\theta})$ and $L_n(\hat{\theta})$ are of the order $O_p(1/n)$.*

Non-Donsker cases

A marginal of U_n on finite points in C converges to a normal distribution with covariance $E_P[v_i v_j]$. The components are independent if the covariance is zero. If we can find an arbitrary number of "almost independent" variables in C, the supremum of U_n takes an arbitrary large value. Hartigan [8] applied this idea to a normal mixture, calculating the covariance explicitly. We extend it to the following

Theorem 3. *Let a statistical model $S = \{f(z; (\boldsymbol{\alpha}, \beta))\}$ be locally conic at $f_0(z)$, and $C = \{v_{\boldsymbol{\alpha}}(z) = \frac{\partial f(z;(\boldsymbol{\alpha},0))}{\partial \beta}\}$ be the basis of the tangent cone. Suppose there exists a sequence $\{v_n\}$ in C such that $v_n \to 0$ almost every z, then, for arbitrary $M > 0$*

$$\lim_{n\to\infty} \Pr(\sup_{(\boldsymbol{\alpha},\beta)} nL_n(\boldsymbol{\alpha}, \beta) \leq M) = 0. \tag{7}$$

Proof. From Proposition 2, for any $\varepsilon > 0$ and $K \in \mathbb{N}$, there exist K elements in C such that $|E[v_{\alpha_i} v_{\alpha_j}]| < \varepsilon$. Then, use the same argument as Hartigan [8]. □

Proposition 2. *Let $\{v_n\}$ be a sequence in $L^2(P)$ such that $\|v_n\|_{L^2(P)} = 1$ for all n, and $v_n(x) \to 0$ for almost every x. Then, for arbitrary $n \in \mathbb{N}$ and $\varepsilon > 0$, there exists $\ell_0 \in \mathbb{N}$ such that $E_P|v_n v_\ell| < \varepsilon$ holds for all $\ell \geq \ell_0$.*

Likelihood of multilayer perceptrons

We apply the results in the previous sections to multilayer perceptrons. For simplicity, we discuss only networks without bias terms in the output unit; that is,

$$\varphi(x; \theta) = \textstyle\sum_{j=1}^{H} b_j\, s(a_j x + c_j). \tag{8}$$

We further assume the constant zero as the true function. Similar to the previous discussion, a locally conic parameterization is given by $\psi(z; (\boldsymbol{\alpha}, \beta)) = \beta \sum_{j=1}^{H} \eta_j\, s(a_j x + c_j)$ for $\boldsymbol{\alpha} = (a_j, c_j, \eta_j)$ and $\beta \in \mathbb{R}$. We use the exponential family in the previous section for $r(y|u)$, and assume the variance of the

y is one w.l.o.g. The basis of the tangent cone C consists of the functions $v_{\alpha}(x,y) = y\sum_{j=1}^{H}\eta_j s(a_j x + c_j)/\|\sum_{j=1}^{H}\eta_j s(a_j x + c_j)\|_{L^2(Q)}$. It is easy to see that C includes a sequence converging to 0 almost everywhere, if $H \geq 2$. In fact, we can make such a sequence by taking $y\{s(a_1 x + c) - s(a_2 x - c)\}/\|s(a_1 x + c) - s(a_2 x - c)\|$, and making $c \to 0$ and $a_1, a_2 \to \infty$. Thus, Theorem 3 gives

Theorem 4. *Assume the true function is constant zero, and the noise model $r(y|u)$ is $r(y|u) = \exp(y\kappa(u) + \tau(y) - \zeta(u))$ satisfying $\kappa'(u) \neq 0$ and $\int yr(y|u)dy = u$. For the multilayer perceptron model (8) with $H \geq 2$, we have*

$$\sup_{\theta} L_n(\theta) \gtrsim O_p(1/n). \tag{9}$$

We can derive a tighter lower bound by counting almost independent variables.

Theorem 5. *Under the same assumptions as in Theorem 4, we have*

$$\sup_{\theta} L_n(\theta) \gtrsim O_p\left(\tfrac{\log n}{n}\right). \tag{10}$$

We omit the proof. The order $O_p(\log n/n)$ has been obtained by Hagiwara et al. [2] under the assumption of additive Gaussian noise. Our approach can be applied to various noise models, including binary output models.

The behavior of $\sup_{\theta} L_n(\theta)$ deeply depends on the functional property of the tangent cone C. If the multilayer perceptron has only one hidden unit, the behavior is totally different. We can prove that C is Donsker, and obtain the following

Theorem 6. *Under the same assumptions as in Theorem 4, for the multilayer perceptron (8) with 1 hidden unit, we have $D(\hat{\theta}) = L_n(\hat{\theta}) + o_p(1/n) = O_p(1/n)$.*

Remark: We have omitted many of the proofs, which will be given in a forthcoming paper.

REFERENCES

1. Dacunha-Castelle, D. and Gassiat, E., *ESAIM Prob. and Statist.*, **1**, 285–317 (1997).
2. Hagiwara, K., Kuno, K., and Usui, S., "On the problem in model selection of neural network regression in overrealizable scenario," in *Proc. IJCNN2000*, Incontrol Productions, California, 2000, vol. VI, pp.461–466.
3. Sussmann, H.J., *Neural Networks* **5**, 589–593 (1992).
4. Fukumizu, K., *Neural Networks*, **9**, 871–879 (1996).
5. Fukumizu, K. and Amari, S., *Neural Networks*, **13**, 317–327 (2000).
6. Pistone, G. and Sempi, C., *Ann. Statist.*, **23**, 1543–1561 (1995).
7. Hartigan, J.A., "A failure of likelihood asymptotics for normal mixtures," in *Proc. Berkeley Conf. in Honor of Jerzy Neyman and Jack Kiefer*, vol.II, 1985, pp.807–810.
8. Hartigan, J.A., "A failure of likelihood asymptotics for normal mixtures," in *Proc. Berkeley Conf. in Honor of Jerzy Neyman and Jack Kiefer*, edited by L.M. Le Cam and E.A. Olshen, Wadsworth, California, 1985, vol.II, pp.807–810.
9. Van der Vaart, A.W. and Wellner, J.A., *Weak convergence and empirical processes.* Springer Verlag, New York, 1996.

Information Geometry and the Quantum Estimation Problem: The Phase-Space Connection

Demetris P.K. Ghikas[1]

Dept. of Mathematics, King's College, University of London, Strand London WC2R 2LS, UK

Abstract. The tools of classical information geometry have not been fully extended to the estimation problem in quantum mechanics. We propose a phase-space definition of information geometry as an intermediate structure between the classical and quantum ones. We present the first results on a particular example and discuss the possibilities and open problems.

INTRODUCTION AND MAIN RESULTS

In the classical parameter estimation theory one is given the data of an experiment and the problem is to determine the parameters of the assumed model. Though the precise form of the model (be it parametric or non-parametric) is not known, its objective existence is not questioned. Furthermore, in general, for this stage of statistical analysis the knowledge of the experimental setup for the acquisition of the data is not important. On the contrary in the quantum estimation problem, the very meaning of the a priori distribution is part of the foundation of the theory of quantum mechanics. Not only the objective existence of a state function is a delicate question, but furthermore the quantum mechanical measurement, that is the collection of data to be analysed, needs an extremely careful approach. We may not touch on these fine questions here. We simply want to stress that any extension of classical information concepts to quantum mechanics is bound to involve difficult mathematical tools and a very careful treatment of the relevant physical concepts. Quantum information geometry is a serious attempt to extend the classical geometry to the non-commutative environment of quantum phenomena. There exists already a relatively large literature on this extension, but there is much to be done, especially on the subject of uniqueness and the area of asymptotic estimation theory. We give here a very short review of classical and quantum estimation problems.

[1] Address from 1st August 2000: Dept. of Physics, University of Patras, Patras 26110, Greece e-mail: ghikas@physics.upatras.gr

CP553, *Disordered and Complex Systems*, edited by P. Sollich, et al.

Classical Estimation Problem: The C-R inequality is $V_\xi(\hat{\xi}) \geq G^{-1}(\xi)$, i.e. $V_\xi - G^{-1} \geq 0$ is a positive semi-definite matrix where G_ξ is the *Fisher information matrix* and $V_\xi(\hat{\xi})$ is the m.s.e. of the unbiased estimator $\hat{\xi}$. If $V_\xi(\hat{\xi}) = G(\xi)^{-1}\ \forall\xi$ then $\hat{\xi}$ is an *efficient estimator*. It is proved that the maximum likelihood estimator is *asymptotically efficient*.

Quantum State Reconstruction: In the *estimation problem* our concern is the estimation of certain parameters of the quantum state which completely or partially characterize it. In the *reconstruction problem* the objective is to reconstruct the state without assuming a particular parametrization but considering the quantum system fully characterized by a set of observables. A very nice introduction can be found in [12].

• To completely reconstruct the quantum state of a system we need an:

accurate	measurement of a
complete	set of observables (the quorum) on an
infinite	set of
identically	prepared systems.

Given the *identically* prepared set we have the possibilities :

• a. *Accurate measurement of the quorum*

1. A *single* observable is measured. If ψ_A is an eigenstate of A the distribution $w_\psi(A) = |\ (\psi_A, \psi)\ |^2$ *cannot* reproduce ψ. This is related to Pauli's problem. In the optical homodyne tomography practice for the harmonic oscilator one can instead measure the so called *quadrature* observable $\hat{x}_\theta = \hat{q}\cos\theta + \hat{p}\sin\theta$. This offers an infinity of distribution functions which are used to reconstruct the Wigner function.
2. Measure *conjugate* observables (like $\hat{q}, \hat{p}$). Due to non-commutativity this introduces quantum noise. In this case modeling the measuring device as a filter (quantum ruler) we may obtain the *operational probability distributions* which give smoothed Wigner function (Husimi distributions).

• b. *Accurate* measurement of a *subset* of observables (reduced observational level). This gives a density operator ρ but it is *non unique.* To achieve uniqueness a selection principle must be employed like *Jayne's maximum entropy principle*, which is equivalent to assuming unbiased prior knowledge of the quantum state.

• c. *Finite* number of repeated measurements. Here we need *a priori* conditioning and selection of best strategies. This is achieved with *Bayesian* techniques and appropriate cost functions.

• d. *Finite* number but many *different* experimental possibilities. Here, given the *a priori* inaccuracies due to the finiteness of the number of measurements, we may improve our results by *optimally* choosing the measurements.

Quantum State Estimation: We have a *finite* ensemble of identically prepared systems for *specific* parameters to measure, selecting *optimal, minimal* measurements and using maximum likelihood estimators to quantify the procedure. We need *criteria* of accuracy which are mainly of the form of *Fisher information, Kullback divergences* and *fidelity functions.*

Conceptual Framework: The above (crash) introduction to the classical and quantum estimation problem was meant to present the conceptual and practical environment of a highly non-academic problem, that of describing concrete quantum states in real systems. We state here the conceptual framework of our understanding of the problem and the proposed possible route for new practical techniques.
• There is a *need* for refinement in the quantum estimation analysis.
• There exist advanced geometrical tools in classical statistical geometry.
• There are serious mathematical difficulties for the full extension of these tools to Quantum Mechanics.
• Proposal for an *intermediate* approach through "phase-space information manifolds".

QUANTUM INFORMATION GEOMETRY

The extension of information theory to quantum mechanics has a long history, of which we highlight:
• Need for information theory in quantum mechanics, von Neuman entropy, projective valued measures (PVM). [4]
• Generalization to positive operator valued measures (POVM).
• Relative entropy techniques. [3]
• Geometry of quantum states and statistical distances. [5]
• Information geometry. [1,2]
In the classical case the exponential and mixture families play a crucial role in the development of estimation theory. Furthermore the special tools needed for the asymptotic analysis are based on these families. This fact is connected with the properties of the related exponential and mixture parametrization of the information manifold. In the quantum case the same type of parametrization can be carried out and some quantum estimation results have been obtained. Nevertheless, a higher asymptotic analysis has not been established yet. Furthermore, the extension to quantum mechanics of the classical information techniques is still restricted by some crucial facts, namely [6–9]:
• non-uniqueness of the structure
• definition dependent estimation
• difficulties due to delicate problems when $\dim \mathcal{H} = \infty$.

SOME SELECTED RESULTS

There is a vast and fast increasing literature on quantum information in general and on quantum estimation in particular. There are three main lines of development: Construction of POVM, estimation and reconstruction. We selectively give two representative results for the last two from the literature.

Estimation [11]: There is much information contained in this paper. We select just a few key points which are relevant for our proposal.

• A new CR inequality is established.
• One can "trade" information about one parameter for that about another.
Some details: Consider a finite N and $\rho^N = \rho \otimes \cdots \otimes \rho, M = \{M_\xi\}$ a POVM, i.e. $M_\xi \geq 0, \sum_\xi M_\xi = I$. Let $p(\xi \mid \theta) = Tr[\rho^N(\theta)M_\xi)]$ be the probability of a measurement M_ξ. Let $V_{ij}^N(\theta)$ be the covariance of the estimator and $I_{ij}^N(M,\theta)$ the classical Fisher matrix. The classical C-R inequality says $V^N \geq (I^N)^{-1}$. Then the following questions are posed: 1. Is there a simple bound for I^N? 2. Is this bound asymptotically valid for biased estimators? 3. Can this bound be attained for N large? For θ 1-dimensional we have $\partial_\theta \rho = \frac{1}{2}(\lambda_\theta \rho + \rho \lambda_\theta)$, $I_{\theta\theta}^N(M,\theta) \leq NTr(\rho\lambda_\theta\lambda_\theta)$, and the bound can be attained. For the multiparameter case for the Helstrom information matrix $H_{ij} = \frac{1}{2}Tr[\rho(\lambda_i\lambda_j + \lambda_j\lambda_i)]$ we have $V^N(M,\theta) \leq NH(\theta)$ and the bound is *not* in general achievable. Some of the many results of the paper are:
• For pure states $Tr[H^{-1}(\theta)I^N(M,\theta)] \leq (d-1)N$, $\quad(\star)$
• For mixed states $(\star)$ is satisfied for separate measurements.
• Non-additivity of Fisher information, i.e., collective measurements do *not* satisfy the condition $(\star)$.
• $\int d\theta\lambda(\theta)Tr[C(\theta)V^N(\theta)] \geq \int d\theta\lambda(\theta)Tr[H^{-1/2}C(\theta)H^{-1/2}(\theta)]^2 - \frac{\alpha}{N^2}$ where $\lambda(\theta)$ is a prior distribution and $C(\theta)$ a quadratic cost function.
We are certainly not doing justice to this paper here, but only wanted to show the type of analysis one performs for the estimation problem, and to point to the existence of asymptotics that one should develop further.

Reconstruction [10]: The whole idea of quantum state reconstruction is based on the phase-space representation of quantum mechanics. This is a very old and vast area, both in terms of abstract and rigorous developments and lately of concrete practical applications. The framework here is a phase-space formulation using coherent states for Lie Groups (Heisenberg-Weyl, SU(2)). For an operator A on $\mathcal{H}$ the Stratonowitz-Weyl correspondence is a map $A \mapsto F_A(\Omega; s)$ where $F_A(\Omega; s)$ is a function on the phase-space (Ω=coherent state). F_A is called the SW symbol of A. We need to construct the SW kernel $\Delta(\Omega; s)$ s.t.

$$F_A(\Omega; s) = Tr[A\Delta(\Omega; s)], \text{ and } A = \int d\mu(\Omega)F_A(\Omega; s)\Delta(\Omega; -s)$$

The construction of $\Delta(\Omega; s)$ has been established using various tools and under various disguises, old and new. We point out the specific resulting function when $A = \rho$, i.e. $F_\rho(\Omega; s)$. We have: $s = 0$ for Wigner functions, $s = -1$ for the Q functions (Husimi) and $s = 1$ for the P functions. The crucial observation is that $s = -1$ is the maximal negative value for which $F_\rho(\Omega; s)$ is *non negative.* This is important for the proposed use of phase-space densities.

Comments:
a. In practice we have at our disposal a finite number N of identically prepared systems and we may employ a finite number n of optimally selected POVM.
b. The estimation depends crucially on n and N. We need control over the asymptotics of the C-R bounds.

c. For the state reconstruction we need measurements of the operational probabilities that are as accurate as possible, given the fact that n and N are finite.

Possibilities:

1. Ordering of importance for the parameters, i.e. a selection of models with the most important parameters to be measured. We may then take into account the rest of the parameters using the asymptotics provided by the information geometric structure.
2. Stochastic parameters and their effect on estimation. These could possibly be treated using the geometry.

INFORMATION GEOMETRIC TOOLS

As has been pointed out in the previous paragraph, in order to control the asymptotics of estimation (and reconstruction) we need higher order statistical analysis. In classical information geometry such tools exist, but their quantum extensions are not yet available [1,2]. In the classical case we embed the statistical model in a bigger exponential family and then the geometry of the embedding provides control over the bounds in terms of the appropriate curvatures. From this point of view we see the possibility of selecting important parameters as a selection of the model to be embedded and the treatment of the rest as playing the role of ancillary parameters. In this framework one could treat the stochastic parameters as nuisance parameters. Unfortunately, in the case of quantum manifolds it is not clear (to us) how this embedding can be done in a unique way. Our proposal is to use phase-space information geometry based on quasi probability distributions in the hope that this can provide a better control of the problem (hopefully unique).

Classical asymptotic analysis [1,2]: Let $H_M^{(e)}, H_A^{(m)}$ be the embedding curvatures of $\mathcal{M}$, the submanifold of the model and $\mathcal{A}$ the ancillary submanifold with the indicated connections., and $\Gamma_M^{(m)}$ the m-curvature of $\mathcal{M}$. We have: The asymptotic m.s.e. of a bias-corrected efficient estimator depends in a non trivial way on these geometric quantities. We have to comment here that one of the problems is the non-exponentiality of a general model. This is a current problem and some deep results have already been obtained.

Phase-space possibilities: The phase-space representation of quantum mechanics has been very fruitful in producing alternative approaches to both fundamental questions and concrete practical applications. It is a picture somehow between the quantum system and its classical relative. It seems plausible that a phase-space information geometry might play a similar role. We are currently investigating this possibility. We summarize our approach as follows: 1. Construct appropriate phase-space representation. 2. Select a positive Q.P. density. 3. Construct a Fisher type information manifold using logarithmic derivatives. 4. Use this geometry to construct C-R asymptotic bounds. 5. Use SW correspondence to "invert" the logarithmic derivatives and compare with the choices of quantum information geometry.

A phase-space exercise: Let $\mid z, \xi\rangle = D(z)S(\xi) \mid 0\rangle$ be a squeezed state, where $D(z)$, $S(\xi)$ are the displacement and squeezing operators and $\mid 0\rangle$ the Fock vacuum. Then the Wigner function has the form:

$$\rho(q,p) = C \exp[-\alpha(q - Az_1)^2 - \beta(p/\hbar - Bz_2)^2 - \gamma(q - Az_1)(p/\hbar - Bz_2)]$$

where q, p are the phase-space coordinates, $z = z_1 + z_2$ the coherent state parameters and α, β, γ, C are constants. We have computed the classical Fisher matrix for this density function and found exactly the form of the quantum Fisher matrix of Fujiwara and Nagaoka [13] (apart from a puzzling factor of 2).

ACKNOWLEDGMENTS

I would like to thank the Mathematics Department of King's College, University of London for their kind hospitality during the academic year 1999-2000 and especially Prof. R.F. Streater for his long, deep and exciting private seminars on information geometry.

REFERENCES

1. Amari S.-I., *Differential Geometric Methods in Statistics*, Lecture Notes in Statistics, vol 28, Springer, New York (1985).
2. Marray M. K., Rice J. M., *Differential Geometry and Statistics*, Chapman and Hall, London (1993).
3. Ohya M., Petz D.,*Quantum Entropy and Its Use*, Springer-Verlag, Berlin (1993).
4. Streater R. F., *Journ. Math. Phys.* 3556-3603 (2000).
5. Uhlmann A., *Rep. Math. Phys.* **33**, 253-263 (1993).
6. Petz D., *Jour. Math. Phys.* **35**, 780-795 (1994).
7. Petz D., Sudar C.: *Journ. Math. Phys.* **37**, 2662-2673 (1996).
8. Hasegawa H., *Rep. Math. Phys.* **33**, 87-93 (1993) and **39**, 49-68 (1997).
9. Grasselli M. R., Streater R. F., "Uniqueness of the Fisher Metric in Finite Dimensional Quantum Information Geometry", preprint math-ph/9910031.
10. Bril C., Mann A., "Phase space formulation of Quantum Mechanics and quantum state reconstruction for physical systems", preprint quant-ph/9809052.
11. Gill R. D., Massar S., "State estimation for large ensembles", preprint quant-ph/9902063.
12. Buzek V., Drobny G., Derka R., Adam G., Wiedemann H., "Quantum State reconstruction for incomplete date", preprint quant-phys/9805020.
13. Fujiwara A., Nagaoka H., *Journ. Math. Phys.* **40**, 4227-4239 (1999).

Monotone Metrics on Statistical Manifolds of Density Matrices by Geometry of Non-Commutative L^2-Spaces

Paolo Gibilisco*, Tommaso Isola†

*Dipartimento di Matematica, Politecnico di Torino, Corso Duca degli Abruzzi 24, I–10129 Torino, Italy, and Centro Vito Volterra, Università di Roma "Tor Vergata". E-mail: gibilisco@polito.it.

†Dipartimento di Matematica, Università di Roma "Tor Vergata", via della Ricerca Scientifica, I–00133 Roma, Italy. Email: isola@mat.uniroma2.it

Abstract. Using an integral decomposition of non-commutative monotone metrics we show that each monotone metric described by the Petz classification theorem is related to the geometry of a suitable non-commutative L^2-space. This exactly reproduces and generalizes the commutative case where the unique monotone metric (Chentsov theorem about Fisher-Rao metric) is classically related to the commutative L^2-geometry.

INTRODUCTION

The concept of monotone metric for parametric statistical manifolds has been introduced by Chentsov [3,4] and further developed by Petz [14]. There are two fundamental results: i) the Chentsov uniqueness theorem [2,3]; ii) the Petz classification theorem [14]. The first theorem says that in the commutative case there exists a unique monotone metric (up to a scalar factor) and that this metric coincides with the well-known Fisher-Rao metric. In the non-commutative case the situation is, as usual, more complicated and richer. This means that there is no uniqueness and that we have an infinite family of different monotone metrics. The classification theorem by Petz shows that there is a natural bijection between the family of monotone metrics and the family of operator monotone functions.
It is well-known that the Fisher-Rao metric can be related to the geometry of commutative L^2-spaces [5]. The purpose of this paper is to answer the following question: which non-commutative monotone metrics arise from the geometry of non-commutative L^2-spaces (following the line of the commutative case)? Such a question is relevant for at least three reasons: i) it is natural to ask which features of the commutative case survive in the non-commutative one; ii) to interpret some

CP553, *Disordered and Complex Systems*, edited by P. Sollich, et al.

norms and scalar products as L^2-norms and scalar products opened the way to the recently established non-parametric version of information geometry [6,16,17]; iii) it has been proved [6,8] in the commutative case that α-geometries for $\alpha \in (-1,1)$ are related to the geometry of L^p-spaces where $p = \frac{2}{1-\alpha}$. The classic work of Amari shows that the Riemannian Fisher-Rao metric induces the 0-connection. It is reasonable therefore to hope that one may associate, using suitable non-commutative L^p-spaces, a family of α-connections to those non-commutative monotone metrics that are associated to L^2-type metrics [8].
The purpose of this note is to show that there is a general positive answer, given by Theorem 12, to the question formulated above. This means the following. Let $\mathcal{P}_n$ be the probability simplex in $\mathbf{R}^n$ and $T_p\mathcal{P}_n$ be the tangent space at $p \in \mathcal{P}_n$, and let us identify $\mathbf{R}^n$ with $L^2(m)$, where m is the counting measure on $\{1,\ldots,n\}$. If one defines $M_p(v)_i := p_i v_i$, $p \in \mathcal{P}_n$, $v \in T_p\mathcal{P}_n$, then the Chentsov Theorem can be rephrased by saying that "each" monotone metric turns into an isometry the linear map $v \in T_p\mathcal{P}_n \to M_p^{-1/2}(v) \in L^2(m)$, and that this property characterises monotone metrics.
Now let $\mathcal{D}_n$ be the manifold of invertible density matrices, so that $T_\rho\mathcal{D}_n$, $\rho \in \mathcal{D}_n$, is the space of hermitian, traceless matrices, and denote by $L^2(\tau)$ the Hilbert space of all n by n matrices endowed with the scalar product given by the normalised trace τ. Define $L_\rho(A) := \rho A$, $R_\rho(A) := A\rho$, and for any symmetric measure μ on $[0,1]$ define $M_{\rho,\mu}^{-1/2} : T_\rho\mathcal{D}_n \to \mathcal{H}_\mu := L^2([0,1],\mu;L^2(\tau))$, by $M_{\rho,\mu}^{-1/2} := \int_{[0,1]}^{\oplus}((1-s)L_\rho + sR_\rho)^{-1/2}d\mu(s)$. Then Theorem 12 shows that each noncommutative monotone metric turns into an isometry the linear map $A \in T_\rho\mathcal{D}_n \to M_{\rho,\mu}^{-1/2}(A) \in \mathcal{H}_\mu$, by a suitable measure μ, and that this property characterises monotone metrics. In this sense $M_{\rho,\mu}^{-1/2}$ appears to be a noncommutative analogue of the division by the square root, and $L^2([0,1],\mu;L^2(\tau))$ an analogue of $L^2(m)$.
In the last section we discuss the possible relevance of this result to the noncommutative theory of α-connections.

THE FISHER-RAO METRIC AND THE CHENTSOV UNIQUENESS THEOREM

Denote by $\mathcal{P}_n = \{p \in \mathbf{R}^n : \sum_{i=1}^n p_i = 1, p_i > 0, i = 1,\ldots,n\}$ the probability simplex in $\mathbf{R}^n$ and by S the sphere of radius 2 in $\mathbf{R}^n$. Define a function $A : \mathcal{P}_n \to S$ by $A(p)_i = 2p_i^{\frac{1}{2}}$ and consider the Riemannian structure that S induces on $\mathcal{P}_n$ by this embedding. Let $p(t)$ be a curve on $\mathcal{P}_n$. We transport this curve on S by the embedding A and determine the induced Riemannian metric. As $\frac{d}{dt}A(p(t))_i = \frac{1}{\sqrt{p_i(t)}}\frac{d}{dt}(p_i(t))$, we get

$$\|\frac{d}{dt}A(p(t))\|^2 = \sum_{i=1}^n(\frac{d}{dt}A(p(t))_i)^2 = \sum_{i=1}^n \frac{1}{p_i(t)}\left(\frac{d}{dt}(p_i(t))\right)^2.$$

So we obtain the well-known Fisher-Rao metric

$$\sum_{i=1}^{n} \frac{1}{p_i} v_i w_i = \sum_{i=1}^{n} \frac{v_i}{\sqrt{p_i}} \frac{w_i}{\sqrt{p_i}}.$$

Now let m be the counting measure on $\{1, \ldots, n\}$, so that $L^2(m)$ can be identified with $\mathbf{R}^n$ with the usual scalar product. Then the Fisher-Rao metric is induced by the linear isomorphism $v \to \frac{v}{\sqrt{p}}$ that identifies the tangent space of $\mathcal{P}_n$ at p, $T_p\mathcal{P}_n = \{v \subset \mathbf{R}^n : \sum v_i = 0\}$, with the tangent space of the unit sphere $S^2(m) \subset L^2(m)$ at the point $\sqrt{p}$, $T_{\sqrt{p}}(S^2(m)) = \{u \in L^2(m) : \langle u, \sqrt{p} \rangle = 0\}$, where $\langle u, v \rangle := \sum_{i=1}^{n} u_i v_i$ denotes the scalar product of $\mathbf{R}^n$.

Recall that a monotone metric is a family $\{g_p : p \in \mathcal{P}_n, n \in \mathbf{N}\}$ of inner products on $T_p\mathcal{P}_n$, such that $g_{Tp}(Tu, Tu) \leq g_p(u, u)$, for any stochastic map $T : \mathbf{R}^n \to \mathbf{R}^k$.

Define $(M_p(v))_i = p_i v_i$, $i = 1, \ldots, n$. We may formulate the

Theorem 1. *Chentsov uniqueness theorem.* There exists a unique (up to a scalar factor) monotone metric on $T_p\mathcal{P}_n$. This metric is the scalar product that turns into an isometry the linear map

$$v \in T_p\mathcal{P}_n \to M_p^{-1/2}(v) \in L^2(m).$$

Proof. Evidently

$$\sum \frac{1}{p_i} v_i w_i = \langle M_p^{-1/2}(v), M_p^{-1/2}(w) \rangle.$$

The proof of monotonicity can be found in [2,3]. ♣

OPERATOR MONOTONE FUNCTIONS AND CHENTSOV-MOROZOVA FUNCTIONS.

Let us recall [1] that a function $f : (0, \infty) \to R$ is called *operator monotone* if for any $n \in N$, any A, $B \in M_n(\mathbf{C})$ such that $0 \leq A \leq B$, the inequalities $0 \leq f(A) \leq f(B)$ hold. By Löwner's results they can be represented in integral form. To make expressions compact, let us introduce the notation

$$\phi(x, t) = \frac{x(1+t)}{x+t}, \quad \text{for} \quad x > 0, \qquad t \geq 0.$$

For fixed $x > 0$ the function $\phi(x, t)$ is bounded and continuous on the extended half–line $[0, \infty]$.

Theorem 2. ([12] p. 208–9) The map $m \to f$, defined by

$$f(x) = \int_{[0,\infty]} \phi(x, t) dm(t), \qquad \text{for} \quad x > 0,$$

establishes an affine isomorphism from the class of positive Radon measures on $[0,\infty]$ onto the class of operator monotone functions.

Remark. In the above representation one has $f(0) = \inf_x f(x) = m(\{0\})$ and $\inf_x \frac{f(x)}{x} = m(\{\infty\})$.

Some other operator monotone functions are associated with a given operator monotone function f ([12] p. 213–4), among them the *transpose* function, $f'(x) := xf(x^{-1})$, and the *dual* function, $f^{\perp}(x) := \frac{x}{f(x)}$. These transformations are involutive, that is $f'' = f$, $f^{\perp\perp} = f$. Moreover f is said to be *symmetric* if $f = f'$.

We need a different representation for the operator monotone functions. It is based on the following result.

Lemma 3. (see [11]) Define $g : [0,1] \to [0,\infty]$ by $g(s) = \frac{s}{1-s}$, $s \in [0,1)$, and $g(1) = \infty$. Let $A \subset [0,1]$ and $B \subset [0,\infty]$ be measurable sets and let μ be a positive Radon measure on $[0,1]$ and m a positive Radon measure on $[0,\infty]$. The formulae $m_\mu(B) := \mu(g^{-1}(B))$, $\mu_m(A) = m(g(A))$, establish a bijection between the class of positive Radon measures on $[0,1]$ and the class of positive Radon measures on $[0,\infty]$. Moreover, if h is an integrable function w.r.t. m, then

$$\int_{[0,\infty]} h(t)dm(t) = \int_{[0,1]} h(g(s))d\mu_m(s).$$

Proposition 4. The map $\mu \to f$, defined by

$$f(x) = \int_{[0,1]} \frac{x}{(1-s)x+s} d\mu(s), \qquad \text{for} \quad x > 0,$$

establishes a bijection between the class of positive Radon measures on $[0,1]$ and the class of operator monotone functions.

Proof.

$$f(x) = \int_{[0,\infty]} \phi(x,t)dm(t) = \int_{[0,1]} \phi(x,g(s))d\mu_m(s) =$$

$$= \int_{[0,1]} \frac{x\left(1+\frac{s}{1-s}\right)}{x+\frac{s}{1-s}} d\mu_m(s) = \int_{[0,1]} \frac{x}{(1-s)x+s} d\mu_m(s).$$

♣

In the above correspondence we write $f = f_\mu$ or $\mu = \mu_f$ to indicate that f_μ is the operator monotone function associated with μ or, conversely, that μ_f is the measure associated with f.

Corollary 5. ([21] p. 474) The map $\mu \mapsto f$, defined by

$$\frac{1}{f(x)} = \int_{[0,1]} \frac{1}{(1-s)x+s} d\mu(s) \qquad \text{for} \quad x > 0$$

establishes a bijection between the class of positive Radon measures on $[0,1]$ and the class of operator monotone functions.
Proof. For each operator monotone function f one has

$$\frac{1}{f(x)} = \frac{1}{x} f^{\perp}(x) = \frac{1}{x} \int_{[0,1]} \frac{x}{(1-s)x+s} \, d\mu_{f^{\perp}}(s) = \int_{[0,1]} \frac{1}{(1-s)x+s} \, d\mu_{f^{\perp}}(s).$$

♣

Definition 6. With each operator monotone function f one associates the so-called *Chentsov–Morozova* function

$$c_f(x,y) := \frac{1}{y f\left(\frac{x}{y}\right)}, \qquad \text{for} \quad x,y>0.$$

Proposition 7. The map $\mu \mapsto c(\cdot,\cdot)$, defined by

$$c(x,y) = \int_{[0,1]} \frac{1}{(1-s)x+sy} \, d\mu(s), \qquad \text{for} \quad x,y>0$$

establishes a bijection between the class of positive Radon measures on $[0,1]$ and the class of Chentsov–Morozova functions.
Proof. By the Corollary 5 we have

$$c_f(x,y) = \frac{1}{y f\left(\frac{x}{y}\right)} = \frac{1}{y} \int_{[0,1]} \frac{1}{(1-s)\frac{x}{y}+s} \, d\mu_{f^{\perp}}(s) = \int_{[0,1]} \frac{1}{(1-s)x+sy} \, d\mu_{f^{\perp}}(s).$$

♣

Note that f is symmetric *iff* c_f (or μ_f) is, *i.e.* it satisfies $c(x,y) = c(y,x)$ (or $d\mu(s) = d\mu(1-s)$).

THE MAIN RESULT

Let τ be the normalised trace on $M_n(\mathbf{C})$, and let $\mathcal{D}_n := \{A \in M_n(\mathbf{C}) : A > 0, \tau(A) = 1\}$ be the set of density matrices. The tangent space to $\mathcal{D}_n$ at $\rho \in \mathcal{D}_n$ can be naturally identified with $\{A \in M_n(\mathbf{C}) : A = A^*, \tau(A) = 0\}$. Similarly to the commutative case, a symmetric monotone metric is a family $\{g_\rho : \rho \in \mathcal{D}_n, n \in \mathbf{N}\}$ of inner products on $T_\rho\mathcal{D}_n$, such that $\rho \in \mathcal{D}_n \to g_\rho(A,A) \in [0,\infty)$ is continuous, for any $A \in T_\rho\mathcal{D}_n$, and $g_{T\rho}(TA,TA) \leq g_\rho(A,A)$, for any stochastic (i.e. completely positive, trace preserving) map $T : M_n(\mathbf{C}) \to M_k(\mathbf{C})$.

Theorem 8. *Petz classification theorem.* There exists a bijective correspondence between symmetric monotone metrics and symmetric operator monotone functions

$f : (0, \infty) \to (0, \infty)$, which is given by $g_\rho(A, B) = \tau(Ac_f(L_\rho, R_\rho)(B))$, for A, $B \in T_\rho \mathcal{D}_n$, where c_f is the CM-function associated with f.

We want to give a different description of the Petz classification theorem. So let us start with some definitions.

Definition 9. Denote by $L^2(\tau)$ the vector space $M_n(\mathbf{C})$ endowed with the inner product $\langle A, B \rangle := \tau(A^* B)$. For any $\rho \in \mathcal{D}_n$, $s \in [0, 1]$, define the operators $L_\rho(A) = \rho A$, $R_\rho(A) = A\rho$, and $M_{\rho,s} := (1 - s)L_\rho + sR_\rho$. Then L_ρ, R_ρ, $M_{\rho,s}$ are positive invertible linear operators on $L^2(\tau)$.

Definition 10. Set $\mathcal{H}_s := L^2(\tau)$, $s \in [0, 1]$, and, for any symmetric positive Radon measure μ on $[0, 1]$, set

$$\mathcal{H}_\mu := \int_{[0,1]}^{\oplus} \mathcal{H}_s d\mu(s) \cong L^2([0, 1], d\mu; L^2(\tau)) \cong L^2([0, 1], d\mu) \otimes L^2(\tau),$$

and $M_{\rho,\mu} := \int_{[0,1]}^{\oplus} M_{\rho,s} d\mu(s)$.

Therefore $\mathcal{H}_\mu$ is endowed with the inner product $\langle A, B \rangle := \int_0^1 \langle A(s), B(s) \rangle d\mu(s)$, if $A : s \in [0, 1] \to A(s) \in L^2(\tau)$, and analogously for B.

Definition 11. Let μ be a symmetric positive Radon measure on $[0, 1]$. The μ-metric on $T_\rho \mathcal{D}_n$, denoted by $\langle \cdot, \cdot \rangle_{\rho,\mu}$, is the inner product on $T_\rho \mathcal{D}_n$ which turns into an isometry the linear map

$$A \in T_\rho \mathcal{D}_n \to M_{\rho,\mu}^{-1/2}(A) \in \mathcal{H}_\mu,$$

where $M_{\rho,\mu}^{-1/2} = \int_{[0,1]}^{\oplus} M_{\rho,s}^{-1/2} d\mu(s)$.

Theorem 12. The family of μ-metrics coincides with the family of symmetric monotone metrics classified by the Petz theorem.

Proof. By the Petz theorem each monotone metric has the form $g_\rho(A, B) = \tau(Ac(L_\rho, R_\rho)(B))$. Therefore we get the conclusion by the following calculation

$$\langle A, B \rangle_{\rho,\mu} = \int_{[0,1]} \langle M_{\rho,s}^{-1/2}(A), M_{\rho,s}^{-1/2}(B) \rangle d\mu(s) =$$

$$= \int_{[0,1]} \tau(A \cdot M_{\rho,s}^{-1}(B)) d\mu(s) = \int_{[0,1]} \tau(A((1 - s)L_\rho + sR_\rho)^{-1}(B)) d\mu(s)$$

$$= \tau\left(A\left(\int_{[0,1]} ((1 - s)L_\rho + sR_\rho)^{-1} d\mu(s)\right)(B)\right) = \tau(Ac_\mu(L_\rho, R_\rho)(B)).$$

♣

The integral decomposition on which Theorem 12 is based has been sketched by Uhlmann in [21].

A DIFFERENT APPROACH

One could also follow a different approach, choosing as a noncommutative analogue of M_p a different interpolation, namely $\widetilde{M}_{\rho,s} := L_\rho^{1-s} R_\rho^s$, for $s \in [0,1]$, as suggested by one of the forms of the BKM monotone metric. We need some preliminaries.

Proposition 13. Let ν be a symmetric positive Radon measure on $[0,1]$. The formula

$$f^\nu(x) := \int_{[0,1]} x^t \mathrm{d}\nu(t)$$

defines a map from the family of positive Radon measures on [0,1] to the class of operator monotone functions. The map is not surjective.

Proof. It is well-known [11] that $f^\nu(x) = \int_{[0,1]} x^t d\beta(t)$, where $\beta : [0,1] \to [0,\infty)$, is increasing and left-continuous, and $\beta(0) = 0$. Besides, it is easy to prove that there are $\beta_n : [0,1] \to [0,\infty)$ increasing, left-continuous, and piecewise constant functions, such that $\beta_n \to \beta$ uniformly in $[0,1]$. As for any fixed $x \in [0,\infty)$, the function $t \in [0,1] \to x^t \in [0,\infty)$ is continuous and bounded, by means of Helly's theorem we get $f^\nu(x) = \lim_{n\to\infty} f_n(x)$, where $f_n(x) := \int_0^1 x^t d\beta_n(t)$. Moreover, if $k \in \mathbf{N}$, $A = A^* \in M_k(\mathbf{C})$ with spectrum contained in $[0,1]$, then $f^\nu(A) = \lim_{n\to\infty} f_n(A)$. Indeed, if $A = \sum_{i=1}^k \lambda_i e_i$ is its spectral decomposition, then $f^\nu(A) = \sum_{i=1}^k f^\nu(\lambda_i)e_i = \sum_{i=1}^k \lim_{n\to\infty} f_n(\lambda_i)e_i = \lim_{n\to\infty} f_n(A)$. Now it is easy to see that f_n is an operator monotone function, being a linear combination with positive coefficients of functions x^{t_i}, which are operator monotone [1]. Finally if $k \in \mathbf{N}$, A, $B \in M_k(\mathbf{C})$, are such that $0 \le A \le B$ we get $f^\nu(A) = \lim_{n\to\infty} f_n(A) \le \lim_{n\to\infty} f_n(B) = f^\nu(B)$, which proves that f^ν is operator monotone.
As for the last statement, the function $\frac{2x}{x+1}$, which is operator monotone [1] and gives the largest monotone metric, is not in the range of the map, because, if ν is not a multiple of the Dirac measure at 0, any f^ν is such that $\lim_{x\to\infty} f^\nu(x) = \infty$, otherwise f^ν is constant. ♣

Corollary 14. The formula

$$c^\nu(x,y) := \int_{[0,1]} (x^{1-t}y^t)^{-1} \mathrm{d}\nu(t)$$

defines a map from the family of positive Radon measures on [0,1] to the class of CM-functions. The map is not surjective.

Definition 15. Let ν be a symmetric positive Radon measure on $[0,1]$. The ν-metric on $T_\rho\mathcal{D}_n$, denoted by $\langle\cdot,\cdot\rangle_\rho^\nu$, is the inner product on $T_\rho\mathcal{D}_n$ which turns into an isometry the linear map

$$A \in T_\rho\mathcal{D}_n \to \widetilde{M}_{\rho,\nu}^{-1/2}(A) \in \mathcal{H}_\nu,$$

where $\widetilde{M}_{\rho,\nu}^{-1/2} := \int_{[0,1]}^{\oplus} M_{\rho,s}^{-1/2} d\nu(s)$.

Theorem 16. The family of ν-metrics is a proper subset of the family of monotone metrics classified by the Petz theorem.
Proof. By definition

$$\langle A, B\rangle_\rho^\nu = \int_{[0,1]} \langle \widetilde{M}_{\rho,s}^{-1/2}(A), \widetilde{M}_{\rho,s}^{-1/2}(B)\rangle d\nu(s) =$$

$$= \int_{[0,1]} \tau(A \cdot \widetilde{M}_{\rho,s}^{-1}(B)) d\nu(s) = \int_{[0,1]} \tau(A(L_\rho^{1-s} R_\rho^s)^{-1}(B)) d\nu(s)$$

$$= \tau\left(A\left(\int_{[0,1]} (L_\rho^{1-s} R_\rho^s)^{-1} d\nu(s)\right)(B)\right) = \tau(Ac^\nu(L_\rho, R_\rho)(B)).$$

Therefore the conclusion follows from the previous results. ♣

UNIFORMLY CONVEX BANACH SPACES

The purpose of this section is to review some results on the geometry of uniformly convex Banach spaces, needed in the sequel. We refer to [8] for full proofs and consider only real Banach spaces. Denote by $\widetilde{X}$ the dual space of X and by S^X the unit sphere of X. If $L \in \widetilde{X}$ and $x \in X$ we write $\langle L, x\rangle = L(x)$.

Definition 17. We say that x is *orthogonal* to y, and denote it by $x \perp y$, if $\|x\| \leq \|x + \lambda y\|$, for any $\lambda \in \mathbf{R}$. Moreover, if $A \subset X$, $x \perp A$ means $x \perp y$, for any $y \in A$.

Definition 18. The *duality mapping* $J : X \to \text{Subsets}(\widetilde{X})$ is defined by $J(x) := \{v \in \widetilde{X} : \langle v, x\rangle = \|x\|^2 = \|v\|^2\}$. We say that X has the *duality map property* if J is single-valued. In this case we set $\widetilde{x} := J(x)$.

Definition 19. We say that X has the *projection property* if for any closed convex $M \subset X$ and any $x \in X$ there is a unique $m \in M$ s.t. $\|x - m\| = \inf\{\|x - z\| : z \in M\} \equiv d(x, M)$. In this case we define $\pi_M(x) := m$.

Definition 20. X is *uniformly convex* if for any $\varepsilon > 0$ there is $\delta > 0$ s.t. $x, y \in S^X$ and $\|\frac{x+y}{2}\| > 1 - \delta$ implies $\|x - y\| < \varepsilon$.

Proposition 21. Let X and $\widetilde{X}$ be uniformly convex Banach spaces. Then
i) X has the projection property.
ii) X has the duality map property.
iii) $x \perp \ker(\widetilde{x})$.
iv) If $M := \ker(\widetilde{x})$, then $\pi_M(v) = v - \frac{\langle \widetilde{x}, v\rangle}{\langle \widetilde{x}, x\rangle} x$.

Now recall that if $\mathcal{M}$ is a Banach manifold and $\mathcal{N} \subset \mathcal{M}$ is a submanifold, then for any $p \in \mathcal{N}$ there is a splitting of the tangent space $T_p\mathcal{M} = T_p\mathcal{N} \oplus V$ and a projection operator $\pi_p : T_p\mathcal{M} \to T_p\mathcal{N}$. Moreover if there is a connection ∇ on $\mathcal{M}$, one gets a connection ∇' on the submanifold $\mathcal{N}$, by setting $\nabla' := \pi \circ \nabla$.

Proposition 22. Let X, $\widetilde{X}$ be uniformly convex Banach spaces. Then
i) S^X is a Banach submanifold of X.
ii) T_xS^X, the tangent space to S^X at $x \in S^X$, can be identified with $\ker(\widetilde{x})$.
iii) The projection operator $\pi_x : T_xX \to T_xS^X$ is given by $\pi_x(v) = v - \langle \widetilde{x}, v\rangle x$.

Using this projection, the trivial connection on X induces a connection on S^X, which we call the *natural connection* on S^X.

We may rephrase the content of this section by saying that if X, $\widetilde{X}$ are uniformly convex then X is "almost an Hilbert space". Note that if X is an Hilbert space, then the natural connection on S^X is just the Levi-Civita connection on the sphere.

α-CONNECTIONS FOR COMMUTATIVE STATISTICAL MANIFOLDS

In this section we summarise some of the results of [6] in the light of the abstract setting of the previous section. Let $(X, \mathcal{A}, \mu)$ be a measure space. We give the following

Definition 23. If $\alpha \in (-1, 1)$, set $p := \frac{2}{1-\alpha}$. $L^p_{\mathbf{R}} \equiv L^p_{\mathbf{R}}(X, \mathcal{A}, \mu) := \{u : X \to \mathbf{R} : u \text{ is } \mathcal{A}\text{-measurable}, \int_X |u|^p d\mu < \infty\}$, for $p \in [1, \infty)$. The unit sphere is denoted by $S^p := \{f \in L^p_{\mathbf{R}} : \|u\|_p = 1\}$. $\mathcal{P}_\mu := \{\rho \in L^1_{\mathbf{R}} : \rho > 0, \int_X \rho = 1\}$. For any $\rho \in \mathcal{P}_\mu$ we set $\mathcal{F}^\alpha_\rho \equiv L^p_0(\rho) := \{u \in L^p_{\mathbf{R}}(X, \mathcal{A}, \rho\mu) : \int_X u\rho d\mu = 0\}$. If $p > 1$ we define $\widetilde{p}$ by $\frac{1}{p} + \frac{1}{\widetilde{p}} = 1$.

A calculation shows that the duality map is given by $u \in L^p_{\mathbf{R}} \to \widetilde{u} := \|u\|_p^{2-p} \mathrm{sgn} u |u|^{\frac{p}{\widetilde{p}}} \in L^{\widetilde{p}}_{\mathbf{R}}$. Therefore, if $\rho \in \mathcal{P}_\mu$, we have that $\rho^{1/p} \in S^p$ and $\widetilde{\rho^{1/p}} = \rho^{1/\widetilde{p}} \in S^{\widetilde{p}}$. The spaces $L^p_{\mathbf{R}}$ are uniformly convex, so the results of the previous section are applicable. For the tangent space of S^p at $\rho^{1/p}$ we have $T_{\rho^{1/p}}S^p = \{u \in L^p_{\mathbf{R}} : \int u\rho^{1/\widetilde{p}} d\mu = 0\}$. We denote by ∇^p the natural connection on S^p induced by the trivial connection on $L^p_{\mathbf{R}}$. Observe that the isometric isomorphism $I^p_\rho : u \in L^p_{\mathbf{R}}(X, \mathcal{A}, \mu) \to u\rho^{-1/p} \in L^p_{\mathbf{R}}(X, \mathcal{A}, \rho\mu)$ sets up a bijection between $T_{\rho^{1/p}}S^p$ and $L^p_0(\rho)$.

Let $\mathcal{N} \subset \mathcal{P}_\mu$ be a statistical model, equipped with a structure of a differential manifold. Consider the bundle-connection pair on S^p given by the tangent bundle and the natural connection (TS^p, ∇^p). Making use of the Amari embedding $A^\alpha : \rho \in \mathcal{N} \to \rho^{1/p} \in S^p$, we may construct the pull-back $((A^\alpha)^*TS^p, (A^\alpha)^*\nabla^p)$ of the bundle-connection pair (TS^p, ∇^p) to $\mathcal{N}$. This means that the fibre over $\rho \in \mathcal{N}$ of the pull-back bundle is given by $T_{\rho^{1/p}}S^p$. Consider now $\mathcal{F}^\alpha := \cup_{\rho \in \mathcal{N}} \mathcal{F}^\alpha_\rho$. Using the family of isomorphisms I^p_ρ, $\rho \in \mathcal{N}$, it is possible to identify $\mathcal{F}^\alpha$ with the pull-back bundle $(A^\alpha)^*TS^p$. One can also transfer the pull-back connection $(A^\alpha)^*\nabla^p$ using this isomorphism. We denote by ∇^α this last connection on the bundle $\mathcal{F}^\alpha$.

Theorem 24. [6] Consider the bundle-connection pair $(\mathcal{F}^\alpha, \nabla^\alpha)$, $\alpha \in (-1, 1)$, on the statistical manifold $\mathcal{N}$. Then ∇^α coincides with the Amari–Chentsov α-connection.
Proof. One obtains

$$\nabla^\alpha = \frac{1+\alpha}{2}\nabla^e + \frac{1-\alpha}{2}\nabla^m \tag{1}$$

where ∇^m and ∇^e are the usual mixture and exponential connections defined by parallel transport on the mixture and exponential bundles $L_0^{x\log x}$ and $L_0^{\exp}$ (see [6] for details). ♣

It is useful to emphasize the new aspects that this theorem introduces into information geometry. First of all, it solves the longstanding problem of an infinite dimensional theory for α-geometries (note that we may discuss orthogonality, projections, etc. also in a non-Riemannian, non-Hilbertian setting). Moreover α-connections appear as L^p-connections in a disguised form (a new result even in the parametric case). Following this line of thought we want to stress that equality (1) should be seen as a theorem and not as a definition. In this sense the parametric case could be seriously misleading: indeed the α-connections are not defined on the tangent space, in general, but on a suitable α-bundle (this point is still overlooked even in some recent papers). In addition one should note that the problem of different geodesics intersecting at right angles cannot be solved naively. In general these geodesics will be on two different manifolds (the target manifolds of different embeddings of the densities) such that a duality pairing exists between the two tangent bundles. A theory of this type has been outlined in [7] and this can probably also be the right approach in the non-commutative setting (see the work of Streater [18,19] where the use of +1 and -1 geodesics is of great importance in the theory of statistical dynamics). But probably the most important aspect is that one can see the whole construction from an abstract point of view (that is for uniformly convex spaces) so that this kind of family of dual geometries should appear whenever one has a family of L^p-type spaces. We have discussed this approach in a previous paper [8] regarding a non-commutative non-parametric generalisation of the α-connections.

NORMS OF L^p-TYPE AND α-CONNECTIONS ASSOCIATED TO MONOTONE METRICS

A general approach to noncommutative α-connections is still missing, even though a number of different points of view exist [8–10,13]. But now Theorem 12 shows that each monotone metric can be obtained by an L^2 scalar product.

Moreover, motivated by Theorem 24 and by the considerations of the previous section, we suggest that one should try to construct α-geometries associated with an arbitrary monotone metric by the construction of an L^p-norm associated with that monotone metric. What follows is a tentative first step in that direction.

Let $(E, \|\cdot\|)$ be a Banach space and denote by $\mathcal{L}(E)$ the set of continuous linear operators on E, and by $GL(E)$ the subset of the invertible ones. If $T \in \mathcal{L}(E)$, we may define a new Banach space $(E, \|\cdot\|_T)$, where $\|v\|_T := \|Tv\|$. Moreover, if $T \in GL(E)$, then $T^{-1} : (E, \|\cdot\|) \to (E, \|\cdot\|_T)$ is an isometric isomorphism.
Now let $L^p(\tau)$ be the matrix von Neumann-Schatten class, that is $M_n(\mathbf{C})$ endowed with the norm $\|A\|_p := \tau(|A|^p)^{1/p}$, and consider $T := M_{\rho,s}^{1/p} \in \mathcal{L}(L^p(\tau))$. Therefore, we may consider the norm $\|A\|_T := \|T(A)\|_p = (\tau(|M_{\rho,s}^{1/p}(A)|^p))^{1/p}$, for $A \in M_n(R)$. Analogously, if we set $\widetilde{T} := \widetilde{M}_{\rho,s}^{1/p}$, we have the norms $\|A\|_{\widetilde{T}} := (\tau(|\widetilde{M}_{\rho,s}^{1/p}(A)|^p))^{1/p}$. So we may define the Banach spaces $L^p(\rho)_s := (M_n(R), \|\cdot\|_T)$, and $L^p(\rho)^s := (M_n(R), \|\cdot\|_{\widetilde{T}})$. The latter spaces are the matrix version of the spaces introduced by Trunov and Zolotarev [20,22] and studied by several authors.
In the construction of commutative α-connections the isomorphism $u \in L^p(\rho) \to u\rho^{1/p} \in L^p(\tau)$ is fundamental; it allows one to identify $T_{\rho^{1/p}}S^p$ with the space $L^p_0(\rho)$ of p-integrable ρ-centred random variables. The operator $A \in L^p(\rho)_s \to M_{\rho,s}^{1/p}(A) \in L^p(\tau)$ could play the same role. If $p = 2$, we may identify $L^2(\rho)_\mu := \int_{[0,1]}^{\oplus} L^2(\rho)_s d\mu(s)$ with $L^2([0,1], d\mu) \otimes L^2(\tau)$, by means of the operator $\int_{[0,1]}^{\oplus} M_{\rho,s}^{1/2} d\mu(s)$. For example the proof of Theorem 12 can be reformulated using $M_{\rho,\mu}^{-1}$ instead of $M_{\rho,\mu}^{-1/2}$ and $L^2(\rho)_\mu$ instead of $L^2([0,1], d\mu) \otimes L^2(\tau)$. In a similar way one may consider $L^2(\rho)^\nu := \int_{[0,1]}^{\oplus} L^2(\rho)^s d\nu(s)$ (this kind of inner product has been introduced by Petz and Toth [15]) and accordingly give a different proof of Theorem 16.
In view of the above considerations, we conjecture that it could be possible to associate with an arbitrary monotone metric a family of α-connections, using a kind of direct integral of the Banach spaces $L^p(\rho)_s$, $s \in [0,1]$, with respect to a positive Radon measure μ on $[0,1]$.

REFERENCES

1. Bhatia R., *Matrix Analysis*. New York: Springer Verlag, 1997.
2. Campbell L. L., An extended Chentsov characterization of the information metric, *Proc. Amer. Math. Soc.* **98**, 135 (1986).
3. Chentsov N. N, *Statistical decision rules and optimal inference.* Transl. Math. Monographs, **53**, Providence: American Math. Soc., 1982.
4. Chentsov N. N, and Morozova E. A., Markov invariant geometry on state manifolds, (in Russian), *Itogi Nauki i Tekhniki* **36**, 69 (1990).
5. Dawid A. P., Further comments on some comments on a paper by Bradley Efron, *Ann. Stat.* **5**, 1249 (1977).
6. Gibilisco P., and Pistone G., Connections on non-parametric statistical manifolds

by Orlicz space geometry, *Infinite Dimensional Analysis, Quantum Probability & Related Fields* **1**, 325 (1998).

7. Gibilisco P., and Pistone G., Analytical and geometrical properties of statistical connections in information geometry, in *Mathematical Theory of Networks and Systems*, eds. A. Beghi, L. Finesso and G. Ricci, Pädova: Il Poligrafo, 1999.
8. Gibilisco P., and Isola T., Connections on statistical manifolds of density operators by geometry of noncommutative L^p-spaces, *Infinite Dimensional Analysis, Quantum Probability & Related Fields* **2**, 169 (1999).
9. Hasegawa H., Non-commutative extension of the Information geometry, Proc. Int. Workshop on *Quantum Communications and Measurement*, Nottingham 1994, eds. V. P. Belavkin, O. Hirota and R. L. Hudson, Plenum Press, 1995.
10. Hasegawa H., and Petz D., Non-commutative extension of Information geometry II, in *Quantum Communication, Computing, and Measurement*, edited by Hirota *et al.*, Plenum Press, New York, 1997.
11. Hewitt E., and Stromberg K., *Real and Abstract Analysis.* New York: Springer Verlag, 1965.
12. Kubo F., and Ando T., Means of positive linear operators, *Math. Ann.* **246**, 205 (1980).
13. Nagaoka H., Differential geometrical aspects of quantum state estimation and relative entropy, Proc. Int. Workshop on *Quantum Communications and Measurement*, Nottingham 1994, eds. V. P. Belavkin, O. Hirota and R. L. Hudson, Plenum Press, 1995.
14. Petz D., Monotone metrics on matrix spaces, *Linear algebra and appl.* **244**, 81 (1996).
15. Petz D., and Toth G., The Bogoliubov inner product in quantum statistics, *Lett. Math. Phys.* **27**, 205 (1993).
16. Pistone G., and Rogantin M. P., The exponential statistical manifold: mean parameters, orthogonality, and space transformation, *Bernoulli* **5**, 721, (1999).
17. Pistone G., and Sempi C., An infinite dimensional geometric structure on the space of all the probability measures equivalent to a given one, *Ann. Stat.* **33**, 1543 (1995).
18. Streater R., *Statistical Dynamics*, London: Imperial College Press, 1995.
19. Streater R., Statistical dynamics and information geometry, in "Geometry and nature", *Contemp. Math.* **203**, p. 117, Amer. Math. Soc., Providence, RI, 1997.
20. Trunov N. V., On a noncommutative analogue of the space L_p, *Iz. VUZ Matematika* (Soviet Mathematics Translation) **23**, 69, (1979).
21. Uhlmann A., Geometric phases and related structures, *Reports on Math. Phys.* **36**, 461, 1995.
22. Zolotarev A. A., L^p spaces with respect to states on a von Neumann algebra and interpolation, *Iz. VUZ Matematika* (Soviet Mathematics Translation) **26**, 36, (1982).

Infinite Dimensional Quantum Information Geometry

Matheus R. Grasselli[1]

Department of Mathematics, King's College London, Strand, London WC2R 2LS, U.K.

Abstract. We present the construction of an infinite dimensional Banach manifold of quantum mechanical states on a Hilbert space $\mathcal{H}$ using different types of small perturbations of a given Hamiltonian H_0. We provide the manifold with a flat connection, called the exponential connection, and comment on the possibility of introducing the dual mixture connection

INTRODUCTION

In finite dimensional quantum information geometry, the set upon which the geometric structures are defined is simply the set of all (invertible) density matrices on a finite dimensional Hilbert space [3]. Already in the definition of the underlying set in infinite dimensions, we need to be slightly more careful and take a more restrictive set than just that of all (invertible) density operators. The reason for this is that, if we modify a given state in the set with a small perturbation, we want the perturbed state to have the same properties as the original one.

Let $\mathcal{C}_p, 0 < p < 1$, denote the set of compact operators $A : \mathcal{H} \mapsto \mathcal{H}$ such that $|A|^p \in \mathcal{C}_1$, where $\mathcal{C}_1$ is the set of trace-class operators on $\mathcal{H}$. Define $\mathcal{C}_{<1} := \bigcup_{0<p<1} \mathcal{C}_p$. We take the underlying set of the quantum information manifold to be $\mathcal{M} = \mathcal{C}_{<1} \cap \Sigma$ where $\Sigma \subseteq \mathcal{C}_1$ denotes the set of density operators. This guarantees that, if $\rho_0 \in \mathcal{M}$, there exists a $\beta_0 < 1$ such that $\rho_0^{\beta_0}$ is a density as well as ρ_0 itself. This set has an affine structure induced from the linear structure of each $\mathcal{C}_p$ in the following way: let $\rho_1 \in \mathcal{C}_{p_1} \cap \Sigma$ and $\rho_2 \in \mathcal{C}_{p_2} \cap \Sigma$; take $p = \max\{p_1, p_2\}$, then $\rho_1, \rho_2 \in \mathcal{C}_p \cap \Sigma$, since $p \leq q$ implies $\mathcal{C}_p \subseteq \mathcal{C}_q$ [4]; define "$\lambda\rho_1 + (1-\lambda)\rho_2, 0 \leq \lambda \leq 1$" as the usual sum of operators in $\mathcal{C}_p$. This is called the (-1)-affine structure. However, this is not the affine structure we use to define a flat connection on the manifold. Instead, we equip $\mathcal{M}$ with an exponential affine structure and then define the natural connection associated with it.

To each $\rho_0 \in \mathcal{C}_{\beta_0} \cap \Sigma$, $\beta_0 < 1$, let $H_0 = -\log \rho_0 + cI \geq I$ be a self-adjoint operator with domain $\mathcal{D}(H_0)$ such that $\rho_0 = Z_0^{-1} e^{-H_0} = e^{-(H_0 + \Psi_0)}$.

[1] Supported by a grant from Capes-Brazil.

CP553, *Disordered and Complex Systems*, edited by P. Sollich, et al.

The idea is to perturb the Hamiltonian H_0, obtaining a new Hamiltonian H_X and then construct a neighbourhood of the point ρ_0 consisting of the perturbed states ρ_X. The perturbations considered are of three different types, and that is the content of the next section.

PERTURBATIONS

The most general class of perturbations we use are the form bounded perturbations. Given a positive self-adjoint operator H with associated form q_H and form domain $Q(H)$, we say that a symmetric quadratic form X (or the symmetric sesquiform obtained from it by polarization) is *q_H-bounded* if

i. $Q(H) \subset Q(X)$ and

ii. there exist positive constants a and b such that $|X(\psi, \psi)| \leq a q_H(\psi, \psi) + b(\psi, \psi)$, for all $\psi \in Q(H)$.

Although form bounded perturbations are of much interest in the study of Schrödinger operators for a variety of quantum systems, they provide very little regularity for the quantum information manifold. In [7], Streater was able to show that the manifold constructed using them has Lipschitz structure. However, more regularity is needed if we want to define a metric on it by, say, the second derivative of the free energy. This led to the idea of looking at the more restrictive case of operator bounded perturbations. Given operators H and X defined on dense domains $\mathcal{D}(H)$ and $\mathcal{D}(X)$ in a Hilbert space $\mathcal{H}$, we say that X is *H-bounded* if

i. $\mathcal{D}(H) \subset \mathcal{D}(X)$ and

ii. there exist positive constants a and b such that $\|X\psi\| \leq a\,\|H\psi\| + b\,\|\psi\|$, for all $\psi \in \mathcal{D}(H)$.

For both form bounded and operator bounded perturbations, the infimum of such a is called the relative bound of X (with respect to H or with respect to q_H, respectively). If $a < 1$, the perturbation is said to be small.

Operator bounded perturbations are also used in the study of Schrödinger operators for quite a broad range of quantum systems [6], with the additional property of providing enough regularity for the manifold $\mathcal{M}$ to have an analytic free energy, for instance.

The following lemma tells us how to characterise operator bounded and form bounded perturbations in terms of certain norms. Before we state it, we need to say what we mean by a form multiplied from both sides by operators. Suppose that X is a quadratic form with domain $Q(X)$ and A, B are operators on $\mathcal{H}$ such that A^* and B are densely defined. Suppose further that $A^* : \mathcal{D}(A^*) \to Q(X)$ and $B : \mathcal{D}(B) \to Q(X)$. Then the expression AXB means the form defined by

$$\phi, \psi \mapsto X(A^*\phi, B\psi), \qquad \phi \in \mathcal{D}(A^*), \quad \psi \in \mathcal{D}(B).$$

Consider now the case where $H_0 \geq I$ is a self-adjoint operator with domain $\mathcal{D}(H_0)$, quadratic form q_0 and form domain $Q_0 = \mathcal{D}(H_0^{1/2})$, and let $R_0 = H_0^{-1}$ be its resolvent at the origin.

Lemma 1 *A symmetric operator $X : \mathcal{D}(H_0) \to \mathcal{H}$ is H_0-bounded if and only if $\|XR_0\| < \infty$. Analogously, a symmetric quadratic form X defined on Q_0 is q_0-bounded if and only if $R_0^{1/2} X R_0^{1/2}$ is a bounded symmetric form defined everywhere. Moreover, if $\left\|R_0^{1/2} X R_0^{1/2}\right\| < \infty$ then the relative bound a of X with respect to q_0 satisfies $a \leq \left\|R_0^{1/2} X R_0^{1/2}\right\|$.*

The set $\mathcal{T}_\omega(0)$ of all H_0-bounded symmetric operators X is a Banach space with norm $\|X\|_\omega(0) := \|XR_0\|$, since the map $A \mapsto AH_0$ from $\mathcal{B}(\mathcal{H})$ onto $\mathcal{T}_\omega(0)$ is an isometry. The set $\mathcal{T}_0(0)$ of all q_0-bounded symmetric forms X is also a Banach space with norm $\|X\|_0(0) := \left\|R_0^{1/2} X R_0^{1/2}\right\|$, since the map $A \mapsto H_0^{1/2} A H_0^{1/2}$ from the set of all bounded self-adjoint operators on $\mathcal{H}$ onto $\mathcal{T}_0(0)$ is again an isometry.

Motivated by Banach space interpolation theory, let us consider, for $\varepsilon \in (0, 1/2)$, the set $\mathcal{T}_\varepsilon(0)$ of all symmetric forms X with $\mathcal{D}(H_0^{\frac{1}{2}-\varepsilon}) \subset Q(X)$ and such that $\|X\|_\varepsilon(0) := \left\|R_0^{\frac{1}{2}+\varepsilon} X R_0^{\frac{1}{2}-\varepsilon}\right\|$ is finite. Then the map $A \mapsto H_0^{\frac{1}{2}-\varepsilon} A H_0^{\frac{1}{2}+\varepsilon}$ is an isometry from the set of all bounded self-adjoint operators on $\mathcal{H}$ onto $\mathcal{T}_\varepsilon(0)$. Hence $\mathcal{T}_\varepsilon(0)$ is a Banach space with the ε-norm $\|\cdot\|_\varepsilon(0)$. Such an X will be called an ε-bounded perturbation and it is shown to interpolate between the extreme cases of form bounded perturbations, for which $\varepsilon = 0$, and operator bounded perturbations, for which $\varepsilon = 1/2$.

Lemma 2 *For fixed symmetric X, $\|X\|_\varepsilon$ is a monotonically increasing function of $\varepsilon \in [0, 1/2]$.*

In the next section, we carry out the programme of using these perturbations to obtain the hoods in the manifold. In what follows, we write $\mathcal{T}_{(\cdot)}(0)$ to indicate that we can use any of the three Banach spaces $\mathcal{T}_\omega(0)$, $\mathcal{T}_0(0)$ or $\mathcal{T}_\varepsilon(0)$.

CONSTRUCTION OF THE MANIFOLD

The two technical tools used in the construction of our manifold are the following.

Theorem 3 (KLMN) *Let H_0 be a positive self-adjoint operator with quadratic form q_0 and form domain Q_0; let X be a q_0-small symmetric quadratic form. Then there exists a unique self-adjoint operator H_X with form domain Q_0 such that*

$$\langle H_X^{1/2}\phi, H_X^{1/2}\psi\rangle = q_0(\phi, \psi) + X(\phi, \psi), \qquad \phi, \psi \in Q_0.$$

Moreover, H_X is bounded below by $-b$.

Lemma 4 (Streater 98) *Let X be a q_0-small with bound $a < 1 - \beta_0$. Denote by H_X the unique operator given by the KLMN theorem. Then $\exp(-\beta H_X)$ is of trace class for all $\beta > \beta_X = \beta_0/(1-a)$.*

The construction of the neighbourhood of ρ_0 goes as follows. In $\mathcal{T}_{(\cdot)}(0)$, take X such that $\|X\|_{(\cdot)}(0) < 1 - \beta_0$. Since $\|X\|_0(0) \leq \|X\|_{(\cdot)}(0) < 1 - \beta_0$, X is also q_0-bounded with bound a_0 less than $1 - \beta_0$. The *KLMN* theorem then tells us that there exists a unique semi-bounded self-adjoint operator H_X with form $q_X = q_0 + X$ and form domain $Q_X = Q_0$. We write $H_X = H_0 + X$ for this operator and consider the state $\rho_X = Z_X^{-1} e^{-(H_0+X)} = e^{-(H_0+X+\Psi_X)}$

Then, from lemma 4, $\rho_X \in \mathcal{C}_{\beta_X} \cap \Sigma$, where $\beta_X = \frac{\beta_0}{1-a_0} < 1$. If we add to H_X a multiple of the identity, we can still have the same state ρ_X, by simply adjusting the partition function Z_X; so we can always assume that, for the perturbed state, we have $H_X \geq I$. We take as a hood $\mathcal{M}_0$ of ρ_0 the set of all such states, that is, $\mathcal{M}_0 = \{\rho_X : \|X\|_{(\cdot)}(0) < 1 - \beta_0\}$.

To give a topology to $\mathcal{M}_0$, we first introduce in $\mathcal{T}_{(\cdot)}(0)$ the equivalence relation $X \sim Y$ iff $X - Y = \alpha I$ for some $\alpha \in \mathbf{R}$, precisely because $\rho_X = \rho_{X+\alpha I}$, as remarked above. We then identify ρ_X in $\mathcal{M}_0$ with the line $\{Y \in \mathcal{T}_{(\cdot)}(0) : Y = X + \alpha I, \alpha \in \mathbf{R}\}$ in $\mathcal{T}_{(\cdot)}(0)/\sim$. This is a bijection from $\mathcal{M}_0$ onto the subset of $\mathcal{T}_{(\cdot)}(0)/\sim$ defined by $\left\{\{X + \alpha I\}_{\alpha \in \mathbf{R}} : \|X\|_{(\cdot)}(0) < 1 - \beta_0\right\}$. The topology in $\mathcal{M}_0$ is then given by transfer of structure. Now that $\mathcal{M}_0$ is a (Hausdorff) topological space, we want to parametrise it by an open set in a Banach space. As in the classical case [5], we choose the Banach subspace of centred variables in $\mathcal{T}_{(\cdot)}(0)$; in our terms, perturbations with zero mean (the 'scores'). The only problem is that, when X is not an operator, it is not immediately clear what its mean in the state ρ_0 should be, let alone the question of whether or not it is finite. To deal with this, define the regularised mean of $X \in \mathcal{T}_{(\cdot)}(0)$ in the state ρ_0 as $\rho_0 \cdot X := \mathrm{Tr}(\rho_0^{\lambda} X \rho_0^{1-\lambda})$, for $0 < \lambda < 1$.

Since $\rho_0 \in \mathcal{C}_{\beta_0} \cap \Sigma$ and X is q_0-bounded, lemma 5 of [7] ensures that $\rho_0 \cdot X$ is finite and independent of λ. It was a shown there that $\rho_0 \cdot X$ is a continuous map from $\mathcal{T}_0(0)$ to $\mathbf{R}$, because its bound contained a factor $\|X\|_0(0)$. Exactly the same proof shows that $\rho_0 \cdot X$ is a continuous map from $\mathcal{T}_{(\cdot)}(0)$ to $\mathbf{R}$. Thus the set $\widehat{\mathcal{T}}_{(\cdot)}(0) := \left\{X \in \mathcal{T}_{(\cdot)}(0) : \rho_0 \cdot X = 0\right\}$ is a closed subspace of $\mathcal{T}_{(\cdot)}(0)$ and so is a Banach space with the norm $\|\cdot\|_{\varepsilon}$ restricted to it. We notice that for the case of operator bounded perturbations, the regularised mean of X coincides with the usual mean $\mathrm{Tr}(\rho_0 X)$.

To each $\rho_X \in \mathcal{M}_0$, consider the point in the line $\{X + \alpha I\}_{\alpha \in \mathbf{R}}$ with $\alpha = -\rho_0 \cdot X$. Write $\widehat{X} = X - \rho_0 \cdot X$ for this point. The map $\rho_X \mapsto \widehat{X}$ is a homeomorphism between $\mathcal{M}_0$ and the open subset of $\widehat{\mathcal{T}}_{(\cdot)}(0)$ defined by $\left\{\widehat{X} : \widehat{X} = X - \rho_0 \cdot X, \|X\|_{(\cdot)} < 1 - \beta_0\right\}$. The map $\rho_X \mapsto \widehat{X}$ is then a global chart for the Banach manifold $\mathcal{M}_0$ modeled by $\widehat{\mathcal{T}}_{(\cdot)}(0)$. The tangent space at ρ_0 is given by $\widehat{\mathcal{T}}_{(\cdot)}(0)$, with the curve $\left\{\rho(\lambda) = Z_{\lambda X}^{-1} e^{-(H_0+\lambda X)}, \lambda \in [-\delta, \delta]\right\}$ having tangent vector

$\widehat{X} = X - \rho_0 \cdot X$.

We extend our manifold by adding new patches compatible with $\mathcal{M}_0$. The idea is to construct a chart around each perturbed state ρ_X as we did around ρ_0. Let $\rho_X \in \mathcal{M}_0$ with Hamiltonian $H_X \geq I$ and consider the Banach space $\mathcal{T}_{(\cdot)}(X)$ of all symmetric forms Y such that the norm $\|Y\|_{(\cdot)}(X)$ is finite, where the expression for this norm is the same as $\|Y\|_{(\cdot)}(0)$ but with all the Hamiltonians replaced by H_X. Then we repeat exactly the same process, namely, take sufficiently small Y (with $\|Y\|_{(\cdot)}(X) < 1 - \beta_X$), obtain from the *KLMN* theorem the Hamiltonian H_{X+Y} with form $q_{X+Y} = q_X + Y = q_0 + X + Y$ and form domain $Q_{X+Y} = Q_X = Q_0$ and take the hood of X to be the set $\mathcal{M}_X$ of all states of the form $\rho_{X+Y} = Z^{-1}_{X+Y} e^{-H_{X+Y}} = Z^{-1}_{X+Y} e^{-(H_0+X+Y)}$. The topology and coordinates for $\mathcal{M}_X$ are then introduced in a completely similar fashion.

We then turn to the union of $\mathcal{M}_0$ and $\mathcal{M}_X$. We need to show that our two previous charts are compatible in the overlapping region $\mathcal{M}_0 \cap \mathcal{M}_X$. For the case of form bounded and operator bounded perturbations, the equivalence of the norms is achieved by a straightforward series of operator identities [7,8]. For the case of ε-bounded perturbations, the argument is more subtle and involves careful consideration of the domains of the operators. Nevertheless, the result still follows [2].

Theorem 5 *$\|\cdot\|_\varepsilon(X)$ and $\|\cdot\|_\varepsilon(0)$ are equivalent norms.*

We can repeat the construction again, starting from any point in $\mathcal{M}_0 \bigcup \mathcal{M}_X$. We then obtain the following definition.

Definition 6 *The information manifold $\mathcal{M}(H_0)$ defined by H_0 consists of all states obtainable in a finite numbers of steps, by extending $\mathcal{M}_0$ as explained above.*

AFFINE GEOMETRY IN $\mathcal{M}(H_0)$

The set $A = \left\{\widehat{X} \in \widehat{\mathcal{T}}_{(\cdot)}(0) : \widehat{X} = X - \rho_0 \cdot X, \|X\|_{(\cdot)}(0) < 1 - \beta_0\right\}$ is a convex subset of the Banach space $\widehat{\mathcal{T}}_{(\cdot)}(0)$ and so has an affine structure coming from its linear structure. We provide $\mathcal{M}_0$ with an affine structure induced from A using the patch $\widehat{X} \mapsto \rho_X$ and call this the canonical or $(+1)$-affine structure. The $(+1)$-convex mixture of ρ_X and ρ_Y in $\mathcal{M}_0$ is then $\rho_{\lambda X+(1-\lambda)Y}$, $(0 \leq \lambda \leq 1)$, which differs from the previously defined (-1)-convex mixture $\lambda\rho_X + (1-\lambda)\rho_Y$.

Given two points ρ_X and ρ_Y in $\mathcal{M}_0$ and their tangent spaces $\widehat{\mathcal{T}}_{(\cdot)}(X)$ and $\widehat{\mathcal{T}}_{(\cdot)}(Y)$, we define the $(+1)$-parallel transport U_L of $(Z - \rho_X \cdot Z) \in \widehat{\mathcal{T}}_{(\cdot)}(X)$ along any continuous path L connecting ρ_X and ρ_Y in the manifold to be the point $(Z - \rho_Y \cdot Z) \in \widehat{\mathcal{T}}_{(\cdot)}(Y)$. Clearly U_L is independent of L by construction, thus the $(+1)$-affine connection is flat. We see that the $(+1)$-parallel transport just moves the representative point in the line $\{Z + \alpha I\}_{\alpha \in \mathbf{R}}$ from one hyperplane to another.

DISCUSSION

The manifold $\mathcal{M}(H_0)$ constructed here does not cover the whole set $\mathcal{M}$ at once. It could not possibly do so, since our small perturbations do not change the domain of the original Hamiltonian H_0, and certainly $\mathcal{M}$ contains states defined by Hamiltonians with plenty of different domains. Also, although we can reach far removed points with a finite number of small perturbations, we cannot move in arbitrary directions. For instance, we cannot reach $X = -H_0$ as the result of our perturbations, since the identity is not an operator of trace class. To cover $\mathcal{M}$ entirely, we have to start at several different points. The whole manifold thus obtained consists of several disconnected parts, pointing towards positive directions with respect to the given Hamiltonians.

Finally, it is clear that $\lambda\rho_X + (1-\lambda)\rho_Y$, for, say, $\rho_X, \rho_Y \in \mathcal{M}_0$, defines a new state in the underlying set $\mathcal{M}$. Nonetheless, we were not yet able to prove that it belongs to the neighbourhood of the original states. This is the main obstacle to defining a mixture connection in our manifold and thence developing a quantum version of Amari's duality theory [1]. One possibility is to change the definition of the neighbourhoods altogether and use, for instance, the condition that states in the same neighbourhood all have finite relative entropy with respect to each other, besides having finite von Neumann entropy.

REFERENCES

1. S.-I. Amari, **Differential Geometric Methods in Statistics**, *Lecture Notes in Statistics*, **28**, Springer-Verlag, New York, 1985.
2. M. R. Grasselli and R. F. Streater, *The Quantum Information Manifold for ε-bounded Forms*, to appear in Rep. Math. Phys., math-ph/9910031.
3. D. Petz and C.Sudar, *Geometries of Quantum States*, J. Math. Phys., **37**, 2662-2673, 1996.
4. A. Pietsch, **Nuclear Locally Convex Spaces**, Springer-Verlag, 1972.
5. G. Pistone and C. Sempi, *An infinite dimensional geometric structure on the space of all probability measures equivalent to a given one*, Ann. Stat., **33**, 1543-1561, 1995.
6. M. Reed and B. Simon, **Methods of Modern Mathematical Physics**, vol. 2, Academic Press, 1975.
7. R. F. Streater, *The Information Manifold for Relatively Bounded Potentials*, to appear in the Bogoliubov Memorial Volume, ed. A. A. Slavnov, Stecklov Institute.
8. R. F. Streater, *The Analytic Quantum Information Manifold*, to appear in **Stochastic Processes, Physics and Geometry**, eds. F. Gesztesy, S. Paycha and H. Holden, Canad. Math. Soc.

Dualistic Properties of the Manifold of Quantum States

Anna Jenčová

Mathematical Institute, Slovak Academy of Sciences
Štefánikova 49, SK-814 73 Bratislava, Slovakia

Abstract. In the finite dimensional case, we introduce a family of torsion-free affine connections in the manifold of all invertible density matrices. These are the quantum version of Amari's α-connections, based on the α-representation of the tangent space. The dual connections with respect to a monotone metric are, in general, not torsion-free. We compute the torsion and the Riemannian curvature. Finally, geodesics are used to define a divergence.

MONOTONE METRICS

Unless stated otherwise, all proofs can be found in [6].

Let $\mathcal{M}$ denote the differentiable manifold of all n-dimensional complex hermitean matrices and let $\mathcal{M}^+ = \{M \in \mathcal{M} \mid M > 0\}$. Let $\tilde{T}_M$ be the tangent space at M. We introduce a Riemannian structure in $\mathcal{M}^+$, defining an inner product in $\tilde{T}_M$ by $\lambda_M(X,Y) = \mathrm{Tr} X J_M(Y)$ where J_M is a suitable superoperator on matrices. The state space of an n-level quantum system can be identified with the submanifold $\mathcal{D} = \{D \in \mathcal{M}^+ \mid \mathrm{Tr} D = 1\}$. The tangent space $T_D \subset \tilde{T}_D$ is the real N-dimensional vector space of all self-adjoint traceless matrices. We require that λ is monotone, in the sense that if T is a stochastic map, then

$$\lambda_{T(M)}(T(X), T(X)) \leq \lambda_M(X,X), \ \forall M \in \mathcal{M}^+, X \in \mathcal{M}$$

As was proved in [8], this is true iff J_M is of the form

$$J_M = (R_M^{\frac{1}{2}} f(L_M R_M^{-1}) R_M^{\frac{1}{2}})^{-1} \tag{1}$$

where $f : R^+ \to R$ is an operator monotone function such that $f(t) = t f(t^{-1})$ for every $t > 0$ and $L_M(A) = MA$ and $R_M(A) = AM$, for each matrix A. We also adopt a normalization condition $f(1) = 1$.

Let T_D^* be the cotangent space of $\mathcal{D}$ at D, then T_D^* is the vector space of all observables A with zero mean at D (i.e. $\mathrm{Tr} DA = 0$). It is easy to see that

CP553, *Disordered and Complex Systems*, edited by P. Sollich, et al.

$$T_D^* = \{J_D(H) \mid H \in T_D\} \tag{2}$$

Here are some examples of monotone metrics.

1. Let $J_D(H) = G$, where $GD + DG = 2H$, the corresponding metric is called the metric of the symmetric logarithmic derivative, see [5,9].

2. If $J_D(H) = \frac{d}{dt} \log(D + tH)|_{t=0}$, we obtain the well-known Kubo-Mori metric.

3. Let $J_D(H) = \frac{1}{2}(D^{-1}H + HD^{-1})$. The corresponding metric is the metric of the right logarithmic derivative, see [5,10].

For more about monotone metrics and their use see [9,10,4].

THE α-REPRESENTATION

Let $g : R \to R$ be a smooth (strictly) monotone function. We define an operator $L_g[M] : \tilde{T}_M \to \tilde{T}_{g(M)}$ by (see [6])

$$L_g[M](H) = \frac{d}{ds} g(M + sH)|_{s=0}$$

In what follows, we omit the indication of the point in square brackets if no confusion is possible. The vector space

$$T_D^g = \{L_g(H) \mid H \in T_D\}$$

will be called the g-representation of the tangent space T_D. The corresponding inner product in T_D^g is

$$\lambda_D^g(G_1, G_2) = \lambda_D(L_g^{-1}(G_1), L_g^{-1}(G_2)) = \mathrm{Tr} G_1 K_g(G_2)$$

where $K_g = L_g^{-1} J_D L_g^{-1}$. The g-representation of the cotangent space is given by

$$T_D^{g*} = \{K_g(G) \mid G \in T_D^g\} = \{L_{g^{-1}} J_D(H) \mid H \in T_D\}$$

Clearly, if g is the identity function, we obtain the usual tangent and cotangent spaces T_D and T_D^*.

The quantum analogue of Amari's α-representations [1] is obtained if we put

$$g(x) = g_\alpha(x) = \begin{cases} \frac{2}{1-\alpha} x^{\frac{1-\alpha}{2}} & \alpha \neq 1 \\ \log x & \alpha = 1 \end{cases}$$

We have $T_D^{\alpha *} = T_D^{-\alpha}$ for each α so that K_α is an isomorphism $K_\alpha : T_D^\alpha \to T_D^{-\alpha}$. We have

$$K_\alpha^{-1} = K_{-\alpha} \iff J_D = J_\alpha := L_{-\alpha} L_\alpha = J_{-\alpha}.$$

The family of metrics J_α was studied in [4] and it was shown that these are monotone for $\alpha \in [-3, 3]$. For example, the Kubo-Mori metric is obtained if $\alpha = \pm 1$ and the right logarithmic derivative metric if $\alpha = \pm 3$.

THE AFFINE CONNECTIONS, TORSION AND CURVATURE

In this section, we define the α-connections in $\mathcal{M}^+$ and $\mathcal{D}$ and investigate the important notions of duality, stated in [1], namely the dual connections and the existence of dual affine coordinate systems.

Let $g : R \to R$ be a smooth strictly monotone function and let $M, M' \in \mathcal{M}^+$. Then there is an isomorphism $\tilde{T}_M \to \tilde{T}_{M'}$ given by $H \mapsto L_g^{-1}[M']L_g[M](H)$. This isomorphism induces an affine connection on $\mathcal{M}^+$. Let us denote the corresponding covariant derivative by $\tilde{\nabla}^g$.

Let $x_1, \ldots, x_{N+1}$ be a coordinate system in $\mathcal{M}^+$. Let us denote $\partial_i = \frac{\partial}{\partial x_i}$ and let $\tilde{H}_i = \partial_i M(x)$, $\tilde{G}_i = L_g(\tilde{H}_i) = \partial_i g(M(x))$, $i = 1, \ldots, N+1$. Then

$$L_g(\tilde{\nabla}^g_{\tilde{H}_j} \tilde{H}_i) = \partial_i \partial_j g(M(x))$$

Hence, the coefficients of the affine connection are

$$\tilde{\Gamma}^g_{ijk}(x) = \lambda_x(\tilde{\nabla}^g_{\tilde{H}_i} \tilde{H}_j, \tilde{H}_k) = \mathrm{Tr}\partial_i\partial_j g(M(x))K_g(\tilde{G}_k), \quad i, j, k = 1, \ldots, N+1$$

From this, it follows that this connection is torsion-free.

If we use the functions g_α, we obtain a one-parameter family of torsion-free connections $\tilde{\nabla}^\alpha$ which is the quantum analogue of the α-connections defined by Amari [1]. But, unlike the classical case, the connections $\tilde{\nabla}^\alpha$ and $\tilde{\nabla}^{-\alpha}$ are not dual in general.

To obtain the dual connection, consider the isomorphism $\tilde{T}_M \to \tilde{T}_{M'}$, given by $H \mapsto J_{M'}^{-1} L_g[M'] L_g^{-1}[M] J_M(H)$. The induced affine connection on $\mathcal{M}^+$, resp. the covariant derivative will be denoted by $\tilde{\nabla}^{g*}$. We have

$$L_g^{-1}[M] J_M(\tilde{\nabla}^{g*}_{\tilde{H}_i} \tilde{H}_j) = \partial_i K_g(\tilde{G}_j(x))$$

The coefficients of this connection are

$$\tilde{\Gamma}^{g*}_{ijk} = \lambda(\tilde{\nabla}^{g*}_{\tilde{H}_i} \tilde{H}_j, \tilde{H}_k) = \mathrm{Tr}\partial_i K_g(\tilde{G}_j)\tilde{G}_k \tag{3}$$

It is easy to prove that $\tilde{\nabla}^g$ and $\tilde{\nabla}^{g*}$ are dual.

The components of the torsion tensor are

$$\tilde{S}^{g*}_{ijk} = \tilde{\Gamma}^{g*}_{ijk} - \tilde{\Gamma}^{g*}_{jik} = \mathrm{Tr}\{\partial_i K_g(\tilde{G}_j(x)) - \partial_j K_g(\tilde{G}_i(x))\}\tilde{G}_k$$

so that this connection is torsion-free iff $\partial_i K_g(\tilde{G}_j) = \partial_j K_g(\tilde{G}_i)$, $\forall i, j = 1, \ldots, N+1$. Obviously, this is not always the case.

Consider now $\mathcal{D}$ as an N-dimensional submanifold in $\mathcal{M}$ and let $t_1, \ldots, t_N$ be a coordinate system in $\mathcal{D}$. As there is no danger of confusion, we use the symbol ∂_i also for $\frac{\partial}{\partial t_i}$. Let $H_i = \partial_i D(t)$, $G_i = \partial_i g(D(t))$, $i = 1, \ldots, N$. The affine structure in $\mathcal{D}$ is obtained by projecting the above affine connections orthogonally onto $\mathcal{D}$. The covariant derivative is given by

$$L_g(\nabla^g_{H_j} H_i) = \partial_i\partial_j g(D_t) - g'(D_t)D_t \mathrm{Tr}(g'(D_t))^{-1}\partial_i\partial_j g(D_t)$$
$$L_g^{-1} J_D(\nabla^{g*}_{H_i} H_j) = \partial_i K_g(G_j(t)) - (g'(D_t))^{-1}\mathrm{Tr} g'(D_t)D_t\partial_i K_g(G_j(t))$$

and the coefficients are

$$\Gamma^g_{ijk}(t) = \mathrm{Tr}K_g(G_k)\partial_i\partial_j g(D_t) \quad \Gamma^{g*}_{ijk}(t) = \mathrm{Tr}G_k\partial_i K_g(G_j(t))$$

Proposition 1 *Let* $\alpha \in [-3, 3]$. *The following statements are equivalent:*

(i) $\tilde{\nabla}^{\alpha*}$ *is torsion-free*

(ii) $\nabla^{\alpha*}$ *is torsion-free*

(iii) $J_D = J_\alpha = L_{-\alpha}L_\alpha$

Proof. The cases (i) $\Rightarrow$ (ii) and (iii)$\Rightarrow$ (i) are rather obvious. We prove (ii)$\Rightarrow$ (iii): Let us consider the natural affine parametrization in $\mathcal{D}$

$$D_t = D_0 + \sum_i t_i H_i \quad t \in \Theta \subset R^N \quad \mathrm{Tr}H_i = 0,\ i = 1, \ldots, N$$

Let $\nabla^{\alpha*}$ be torsion-free, then

$$\mathrm{Tr}\left(\partial_i L_\alpha^{-1} J_D(H_j) - \partial_j L_\alpha^{-1} J_D(H_i)\right) L_\alpha(H_k) = 0 \ i, j, k = 1, \ldots, N$$

Clearly, $D \in \tilde{T}_D$, $D \perp T_D$ for each $D \in \mathcal{D}$ and

$$\mathrm{Tr}\partial_i\{L_\alpha^{-1} J_D(H_j)\}L_\alpha(D) = \partial_i \mathrm{Tr} L_\alpha^{-1} J_D(H_j) L_\alpha(D) - \frac{1-\alpha}{2}\mathrm{Tr} L_\alpha J_D(H_j)\partial_i g_\alpha(D) =$$
$$= -\frac{1-\alpha}{2}\mathrm{Tr} J_D(H_j)H_i = \mathrm{Tr}\partial_j\{L_\alpha^{-1} J_D(H_i)\}L_\alpha(D)$$

It follows that $\partial_i L_\alpha^{-1} J_D(H_j) = \partial_j L_\alpha^{-1} J_D(H_i)$, for $i, j = 1, \ldots, N$. This implies the existence of a potential function $\Phi : \Theta \to \mathcal{M}$, such that $\frac{d}{dt}\Phi(D + tH) = L_\alpha^{-1}[D + tH]J_{D+tH}(H)$. Let $\alpha \neq -1$. If $DH = HD$, then $L_\alpha^{-1}[D + tH]J_{D+tH}(H) = (D + tH)^{\frac{\alpha-1}{2}} H$. Let $H = D - \frac{1}{n}$. Compute

$$\Phi(D) - \Phi(\frac{1}{n}) = \int_0^1 \frac{d}{dt}\Phi(\frac{1}{n} + tH)dt = \int_0^1 (\frac{1}{n} + tH)^{\frac{\alpha-1}{2}} H dt = g_{-\alpha}(D) - g_\alpha(\frac{1}{n})$$

It follows that $L_\alpha^{-1} J_D(H) = L_{-\alpha}(H)$. The case $\alpha = -1$ is proved similarly.

Let us now compute the Riemannian curvature. Let $\tilde{R}^g$ ($\tilde{R}^{g*}$) be the Riemannian curvature tensor of $\tilde{\nabla}^g$ ($\tilde{\nabla}^{g*}$) in $\mathcal{M}^+$, then $\tilde{R}^g = \tilde{R}^{g*} = 0$. The curvature tensor R^g in $\mathcal{D}$ is equal to

$$R^g_{ijkl} = H^g_{jk1}H^{g*}_{il1} - H^g_{ik1}H^{g*}_{jl1}$$

where H^g_{ij1}, resp. H^{g*}_{ij1}, is the Euler-Shouten imbedding curvature. If $g = g_\alpha$,

$$R^\alpha_{ijkl}(D) = \frac{1-\alpha^2}{4}\{\mathrm{Tr}H_j J_\alpha(H_k)\mathrm{Tr}H_i J_D(H_l) - \mathrm{Tr}H_i J_\alpha(H_k)\mathrm{Tr}H_j J_D(H_l)\}$$

We see that $R^\alpha = 0$ iff $\alpha = \pm 1$. The dual curvature $R^{\alpha *}$ is then computed from [2]

$$\tilde{R}(X,Y,Z,W) = -\tilde{R}^*(X,Y,W,Z) \tag{4}$$

We see that, as $\tilde{\nabla}^\alpha$ and $\nabla^{\pm 1}$ connections are flat, there exists an affine coordinate system. On the other hand, no dual coordinate system and hence no potential function exists, unless $J_D = J_\alpha$. It means that, in general, we cannot use Amari's theory to define a divergence.

GEODESICS AND DIVERGENCE FUNCTIONS

As one dimensional submanifolds are always torsion-free, a divergence exists for each $\nabla^{\pm 1*}$ and $\tilde{\nabla}^{\alpha *}$-geodesic. As suggested in [7], we use these to define a divergence in $\mathcal{D}$ $(\mathcal{M}^+)$.

For each α, a $\tilde{\nabla}^{\alpha *}$-geodesic is a solution of

$$L_\alpha^{-1} J_{\rho_t}(\dot{\rho}_t) = A \tag{5}$$

where $A \in \mathcal{M}$, it means that each geodesic is determined by the observable A.

Proposition 2 *Let $\tilde{\rho}_t$ be a solution of $L_\alpha^{-1} J_{\rho_t}(\dot{\rho}_t) = A$. Then $\rho_t = \frac{\tilde{\rho}_t}{\mathrm{Tr}\tilde{\rho}_t}$ is a $\nabla^{\alpha *}$ geodesic.*

A divergence measure in $\mathcal{D}$ $(\mathcal{M}^+)$ can be defined as follows. Let $\rho_0, \rho_1 \in \mathcal{M}^+$. Let $\tilde{\rho}_t$ be the unique $\tilde{\nabla}^{\alpha *}$-geodesic connecting them. The α-divergence is $D_\alpha(\rho_0, \rho_1) = D^{\tilde{\rho}}_\alpha(0,1)$.

Proposition 3 *Let $\alpha \neq 1$, $\rho_0, \rho_1 \in \mathcal{M}^+$. Let $\tilde{\rho}_t$ be as above and let A be the corresponding observable. Then*

$$D_\alpha(\rho_0, \rho_1) = \frac{2}{1-\alpha}(\mathrm{Tr}\rho_0 - \mathrm{Tr}\rho_1) + \mathrm{Tr}g_\alpha(\rho_1)A$$

If $\alpha = -1$ and $\rho_0, \rho_1 \in \mathcal{D}$, we may use the $\nabla^{\alpha *}$-geodesic $\rho_t = (\mathrm{Tr}\tilde{\rho}_t)^{-1}\tilde{\rho}_t$ to define the (-1)- divergence. In this way we obtain the restriction of D_α to $\mathcal{D}$.

Examples. 1. Let $J_D = J_\alpha$ for some $\alpha \in [-3, 3]$. In this case we have the same situation as in the classical case. Thus for $\alpha = \pm 1$, the divergence is given by

$$D(\rho_0, \rho_1) = \mathrm{Tr}\rho_1(\log \rho_1 - \log \rho_0)$$

which is the relative entropy.

For $\alpha \neq \pm 1$, $D_\alpha(\rho_0, \rho_1) = \mathrm{Tr} g_\alpha(\rho_1)(g_{-\alpha}(\rho_1) - g_{-\alpha}(\rho_0))$ This α-divergence was defined also in [3]. It is easy to see that the above divergence functions are the same as those from Proposition 3.

2. [7] Let $\alpha = -1$ and let λ be the metric of the symmetric logarithmic derivative. Then

$$\rho_t = \exp\{\frac{1}{2}(tA - \psi(t))\}\rho_0 \exp\{\frac{1}{2}(tA - \psi(t))\}$$

where $\psi(t) = \log \mathrm{Tr}\rho_0 \exp tA$ is a ∇^{-1*}-geodesic. The divergence is

$$D_{-1}(\rho_0, \rho_1) = 2\mathrm{Tr}\rho_1 \log \rho_0^{-\frac{1}{2}}(\rho_0^{\frac{1}{2}}\rho_1\rho_0^{\frac{1}{2}})^{\frac{1}{2}}\rho_0^{-\frac{1}{2}}$$

which coincides with the relative entropy if ρ_0 and ρ_1 commute.

3. Let $\alpha = -1$ and let λ be the metric of the right logarithmic derivative. Let Q_F be the linear operator given by $Q_F^{-1}(A) = \frac{1}{2}(F^{-\frac{1}{2}}AF^{\frac{1}{2}} + F^{\frac{1}{2}}AF^{-\frac{1}{2}})$, then

$$\rho_t = \rho^{\frac{1}{2}} \exp\{(t-1)Q_\rho(A) - \psi(t)\}\rho^{\frac{1}{2}}$$

with $\psi(t) = \log \mathrm{Tr}\rho \exp\{(t-1)Q_\rho(A)\}$ is a ∇^{-1*}-geodesic. The divergence is given by

$$D_{-1}(\rho_0, \rho_1) = \mathrm{Tr}\rho_1 A = \mathrm{Tr}\rho_1 \log \rho_1^{\frac{1}{2}}\rho_0^{-1}\rho_1^{\frac{1}{2}}$$

which is another version of the relative entropy.

REFERENCES

1. Amari, S. *Differential-geometrical methods in statistics*, Lecture Notes in Statistics 28 (1985) Springer-Verlag
2. Amari, S., Barndorff-Nielsen, O.E., Kass, R.E., Lauritzen, S.L., Rao, C.R. *Differential geometry in statistical inference.* IMS Lecture notes-Monograph series 10 (1987)
3. Hasegawa, H. *Rep. Math. Phys.* 33 (1993) 87–93
4. Hasegawa, H. and Petz, D. Non-commutative extension of information geometry II, in *Quantum Communication, Computing and Measurement*, eds. Hirota et al., Plenum Press, New York (1997)
5. Holevo, A.S. *Probabilistic and statistical aspects of quantum theory*, North-Holland, Amsterdam, 1982
6. Jenčová, A. Geometry of quantum states: dual connections and divergence functions *Rep. Math. Phys.* (to appear)
7. Nagaoka, H. Differential geometrical aspects of quantum state estimation and relative entropy, in *Quantum Communication, Computing and Measurement*, eds. Hirota et al., Plenum Press, New York (1994)
8. Petz, D. and Sudár, Cs. *J. Math. Phys.* **37** (1996) 2662–2673
9. Petz, D. Information geometry of quantum states, in *Quantum Probability Communications*, vol. 10, eds. Hudson, R.L. et al, World Scientific, 135–158
10. Petz, D. *Linear Algebra Appl.* 244 (1996) 81–96

On Real Hilbertian Info-Manifolds

G. Burdet*, Ph. Combe*, H. Nencka†

*Centre de Physique Théorique, CNRS-Luminy, Case 907,
F 13288 Marseille Cedex 9, FRANCE [1]
†Center of Mathematical Sciences, University of Madeira,
9000 Funchal, Madeira - Portugal [2] [3]
E-mail: burdet@cpt.univ-mrs.fr, combe@cpt.univ-mrs.fr, nencka@cpt.univ-mrs.fr

Abstract. The methods of differential geometry applied to probability and statistics open a new domain of investigation of the statistical manifold or information geometry. This framework provides a geometrical description of statistical quantities and leads to a new approach to complex statistical problems. A peculiar feature is that statistical manifolds are naturally associated with a family of affine-metric geometries. In active fields such as information and communication theories, the finite dimensional statistical manifolds approach is not completely satisfactory. In this short note we discuss some possible constructions of a real Hilbertian geometry.

INTRODUCTION

The importance of the curvature in parametric statistical problems was firstly noticed by Efron [9] and Dawid [8] who suggested the introduction of affine-metric connections, not necessarily Riemannian, on the parameter space. Even though the need for a non-parametric approach had already been noticed by Dawid [8], Amari [1] and Chensov [4], most of the literature is still devoted to parametric statistical manifolds (see e.g. [1,3,5]). However, in the 80's Kirillov [13] noticed a probabilistic realization of the infinite dimensional manifold $\mathrm{Diff}_+(S^1)/S^1$ and Friedrich [11] generalized the Kählerian structure of orbits for transformation models [2]. Recently, Gibilesco and Pistone have proposed an Orlicz space geometry for non-parametric exponential model [12,18].

Let $(\Omega, \mathcal{F}, \lambda)$ be a measure space with λ a σ-finite measure and consider the family $\mathcal{P}_\lambda$ of all probabilities ($P \ll \lambda$), absolutely continuous with respect to the measure λ. Then the Radon-Nikodym derivative

---

1) One of us, Ph. C., thanks Prof. R. Streater for his kind invitation to King's College London.
2) Two of us, Ph.C. and H. N., are grateful to G. Pistone for constructive discussions and comments
3) This work is partially supported by the projects plurianual and Praxis XXI, Portugal.

CP553, *Disordered and Complex Systems*, edited by P. Sollich, et al.

$$\rho = \frac{dP}{d\lambda} > 0 \quad \lambda - a.s., \tag{1}$$

exists and the family $\mathcal{P}_\lambda$ can be realized in terms of λ-measurable, normalized, positive random variables:

$$\mathcal{L}_\lambda = \left\{ \rho \in L^1(\Omega, \lambda)\,;\, \rho > 0 \ \lambda - a.s.\,,\ \int_\Omega \rho\, d\lambda = 1 \right\}. \tag{2}$$

The metric space $\mathcal{L}_\lambda$ can be partitioned into convex subsets of equivalent probabilities ($P \equiv P'$), i.e. probabilities with the same null-space. Let us denote by $\mathcal{C}_{\rho_0}$ any one of the equivalent classes characterised by the choice of the representative $\rho_0 = dP_0/d\lambda$. Moreover, the mapping $\rho \to b = \sqrt{\rho}$ defines a one to one correspondance between $\mathcal{L}_\lambda$ and $\mathcal{R}_\lambda$,

$$\mathcal{R}_\lambda = \left\{ b \in L^2(\Omega, \lambda)\,;\, b > 0 \ \lambda - a.s.\,,\ \int_\Omega b^2\, d\lambda = 1 \right\}. \tag{3}$$

A natural real Hilbertian distance on the positive cone $L^2_+(\Omega, \lambda) = \{b \in L^2(\Omega, \lambda)\,;\, b > 0 \ \lambda - a.s.\}$ is given by the Hellinger one,

$$d_H(M_1, M_2) = 2\left[\int_\Omega \left(\sqrt{m_1} - \sqrt{m_2}\right)^2 d\lambda\right]^{\frac{1}{2}} = d_H(m_1, m_2);\ m_i = \frac{dM_i}{d\lambda},\ i = 1.2 \tag{4}$$

between two (positive) measures M_1 and M_2, up to a factor. This distance, in the L^2-cone, is a geometrical one since it is the length of the line $\sqrt{m(t)} = (1-t)\sqrt{m_1} + t\sqrt{m_2}$, which is just the geodesic between $\sqrt{m_1}$ and $\sqrt{m_2}$. However, the restriction of the Hellinger distance to the subspace $\mathcal{R}_\lambda$ is not geometrical. Probabilities are associated with points in the intersection of L^2_+ with the unit L^2-sphere and only the points $\sqrt{\rho_1}$ and $\sqrt{\rho_2}$ on the line $\sqrt{\rho(t)} = (1-t)\sqrt{\rho_1} + t\sqrt{\rho_2}$ define probabilities.

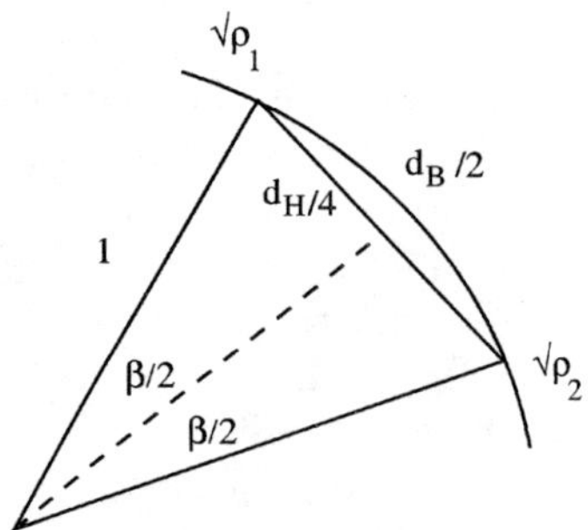

FIGURE 1. Hellinger and Bhattacharya distance.

Intuitively, in this scheme a geodesic connecting $\sqrt{\rho_1}$ and $\sqrt{\rho_2}$ in $\mathcal{R}_\lambda$ is a great arc on the unit L^2-sphere, the length being given by the angle β under which the arc is seen from the top of the cone, $\cos\beta = 2\cos^2\beta/2 - 1 = 1 - d_H^2/8 = \int_\Omega \sqrt{\rho_1\, \rho_2}\, d\lambda$.

This angle β defines a geometrical distance between the probabilities P_1 and P_2 with Radon-Nikodym derivative ρ_1 and ρ_2 respectively, proportional to the so-called Bhattacharyya distance

$$d_B(P_1, P_2) = d_B(\rho_1, \rho_2) = 2\beta = 2 \text{ arccos} \int_\Omega \sqrt{\rho_1 \, \rho_2} \, d\lambda. \tag{5}$$

The infinitesimal distance is then given by $d\ell^2 = \mathbb{E}_P[w^2]dt^2$ where $w = d\ln\rho/dt$, $\rho = dP/d\lambda$ and $d_B(P_1, P_2)$ is the integal along the arc of the great circle between $\sqrt{\rho_1}$ and $\sqrt{\rho_2}$ of the metric tensor

$$g_P(w, w') = \mathbb{E}_P[ww'], \tag{6}$$

where $w = d\ln\gamma/dt(0)$ and $w' = d\ln\gamma'/dt'(0)$ are the tangent vectors at the curves γ and γ' respectively at $\gamma(0) = \gamma'(0) = \rho,$. Tangent vectors at ρ satisfy

$$\mathbb{E}_P[w] = 0. \tag{7}$$

COVARIANT DERIVATIVE

The fact that any curve $s \to (\frac{a}{2}s + \sqrt{(m)})^2 \in \mathcal{L}_\lambda$ with $\mathbb{E}_P[a] = 0$ defines (by the mapping $\rho \to \sqrt{\rho}$) a geodesic in $L^2_+(\Omega, \lambda)$ with affine parameter s, suggests that one should study the following curve

$$\gamma(t) = N\left(\frac{a}{2}s(t) + \sqrt{\rho}\right)^2, \quad N = \int_\Omega \left(\frac{a}{2}s(t) + \sqrt{\rho}\right)^2 d\lambda, \quad \mathbb{E}_P[a] = 0, \tag{8}$$

with

$$s(t) = \frac{1}{2}\left(1 - \frac{\tan\left(\frac{\beta}{2}(1-4t)\right)}{\tan\frac{\beta}{2}}\right), \tag{9}$$

where from Eq. (5),

$$\tan\frac{\beta}{2} = \frac{d_H}{\sqrt{16 - d_H^2}}, \quad d_H^2 = 4\int_\Omega \left(\sqrt{\gamma(1)} - \sqrt{\gamma(0)}\right)^2 d\lambda$$

Some computation shows that the metric defined by Eq. (6) on the tangent vector $w(t) = d\ln\gamma/dt(t)$ to this curve at $\gamma(t)$ is

$$g(w, w)|_t = \mathbb{E}_{\gamma(t)}[w(t)^2] = d_B^2(\gamma(0), \gamma(1)), \tag{10}$$

i.e., a constant along the curve. Then the curve $t \to \gamma(t)$ with affine parameter t minimizes the distance between $\gamma(0)$ and $\gamma(1)$,

$$\frac{d}{dt}\mathbb{E}_{\gamma(t)}[w(t)^2] = 0. \tag{11}$$

Let us define $L_0^2(\Omega, \rho d\lambda) = \{u \in L^2(\Omega, \rho d\lambda),\ \mathbb{E}_\rho[u] = 0\}$ and consider the transport $\mathcal{T}_{t',t} : L_0^2(\Omega, \gamma(t')d\lambda) \to L_0^2(\Omega, \gamma(t)d\lambda)$ along the curve γ, defined by the one to one mapping

$$\mathcal{T}_{t',t}(u) = u\sqrt{\frac{\gamma(t')}{\gamma(t)}} - \mathbb{E}_{\gamma(t)}\left[u\sqrt{\frac{\gamma(t')}{\gamma(t)}}\right]. \tag{12}$$

The covariant derivative associated with this transport, defined by

$$\nabla_w v|_t := \lim_{h\to 0}\frac{1}{h}\left(v(t+h)\sqrt{\frac{\gamma(t+h)}{\gamma(t)}} - \mathbb{E}_{\gamma(t)}\left[v(t)\sqrt{\frac{\gamma(t+h)}{\gamma(t)}}\right]\right), \tag{13}$$

satisfies

$$\nabla_w w|_t = 0. \tag{14}$$

Hence the curve γ can be interpreted as a geodesic in the probability space.

GEOMETRICAL STRUCTURE

The previous considerations suggest an interpretation of $\bigcup_{\rho\in\mathcal{C}_{\rho_0}} L_0^2(\Omega, \rho d\lambda)$ as a tangent bundle $T\mathcal{C}_{\rho_0}$ with projection $\pi : L_0^2(\Omega, \rho d\lambda) \to \rho$ where $\mathcal{C}_{\rho_0}$ is the equivalent class of ρ_0.

Let $L_0^2(\Omega, \rho_0 d\lambda)$ be chosen as model space. The mapping

$$\rho = f_{\rho_0}(u) = \left(u + \sqrt{1 - \mathbb{E}_{\rho_0}[u^2]}\right)^2 \rho_0, \tag{15}$$

from the unit ball $\mathcal{B}_{\rho_0}(0,1) \in L_0^2(\Omega, \rho_d\lambda) \to \mathcal{C}_{\rho_0}$ is surjective. Indeed,

$$u = \varphi_{\rho_0}(\rho) = \sqrt{\frac{\rho}{\rho_0}} - \mathbb{E}_{\rho_0}\left[\sqrt{\frac{\rho}{\rho_0}}\right]; \quad \forall\rho \in \mathcal{C}_{\rho_0}. \tag{16}$$

However, it is injective only on the subset of $\mathcal{B}_{\rho_0}(0,1)$ satisfying, for example,

$$u + \sqrt{1 - \mathbb{E}_{\rho_0}[u^2]} > 0; \tag{17}$$

we denote this subset by $\mathcal{A}_{\rho_0}$. Hence Eq. (16) does not define a chart in the classical sense [16], but many features of a manifold are present as we have already seen in the previous section. In particular, the change of representative $u' = \varphi_{\rho_0'} \circ f_{\rho_0}(u)$,

$$u' = \left(u + \sqrt{1 - \mathbb{E}_{\rho_0}[u^2]}\right)\sqrt{\frac{\rho_0}{\rho_0'}} - \mathbb{E}_{\rho_0'}\left[\left(u + \sqrt{1 - \mathbb{E}_{\rho_0}[u^2]}\right)\sqrt{\frac{\rho_0}{\rho_0'}}\right] \tag{18}$$

preserves the positivity of the subset, $u' + \sqrt{1 - \mathbb{E}_{\rho_0'}[u'^2]} > 0$.

We can also define some equivalent mapping to the tangent one by

$$v' = f_{\rho_0 \star}(v) = \frac{1}{2 f_{\rho_0}(u)} \lim_{t \to 0} \frac{f_{\rho_0}(u + tv) - f_{\rho_0}(u)}{t} = \sqrt{\frac{\rho_0}{\rho}}\left(v - \frac{\mathbb{E}_{\rho_0}[uv]}{\sqrt{1 - \mathbb{E}_{\rho_0}[u^2]}}\right) \tag{19}$$

with inverse

$$f_{\rho_0}^{-1}{}_{\star}(v') = v'\sqrt{\frac{\rho}{\rho_0}} - \mathbb{E}_{\rho_0}\left[v'\sqrt{\frac{\rho}{\rho_0}}\right], \quad v' \in \mathcal{A}_\rho, \tag{20}$$

corresponding to the isomorphism in Eq.(12).

In this framework $\overset{0}{\nabla}(= \nabla)$ given by Eq.(13) defines a "Riemannian"-connection in a weak sense

$$g_\rho(\overset{0}{\nabla}_u v, w) = \mathbb{E}_\rho[uvw] + \frac{1}{2} t[uvw] \tag{21}$$

where t is a symmetric 3-covariant tensor

$$t(u, v, w) = \mathbb{E}_\rho[uvw], \quad u, v, w \in L_0^2(\Omega, \rho_d \lambda). \tag{22}$$

A family of "affine"-connections of Amari-Chensov-type [1,4] can be defined through

$$g_\rho(\overset{\alpha}{\nabla}_u v, w) = g_\rho(\overset{0}{\nabla}_u v, w) + \frac{\alpha}{2} t[uvw]. \tag{23}$$

These results, in comparison with the works of Kirilov-Yuriev and Friedrich [13,11], seem to show that classical modelled manifold approach needs more structure on the Radom-Nikodym derivative than measurability.

REFERENCES

1. S.-I. Amari, *Differential Geometrical methods in Statistics*, Lecture Notes in Statistics, vol. 28, Springer, 1985; *Contemporary Mathematics* **203**, 81–95 (1997).
2. O. Barndorff-Nilsen, *Parametric Statistical Models and Likelihood*, Lecture Notes in Statistics, vol. 50, Springer, 1988.
3. G. Burdet, Ph. Combe, H. Nencka, *Progress in Probability* **45**, 87–99 (Birkhauser 1998).
4. N.N. Chensov, *Statistical Decision and Optimal Inference*, (in Russian), Nauka, Moscow, 1972; English translation: AMS, vol. 53, Providence R.I., 1982.

5. Ph. Combe, H. Nencka, *Contemporary Mathematics* **203**, 105-116 (1997).
6. I. Csiszár, *Ann. of Probab.* **3**, 146-158 (1975).
7. I. Csiszár, J. Körner *Information Theory, Academic Press* , (1981).
8. A.P. Dawid, *Ann. Stat.* **3**, 1231-1234 (1975)., *Ann. Stat.* **5**, 1242 (1977).
9. B. Efron, *Ann. Stat.* **3**, 1189-1942 (1975)., *Ann. Stat.* **6**, 362-376 (1975).
10. R.A. Fisher, *Proc. Camb. Phil. Soc.* **22**, 700-725 (1925).
11. T. Friedrich, *Math. Nachr.* **153**, 273-296 (1991).
12. P. Gibilisco, G. Pistone, *Infinite Dimensional Anal. Quantum Prob. Related Topics* **1**, 325-347 (1998).
13. A. A. Kirillov, D. V. Yuriev, *Functional Analysis and its Applications* **21**, 284-294 (1987).
14. S. Kullback, *Information Theory and Statistics*, Wiley, New York, 1959.
15. S. Kullback, R. Leibler, *Ann. Math. Stat.* **22**, 79-86 (1951).
16. S. Lang, *Differential and Riemannian Manifolds*, Springer, 1994.
17. M.K. Murray, J.W. Rice, *Differential Geometry and Statistics*, Chapman & Hall, 1990.
18. G. Pistone, C. Sempi, *Ann. stat.* **23**, 1543-1561 (1995).
19. C.R. Rao, *Bull. Calcutta Math. Soc.* **37**, 81-91 (1945).
20. I. Vajda, *Theory of Statistical Inference and Information*, Kluwer, 1989.

New Ideas in Non-Parametric Estimation

Giovanni Pistone[1]

Politecnico di Torino
Corso Duca degli Abruzzi 24
10129 Torino — Italy

Abstract. The non-parametric case of statistical manifolds poses peculiar technical problems as the basic paradigms, namely the use of the Fisher information as a metric tensor by Rao in 1945, see [12], and the geometry of exponential models introduced by Efron in 1975, see [3], are difficult to realize rigorously in one single geometric structure as advocated by Amari in 1985, see [1]. The present paper discusses some improvement to the basic construction of the non-parametric exponential model related to: the analyticity of the manifold, the topology induced by the use of exponential Orlicz spaces, the use of other parameterizations. The discussion is restricted to the classical (commutative) case. Further developments on connections both in the classical and in the quantum case are treated in the paper presented in this Conference by P. Gibilisco and T. Isola.

INTRODUCTION

One approach to the construction of a statistical manifold in the non-parametric case has been developed in a series of papers by the author and coworkers, see [10], [5], [9]. In such an approach, the set supporting the manifold consists of all probability densities equivalent to a given reference finite measure μ, and the manifold in a neighborhood of a density p is modeled with a subspace of the Orlicz space associated with the Young function $\Phi(x) = \cosh(x) - 1$, with respect to the probability measure $p \cdot \mu$. A recent account of the relevant theory of Orlicz spaces is given in the monograph by Rao and Ren [13]. We denote the Orlicz space with $L^\Phi(p)$. The model Banach space consists of all centered random variables in $L^\Phi(p)$, or in the closure in this space of bounded centered random variables. Here we denote the model space by B_p.

A number of technical problems are still open, especially those connected with the properties of morphisms of such manifolds and with the properties of sub-manifolds, see [9, Sec. 5 and 7]. The difficulty arises from the peculiar properties of the space $L^\Phi(p)$, which is non separable and non reflexive if it is not finite dimensional.

[1] The author gratefully acknowledges EPSRC for financial support of his visit to King's College July 2000.

CP553, *Disordered and Complex Systems*, edited by P. Sollich, et al.

New ideas and speculations concerning further developments and ongoing research are shortly described in [6], while the connection with the algebraic description of statistical models appears in [11].

This paper is devoted to present in some detail a number of non-systematic remarks and new ideas concerning this basic background. This material either improves over the presentation given in the quoted papers, or explores variations on the earlier approach. All the material is the outcome of a series of discussions with Ph. Combe, P. Gibilisco, H. Nencka, R. Streater, H. Wynn. Warm thanks to all of them. We do not touch upon the non-commutative case, which is treated in the paper presented to this Conference by P. Gibilisco and T. Isola.

Our perspective is limited to the estimation problem that we formalize as follows. Let $\mathcal{M}(X, \mathcal{X}, \mu)$ denote the set of probability densities on the measurable space $(X, \mathcal{X})$, equivalent to the reference measure μ. Let $\mathcal{M}_0$ be a statistical model, $\mathcal{M}_0 \subset \mathcal{M}$. An *estimating function* is a Hilbert valued map $U : X \times \mathcal{M}_0$ such that for all $p \in \mathcal{M}_0$ the section U_p is a centred random variable for the probability measure $p \cdot \mu$. For example, a special case of an estimating function is $U(x,p) = G(x) - g(p)$, were G in an unbiased estimator of the function g. Given an estimating function U, an estimator is given by a $\mathcal{M}_0$-valued random variable G, such that $U(x, G(x)) = 0$. If a proper manifold structure is given, the problem can be regarded as an implicit function problem, and properties of the estimator can be rigorously derived.

We use the notations of [9], apart from the following variants. The definition of the Luxemburg norm of the Orlicz space $L^{\Phi}(p)$, $\Phi(x) = \cosh(x) - 1$, used in [9] is annoying for many reasons, for example because the constant random variable 1 has norm $\Phi(1) = \cosh(1) - 1$ according to this definition. Actually the choice of a specific definition of the norm is usually not relevant, because in most cases only a characterization of convergence is used. One could use normalized Young functions, see [13, Sec. 2.4], or the equivalent norm $||u||_p = \inf \{r > 0 | \mathrm{E}_p (\Phi(u/r)) \leq \Phi(1)\}$. Other options are possible: in the present paper the notation $|| \cdot ||_p$ denotes any of them. To avoid the problem of the non-separability of the space $L^{\Phi}(p)$, we suggest here the use of the closure of the space of simple random variables. This space is denoted by $M^{\Phi}(p)$ as in [13, Sec. 3.4].

ANALYTIC FUNCTIONALS

In the construction of exponential models, the key quantity is the normalizing constant (partition function) $M_p(u) = \mathrm{E}_p(\mathrm{e}^u)$. If for some random variable u the normalizing constant is finite, $M_p(u) < +\infty$, then $p = \exp(u - \ln M_p(u))$ is a probability density. To avoid an arbitrary constant, we assume $\mathrm{E}_p(u) = 0$. The smoothness of the mapping $B_p \ni u \mapsto M_p(u)$ is a generalization of the classical smoothness of the Laplace transform $\mathbb{R} \ni t \mapsto M_p(tu)$. From the theory of the Laplace transform we derive the idea to ask not just for the finiteness of the normalizing constant, but also for the finiteness of the Laplace transform of the random variable u on an *open* interval containing 0 and 1. A detailed study of the theory of the Laplace transform in its connections with exponential models was given by G. Letac in his monograph [8].

We consider now two options. The first one is to consider the class of random variables u such that $\mathrm{E}_p\left(\mathrm{e}^{tu}\right)$ is finite *in a neighborhood* of 0; the second one is to ask for it to be finite for *all* real t. By symmetrizing both assumptions we obtain the following two alternative definitions: (1) $\mathrm{E}_p\left(\Phi(tu)\right)$ is finite for *some* positive $t > 0$; (2) $\mathrm{E}_p\left(\Phi(tu)\right)$ is finite for *all* positive $t > 0$. In both cases the unit ball of the Banach space we use to model our manifold is defined by the inequality $\mathrm{E}_p\left(\Phi(tu)\right) \leq 1$. In case (1) we get the Orlicz space $L^{\Phi}(p)$ endowed with the Luxemburg norm. Note that the function Φ is symmetric, $\Phi(x) = \Phi\left(|x|\right)$, and that from $\Phi(x) = \Phi(x^+)(x > 0) + \Phi(x^-)(x < 0)$ it follows the equivalence of u or $|u|$ satisfying (1) or (2). It is easy to prove that in case (2) we get the space $M^{\Phi}(p)$. In fact, if $u_n = u(|u| \leq n)$ is the random variable u truncated at level n, then $u - u_n = u(|u| > n)$ and $\Phi\left((u - u_n)/r\right) = \Phi(u/r)(|u| > n)$, which implies that $u_n \to u$ as $n \to \infty$ in case (2). On the other side, given a sequence of bounded random variables v_n converging with respect to the Luxemburg norm to a random variable u, then the sequence of truncated u_n converges to u and this in turn implies $u \in M^{\Phi}(p)$, see the detailed proof in [13, Cap. 3].

Now we can prove that the functional M_p is analytic on the unit ball of B_p. By Taylor expansion of the exponential we obtain $\left|\sum_{k=0}^{\infty}\left(u^k/k!\right)\right| \leq \sum_{k=0}^{\infty}|u|^k/k! \leq \mathrm{e}^{|u|}$. From $\mathrm{e}^{|x|} \leq 2\cosh(x)$ and the assumption $\mathrm{E}_p\left(\Phi(u)\right) \leq 1$, we have $\mathrm{E}_p\left(|u|^k\right) \leq 4k! < \infty$ and, by bounded convergence, $M_p(u) = \sum_{k=0}^{\infty}\mathrm{E}_p\left(u^k/k!\right)$. As $u \mapsto \mathrm{E}_p\left(u^k\right)$ defines a power operator whose norm is bounded by $4k!$, we can consider the previous series as a power series. If $||u||_p \leq r < 1$, then $\left|E_p\left(u^k\right)\right| \leq 4k!r^k$ and the series is dominated by $\sum 4r^k$. As the convergence is uniform on closed balls of radius $r < 1$, the functional M_p is Frechet-analytic on the open unit ball. Frechet regularity follows by standard arguments.

In fact the most important quantity is the analytic functional generalizing the cumulant generating function $K_p(u) = \ln M_p(u)$ and the related relative information function $K(p|q) = \mathrm{E}_p\left(\ln(p/q)\right)$. We refer to [9] and to [4] for the following properties, where $q = \exp\left(v - K_p(v)\right)$, and Ψ is the conjugate function of Φ whose Orlicz space is $L^{\Psi}(p)$:

1. $K_p(u) = K(p|q)$.
2. $\mathrm{D}K_p(u)v = \mathrm{E}_q(v)$ and $\mathrm{D}^2K_p(u)vw = \mathrm{Cov}_q(v, w)$.
3. If $H_p : L_0^{\Psi}(p) \to \mathbb{R}$ is the conjugate functional of K_p, and $u^\star = q/p - 1$, then $H_p(u^\star) = \mathrm{E}_p\left((1 + u^\star)\ln(1 + u^\star)\right) = K(q|p)$

TOPOLOGY

In the exponential statistical manifold $\mathcal{M}$, the topology is induced via the charts $\mathrm{E}_p : B_p \supset \mathcal{V}_p \ni u \mapsto q = \exp(u - K_p(u)) \cdot p \in \mathcal{U}_p \subset \mathcal{M}$. This topology is metric and can be defined by the convergence of sequences. Because of the relations between moment generating functions, cumulant generating function, relative information functions, it is useful to express the convergence condition explicitly in terms of convergence of the above mentioned functionals.

Let u_n, $n \in \mathbb{N}$, be a sequence in B_p and let $u \in \mathcal{V}_p$. If $u_n \to u$ in B_p as $n \to \infty$, then the sequence will be eventually in $\mathcal{V}_p$. We can assume without loss of generality $u_n \in \mathcal{V}_p$ for all $n \in \mathbb{N}$, so that the corresponding probability measure sequence of probability densities $q_n = \mathrm{E}_p(u_n) = \exp(u_n - K_p(u_n)) \cdot p$ is defined for all n. Let us analyze first in detail the convergence in B_p. The following statements are all equivalent.

1. $v_n \to 0$ in B_p as $n \to \infty$.

2. There exist a positive constant $K > 0$ such that for all positive $t > 0$, $\limsup_{n\to\infty} \mathrm{E}_p\left(\Phi(tv_n)\right) \leq K$.

3. For all positive $t > 0$, $\lim_{n\to\infty} \mathrm{E}_p\left(\Phi(tv_n)\right) = 0$.

The proof is as follows. According to the definition of the norm $||\cdot||_p$ used in [9], the condition $||v_n||_p \leq \epsilon$ is equivalent to $\mathrm{E}_p\left(\Phi\left(\frac{v_n}{\epsilon}\right)\right) \leq 1$, then the convergence $v_n \to 0$ implies, putting $t = \epsilon^{-1} > 0$, the condition $\limsup_{n\to\infty} \mathrm{E}_p\left(\Phi(tv_n)\right) \leq 1$. From the inequality $\Phi(tv_n) \geq t^{2k}\frac{|v_n|^{2k}}{(2k)!}$, $k \geq 1$, we obtain $\limsup_{n\to\infty} \mathrm{E}_p\left(|v_n|^{2k}\right) \leq \epsilon^{2k}(2k)!$, then the convergence of all moments to zero. Now $\Phi(tv_n) \to 0$, $n \to \infty$, in probability and we must show that the sequence of random variables $\Phi(tv_n)$ is uniformly integrable. This follows from the elementary inequality $\Phi(x)^2 \leq \Phi(2x)$. In fact, for all positive $t > 0$, we have $\limsup_{n\to\infty} \mathrm{E}_p\left(\Phi(tv_n)^2\right) \leq \limsup_{n\to\infty} \mathrm{E}_p\left(\Phi(2tv_n)\right) \leq K$.

Note that $\mathrm{E}_p\left(\Phi(u)\right) = \frac{1}{2}\left(M_p(u) + M_p(-u)\right) - 1$. Then condition of convergence is expressed in term of the moment generating function, for example with the following further conditions equivalent to convergence.

4. For all real $t \in \mathbb{R}$, $\lim_{n\to\infty} M_p(tv_n) = 1$

5. For all real $t \in \mathbb{R}$, $\lim_{n\to\infty} K_p(tv_n) = 0$

The next step is to consider the convergence induced on the manifold. We denote by $q_n = \exp\left(v_n - K_p(v_n)\right) \cdot p$ the sequence of densities with coordinates v_n. We consider the one-dimensional exponential models $q_n(t) = \exp\left(tv_n - K_p(tv_n)\right) \cdot p$ and recall (see [9]) that the value of the cumulant functional $K_p(v)$ is equal to the relative information $K(p|q) = \mathrm{E}_p\left(\ln\left(p/q\right)\right)$ of p with respect to the density q whose coordinate is v. Now we can add to the list of equivalent conditions two new items. Condition 6. was introduced in [10] and used to prove the existence on the manifold.

6. For all real $t \in \mathbb{R}$, $\lim_{n\to\infty} \left(\frac{q_n}{p}\right)^t = 1$ in $L^1(p)$.

7. For all real $t \in \mathbb{R}$, $\lim_{n\to\infty} K(p|q_n(t)) = 0$

The proof is as follows. First we compute $(q_n/p)^t = \exp(tv_n - tK_p(v_n)) = (q_n(t)/p)\exp(K_p(tv_n) - tK_p(v_n)) = (q_n(t)/p)\exp(K(p|q_n(t)) - tK_p(v_n))$, and, by taking the expectation, we obtain $\mathrm{E}_p\left((q_n/p)^t\right) = \exp(K_p(tv_n) - tK_p(v_n)) = \exp(K(p|q_n(t)) - tK_p(v_n))$. Using conditions 1. and 5. we obtain the convergence in the form 6. of the quotient and in the form 7. with the information function.

We recall a simple result for convex functions over the reals $\mathbb{R}$. Let ϕ_n be a sequence of non-negative convex functions with proper domain I_n such that for all real $x \in \mathbb{R}$, $\lim_{n\to\infty}\phi_n(x) = 0$. If ϕ_n' is a sub-differential function of ϕ_n, then passing to the limit in the inequality $\phi_n(y) - \phi_n(x) \geq \phi_n'(x)(y-x)$ we obtain $\lim_{n\to\infty}\phi_n'(x) = 0$. By using this remark for the sequence of convex functions of condition 5., that is $\phi_n(t) = K_p(tv_n)$, we have from [9] that $\phi'(t) = DK_p(tv_n)v_n = \mathrm{E}_{q_n(t)}(v_n) = \eta_n(t)$. It is the so-called expectation parameter for the exponential model $q_n(t)$ and we have a new condition for convergence.

8. For all real $t \in \mathbb{R}$, $\lim_{n\to\infty}\mathrm{E}_{q_n(t)}(v_n) = \lim_{n\to\infty}\eta_n(t) = 0$

All conditions 1. to 8. have something interesting to reveal about the meaning of convergence in the manifold we have defined, but none answers the more difficult question on how to construct statistically meaningful distances on the set of probability densities which are equivalent to or at least continuous for the topology on the manifold. More precisely we expect interesting distances to be actually associated with path metric spaces, see [7, Sec. 1.B].

CHARTS AND PARAMETERIZATIONS

One of the technical problems left open by S.-I. Amari in his seminal work on Information Geometry in [1] is the construction of a Riemannian manifold on the space all probability densities on a given space. Formally this can be done by using the Amari embedding $p \mapsto 2\sqrt{p}$ which maps densities into the surface of a sphere in the Hilbert space $L^2(\mu)$. The Riemannian structure comes from the pull-back of the surface of the Hilbert ball. Actually the construction of the proper Riemannian manifold works only in the finite dimensional case, basically because the image of the Amari embedding is a set with empty interior in the relative topology of the sphere. On the other side it is remarkable that such a set is weakly open, that is to say each finite dimensional restriction is open.

A more interesting construction has been given by G. Burdet, Ph. Combe and H. Nencka in [2] and further developed in the paper given in this Conference. The authors give an explicit chart $\phi_p : q \mapsto \sqrt{q/p} - E_p\left(\sqrt{q/p}\right)$ with values in the open unit ball of $L^2(p)$ and show that the function $\psi_p : u \mapsto \left(u + \sqrt{1 - E_p(u^2)}\right)^2 p$ is such that $\psi \circ \phi(q) = q$. This system of charts permits the derivation of the structure of the statistical connections on the Hilbert bundle of the manifold and clearly has the advantage of giving the charts explicitly, without reference to the pull-back of another manifold. Nevertheless, the same problem as for the Amari embedding arises here: in fact the image of the chart ϕ_p is not the full open ball, as it is restricted to random variables bounded below by a constant and the function ψ_p is surjective but not injective.

Apart from the Riemannian one, other manifold structures are of interest. For example in [9] we study the non parametric equivalent of the expectation parameters, arising from the mapping $q \mapsto q/p - 1 \in L_0^{\Psi}(p)$. Again this mapping cannot be a chart because the values are bounded below by -1. It would be interesting to show that non-Riemannian structures (if any) are actually Finsler manifolds.

Many approaches are possible to try to overcome the problem. In [5] and in the paper presented to this Conference by P. Gibilisco, the idea is to give up the construction of the Riemannian manifold, and to consider the Amari embedding as a morphism from the manifold modelled on Orlicz spaces to the Hilbert sphere, and to derive all relevant structures by pull-back of the tangent bundle. It has also been suggested by G. Ben Arous (private communication, 1998) that a manifold with a regularity weaker than Frechet differentiability, such as some kind of white noise calculus, would lead to a better theory.

REFERENCES

1. Amari, S., *Differential-geometrical methods in statistics*, Springer-Verlag, New York-Berlin, 2nd printing 1990, corrected edn. 1985.
2. Burdet, G., Combe, P., and Nencka, H., "Information de Fisher et connections de Chentsov-Amari pour des varietès statistiques non-parametriques," presented to SFDS99 - Grenoble 17 May 1999.
3. Efron, B., *The Annals of Statistics* **3**, 1189–1242 (1975).
4. Ekeland, I., and Temam, R., *Analyse convexe et problèmes variationnels*, Dunod Gauthier–Villars, Paris, 1979.
5. Gibilisco, P., and Pistone, G., *Infinite Dimensional Analysis, Quantum Probability and Related Topics* **1**, 325–347 (1998).
6. Gibilisco, P., and Pistone, G., "Analytical and geometrical properties of statistical connections in information geometry," in *Mathematical Theory of Networks and Systems. Proceedings of MTNS-98: Padova, Italy, July 6-10 1998*, edited by A. Beghi, L. Finesso, and G. Picci, Il Poligrafo, Padova, Italy, 1999, pp. 811–814.
7. Gromov, M., *Metric structures for Riemannian and non-Riemannian spaces*, Birkhäuser Boston Inc., Boston, MA, 1999.
8. Letac, G., *Lectures on Natural Exponential Families and their Variance Functions*, Istituto de Matemática Pura e Aplicada, Rio de Janeiro, 1992.
9. Pistone, G., and Rogantin, M.-P., *Bernoulli* **5**, 721–760 (1999).
10. Pistone, G., and Sempi, C., *The Annals of Statistics* **33**, 1543–1561 (1995).
11. Pistone, G., and Wynn, H. P., *Statistica Sinica* **9**, 1029–1052 (1999).
12. Rao, C. R., *Bulletin of Calcutta Mathematical Society* **37**, 81–89 (1945).
13. Rao, M. M., and Ren, Z. D., *Theory of Orlicz Spaces*, Marcel Dekker Inc., New York, 1991.

The Uniqueness of the Chentsov Metric

M. R. Grasselli[1] and R. F. Streater

Department of Mathematics, King's College London, Strand, WC2R 2LS, U.K.

Abstract. We show that on the space of faithful density matrices, the only monotone metrics, for which the exponential and mixture affine connections are mutually dual, are constant multiples of the Bogoliubov-Kubo-Mori (*BKM*) metric.

INTRODUCTION

Information manifolds are equipped with two natural flat connections: the mixture connection, obtained from the linear structure of trace class operators themselves, and the exponential connection, obtained when combinations of states are performed by adding their logarithms [4]. Following Amari [1,2], we consider duality to be the fundamental structure. Thus, given these two connections, we find in §3 all the Riemannian metrics that make them dual. In §4, we combine this result with Petz's characterisation [9] of monotone metrics to find that the *BKM* metric is, up to a factor, the unique monotone metric with respect to which the exponential and mixture connections are dual.

THE EXPONENTIAL AND MIXTURE CONNECTIONS

Let $\mathcal{H}^N$ be a finite dimensional complex Hilbert space, $\mathcal{A}$ the subset of self-adjoint matrices and $\mathcal{M}$ the set of all invertible density matrices on $\mathcal{H}^N$. Then $\mathcal{A}$ is an N^2-dimensional real vector space and $\mathcal{M}$ can be regarded as an n-dimensional manifold with $n = N^2 - 1$. Defining the 1-embedding of $\mathcal{M}$ into $\mathcal{A}$ as

$$\ell_1 : \mathcal{M} \to \mathcal{A} \qquad\qquad \rho \mapsto \log \rho,$$

we can use the affine structure of $\mathcal{A}$ to obtain an affine structure on $\mathcal{M}$. Since then $\mathcal{M}$ is flat, we can identify $\mathcal{M}$ with its tangent space. At each point $\rho \in \mathcal{M}$, consider the subspace $\mathcal{A}_\rho = \{A \in \mathcal{A} : \mathrm{Tr}(\rho A) = 0\}$ of $\mathcal{A}$, called the space of 'scores'; we define the isomorphism

$$(\ell_1)_{*(\rho)} : T_\rho\mathcal{M} \to \mathcal{A}_\rho \qquad\qquad v \mapsto (\ell_1 \circ \gamma)'(0),$$

[1] Supported by a grant from CAPES-Brazil.

CP553, *Disordered and Complex Systems*, edited by P. Sollich, et al.

where $\gamma : (-\varepsilon, \varepsilon) \to \mathcal{M}$ is a curve in the equivalence class of the tangent vector v. We call this isomorphism the 1-representation of the tangent space $T_\rho\mathcal{M}$. If $(\theta^1, \ldots, \theta^n)$ is a coordinate system for $\mathcal{M}$, then the 1-representation of the basis $\left\{ \frac{\partial}{\partial\theta^1}\Big|_\rho, \ldots, \frac{\partial}{\partial\theta^n}\Big|_\rho \right\}$ of $T_\rho\mathcal{M}$ is $\left\{ \frac{\partial \log\rho}{\partial\theta^1}, \ldots, \frac{\partial\log\rho}{\partial\theta^n} \right\}$. The 1-representation of a vector field X on $\mathcal{M}$ is therefore the $\mathcal{A}$-valued function $(X)^{(1)}$ given by $(X)^{(1)}(\rho) = (\ell_1)_{*(\rho)} X_\rho$.

The exponential- or 1-connection is that got using the 1-embedding and the following parallel transport [4]

$$\tau^{(1)}_{\rho_0,\rho_1} : T_{\rho_0}\mathcal{M} \to T_{\rho_1}\mathcal{M} \qquad v \mapsto (\ell_1)^{-1}_{*(\rho_1)} \left((\ell_1)_{*(\rho_0)} v - \mathrm{Tr}[\rho_1 (\ell_1)_{*(\rho_0)} v] \right).$$

Giving the parallel transport in a 'hood of ρ is equivalent to specifying the covariant derivative. The 1-representation of the 1-covariant derivative is

$$\left(\nabla^{(1)}_{\frac{\partial}{\partial\theta^i}} \frac{\partial}{\partial\theta^j} \right)^{(1)} = \frac{\partial^2 \log\rho}{\partial\theta^i \partial\theta^j} - \mathrm{Tr}\left(\rho \frac{\partial^2 \log\rho}{\partial\theta^i\partial\theta^j} \right) \tag{1}$$

The construction above corresponds to making $\mathcal{M}$ into an affine space and endowing it with the *natural* flat connection induced by the linear structure of the space of scores, which then provide us with an affine coordinate system.

Now let $\mathcal{A}_0$ be the subspace of traceless operators in $\mathcal{A}$. Consider the -1-embedding

$$\ell_{-1} : \mathcal{M} \to \mathcal{A} \qquad \rho \mapsto \rho,$$

and define, at each $\rho \in \mathcal{M}$, the -1-representation of tangent vectors as

$$(\ell_{-1})_{*(\rho)} : T_\rho\mathcal{M} \to \mathcal{A}_0 \qquad v \mapsto (\ell_{-1} \circ \gamma)'(0),$$

where $\gamma : (-\varepsilon, \varepsilon) \to \mathcal{M}$ is again a curve in the equivalence class of the tangent vector v. In coordinates, the -1-representation of the basis $\left\{ \frac{\partial}{\partial\theta^1}\Big|_\rho, \ldots, \frac{\partial}{\partial\theta^n}\Big|_\rho \right\}$ of $T_\rho\mathcal{M}$ is $\left\{ \frac{\partial\rho}{\partial\theta^1}, \ldots, \frac{\partial\rho}{\partial\theta^n} \right\}$. As before, the -1-representation of a vector field X on $\mathcal{M}$ is an $\mathcal{A}_0$-valued function denoted by X^-.

We obtain the mixture or -1-connection by defining the flat parallel transport

$$\tau^{(-1)}_{\rho_0,\rho_1} : T_{\rho_0}\mathcal{M} \to T_{\rho_1}\mathcal{M} \qquad v \mapsto (\ell_{-1})^{-1}_{*(\rho_1)} \left((\ell_{-1})_{*(\rho_0)} v \right).$$

DUALITY AND THE *BKM* METRIC

Two connections ∇ and ∇^* on a Riemannian manifold $(\mathcal{M}, g)$ are dual with respect to g if and only if

$$X g(Y, Z) = g(\nabla_X Y, Z) + g(Y, \nabla^*_X Z), \tag{2}$$

for any vector fields X, Y, Z on $\mathcal{M}$ [1]. Equivalently, if $\tau_{\gamma(t)}$ and $\tau^*_{\gamma(t)}$ are the respective parallel transports along a curve γ on $\mathcal{M}$, then ∇ and ∇^* are dual with respect to g if and only if

$$g(Y,Z) = g\left(\tau_{\gamma(t)}Y, \tau^*_{\gamma(t)}Z\right). \tag{3}$$

Given any connection ∇ on $(\mathcal{M}, g)$, we can always find a unique connection ∇^* such that ∇ and ∇^* are dual with respect to g. On the other hand, given two connections ∇ and ∇^*, we can ask what are the possible Riemannian metrics g with respect to which they are dual. In particular, we want to explore this question for the case of the exponential and mixture connections on a manifold of density matrices.

Another concept of duality [1] is that of dual coordinate systems, regardless of any connection. Two coordinate systems $\theta = (\theta^i)$ and $\eta = (\eta_i)$ on a Riemannian manifold $(\mathcal{M}, g)$ are dual with respect to g if and only if their natural bases for $T_p\mathcal{M}$ are *biorthogonal* at every point $p \in \mathcal{M}$, that is, $g(\partial/\partial\theta^i, \partial/\partial\eta_j) = \delta^i_j$. Equivalently, $\theta = (\theta^i)$ and $\eta = (\eta_i)$ are dual with respect to g if and only if $g_{ij} = \partial\eta_i/\partial\theta^j$ and $g^{ij} = \partial\theta_i/\partial\eta^j$, at every point $p \in \mathcal{M}$, where, as usual, $g^{ij} = (g_{ij})^{-1}$.

The next theorem gives a characterisation of dual coordinate systems in terms of potential functions, thus introducing convexity theory and the related duality with respect to Legendre transforms.

Theorem 4 (Amari, 1985) *When a Riemannian manifold $(\mathcal{M}, g)$ has a pair of dual coordinate systems (θ, η), there exist potential functions $\Psi(\theta)$ and $\Phi(\eta)$ such that*

$$g_{ij}(\theta) = \frac{\partial^2\Psi(\theta)}{\partial\theta^i\partial\theta^j} \quad \textit{and} \quad g^{ij}(\eta) = \frac{\partial^2\Phi(\eta)}{\partial\eta_i\partial\eta_j}.$$

Conversely, when either potential function Ψ or Φ exists from which the metric is derived by differentiating it twice, there exist a pair of dual coordinate systems. The dual coordinate systems and the potential functions are related by the following Legendre transforms

$$\theta^i = \frac{\partial\Phi(\eta)}{\partial\eta_i}, \quad \eta_i = \frac{\partial\Psi(\theta)}{\partial\theta^i}$$

and

$$\Psi(\theta) + \Phi(\eta) - \theta^i\eta_i = 0$$

In contrast to the case of dual connections, dual coordinate systems do not necessarily exist on every Riemannian manifold [1]. When the additional property of flatness is required, the following theorem provides a link between the two concepts of duality. We say that a connection ∇ on manifold $\mathcal{M}$ is flat if $\mathcal{M}$ admits a global ∇-affine coordinate system. This is equivalent to its curvature and torsion both being zero.

Theorem 5 (Amari) *Suppose that ∇ and ∇^* are two flat connections on a manifold $\mathcal{M}$. If they are dual with respect to a Riemannian metric g on $\mathcal{M}$, then there exists a pair (θ, η) of dual coordinate systems such that θ is ∇-affine and η is ∇^*-affine.*

We now consider the uniqueness of the Riemannian metric on $\mathcal{M}$. Using either the 1 or the -1 representation of the tangent bundle $T\mathcal{M}$, we define a Riemannian metric on $\mathcal{M}$ by a smooth assignment of an inner product $\langle \cdot, \cdot \rangle_\rho$ in $\mathcal{A} \subset B(\mathcal{H}^N)$ for each point $\rho \in \mathcal{M}$. If $A^{(+1)}, B^{(+1)}$ and $A^{(-1)}, B^{(-1)}$ are, respectively, the 1 and -1 representations of $A, B \in T_\rho\mathcal{M}$, then the *BKM* metric is

$$g_\rho^B(A, B) = \operatorname{Tr}\left(A^{(-1)}B^{(+1)}\right) = \int_0^\infty \operatorname{Tr}\left[(\rho+\alpha)^{-1}A^-(\rho+\alpha)^{-1}B^-\right] d\alpha. \tag{6}$$

The (1) and (-1) connections are dual with respect to the *BKM* metric [6,5]. A natural question is whether this is the only metric with this property. The next theorem tells us the consequences of duality alone.

Theorem 7 *If the connections $\nabla^{(1)}$ and $\nabla^{(-1)}$ are dual with respect to a Riemannian metric g on $\mathcal{M}$, then there exists a constant $n \times n$ matrix M, such that $(g_\rho)_{ij} = \sum_{k=1}^n M_{ik}(g_\rho^B)_{kj}$.*

Proof: Since the two connections are flat, by theorem 5, there exist dual coordinate systems (θ, η) such that θ is $\nabla^{(1)}$-affine and η is $\nabla^{(-1)}$-affine. Thus, applying theorem 4, there exist a potential function $\Psi(\theta)$ such that

$$g_{ij}(\theta) = \frac{\partial^2\Psi(\theta)}{\partial\theta^i\partial\theta^j} \qquad \text{and} \qquad \eta_i = \frac{\partial\Psi(\theta)}{\partial\theta^i}.$$

Now since θ is $\nabla^{(1)}$-affine, there exist linearly independent operators $\{1, X_1, ..X_n\}$ such that

$$\rho = \exp\left(\theta^1 X_1 + \cdots + \theta^n X_n - \tilde{\Psi}(\theta)\right), \tag{8}$$

where $\tilde{\Psi} = \log \operatorname{Tr} \exp(\theta^i X_i)$ is the free energy. Any such set of operators defines a $\nabla^{(-1)}$-affine coordinate system through the formula $\tilde{\eta}_i = \operatorname{Tr}(\rho X_i)$. Differentiating $\tilde{\Psi}$ with respect to θ^i we obtain

$$\frac{\partial\tilde{\Psi}(\theta)}{\partial\theta^i} = \operatorname{Tr}(\rho X_i) = \tilde{\eta}_i.$$

Thus $\tilde{\eta}_i = \frac{\partial\tilde{\Psi}(\theta)}{\partial\theta^i}$ and $\eta_i = \frac{\partial\Psi(\theta)}{\partial\theta^i}$ are two $\nabla^{(-1)}$-affine coordinate systems, so they must be related by an affine transformation. So there exist an $n \times n$ matrix M and numbers $(a_1, \ldots, a_n)$ such that

$$\eta_i = \sum_{k=1}^n M_{ik}\tilde{\eta}_k + a_i,$$

that is,

$$\frac{\partial\Psi(\theta)}{\partial\theta^i} = \sum_{k=1}^{n} M_{ik}\frac{\partial\tilde{\Psi}(\theta)}{\partial\theta^k} + a_i,$$

and differentiating this equation with respect to θ^j gives

$$g_{ij}(\theta) = \frac{\partial^2\Psi(\theta)}{\partial\theta^i\partial\theta^j} = \sum_{k=1}^{n} M_{ik}\frac{\partial^2\tilde{\Psi}(\theta)}{\partial\theta^j\partial\theta^k} = \sum_{k=1}^{n} M_{ik}g^B_{kj}. \quad \square \tag{9}$$

THE CONDITION OF MONOTONICITY

We say [3] that a metric g on $\mathcal{A}_0$ (the (-1) representation) is monotone iff

$$g_{S\rho}(SA^-, SA^-) \leq g_\rho(A^-, A^-) \tag{10}$$

for every $\rho \in \mathcal{M}$, $A \in T_\rho\mathcal{M}$, and every completely positive, trace preserving map $S : \mathcal{A} \to \mathcal{A}$.

Let $\widehat{\mathcal{M}}$ be the manifold of faithful weights (the positive-definite matrices). We can extend g and g^B from $T\mathcal{M}$ to $T\widehat{\mathcal{M}}$ as follows. At $\rho \in \mathcal{M}$ and $A \in T\widehat{\mathcal{M}}$, put $A = A_0\rho + A^-$, where $A_0 = \mathrm{Tr}\, A$ and $\mathrm{Tr}\, A^- = 0$. Then put

$$\hat{g}_\rho(A, B) = A_0B_0 + g_\rho(A^-, B^-). \tag{11}$$

For g^B this extension coincides with that given by eq. (6). Then, if g is monotone on $T\mathcal{M}$, its extension is monotone on $T\widehat{\mathcal{M}}$: let S be a trace-preserving CP map on $T\widehat{\mathcal{M}}$; then

$$\hat{g}_{S\rho}(SA, SA) = A_0^2 + g_{S\rho}(SA^-, SA^-) \leq A_0^2 + g_\rho(A^-, A^-) = \hat{g}_\rho(A, A).$$

For any metric g on $T\widehat{\mathcal{M}}$, and putting $A^{(-1)} = (A_0, A^-)$, we define the positive operator K_ρ by

$$g_\rho(A, B) = \left\langle A^{(-1)}, K_\rho\left(B^{(-1)}\right)\right\rangle_{HS} = \mathrm{Tr}\left(A^{(-1)}K_\rho\left(B^{(-1)}\right)\right), \tag{12}$$

acting on $T\widehat{\mathcal{M}}$ furnished with the Hilbert-Schmidt scalar product. Petz uses the following operators on $T\widehat{\mathcal{M}}$: $L_\rho A = \rho A$, $R_\rho A = A\rho$, and has [7,9,10] given a characterisation of monotone metrics on matrix spaces in terms of operator monotone functions. He proved

Theorem 13 (Petz, 1996) *A Riemannian metric g on $\widehat{\mathcal{M}}$ is monotone if and only if*

$$K_\rho = \left(R_\rho^{1/2}f(L_\rho R_\rho^{-1})R_\rho^{1/2}\right)^{-1},$$

where K_ρ is defined in (12) and $f : R^+ \to R^+$ is an operator monotone function satisfying $f(t) = tf(t^{-1})$.

In particular, the *BKM* metric is monotone and its corresponding operator monotone function is $f^B(t) = (t-1)/\log t$. Combining this characterisation with theorem (7), we obtain the following uniqueness result.

Theorem 14 *If the connections $\nabla^{(1)}$ and $\nabla^{(-1)}$ are dual with respect to a monotone Riemannian metric g on $\mathcal{M}$, then g is a* constant *multiple of the BKM metric.*

Proof: Let K^g_ρ and K^B_ρ be the operators associated with the monotone metrics g and g^B as in equation (12). Let us extend the matrix M of theorem (7) from $T\mathcal{M}$ to a matrix on $\widehat{\mathcal{M}}$ by $M_{0,k} = 0 = M_{k,0}$, $M_{0,0} = 1$. Then eq. (9) gives the following relation between the extended metrics $\hat{g}$ and $\hat{g}^B$, in the coordinates $(\theta^0, \theta^1, \ldots, \theta^n)$:

$$\hat{g}_\rho(\partial_i, \partial_j) = \sum_{k=0}^{n} \hat{M}_{ik} \hat{g}^B_\rho(\partial_k, \partial_j). \tag{15}$$

In terms of the kernels, this gives $K^g_\rho = \hat{M} K^B_\rho$. But then Petz's formula for K in theorem (13) leads to $\hat{M} = f^g(L_\rho R_\rho^{-1})^{-1} f^B(L_\rho R_\rho^{-1})$, where f^g and f^B are the operator-monotone functions corresponding to g and g^B respectively. Thus the matrix $\hat{M}$ is a certain function of the operators L_ρ and R_ρ, but is independent of the point ρ. We conclude that it must be a constant multiple of the identity matrix. □

REFERENCES

1. S.-I. Amari, **Differential Geometric Methods in Statistics**, *Lecture Notes in Statistics*, **28**, Springer-Verlag, New York, 1985.
2. S.-I. Amari, *Information Geometry*, Contemp. Math., **203**, 1997.
3. N. N. Čencov, **Statistical Decision Rules and Optimal Inferences**, *Translations of Mathematical Monographs*, American Mathematical Society, Providence, 1982.
4. M. R. Grasselli and R. F. Streater, *The Quantum Information Manifold for ε-bounded Forms*, to appear in Rep. Math. Phys., math-ph/9910031.
5. H. Hasegawa, *Exponential and mixture families in quantum statistics: dual structures and unbiased parameter estimation*, Rep. Math. Phys., **39**, 49-68, 1997.
6. H. Nagaoka, *Differential Geometric Aspects of Quantum State Estimation and Relative Entropy*, in **Quantum Communication and Measurement**, eds. V. P. Balavkin, O. Hirota and R. L. Hudson, Plenum Press, 1995.
7. D. Petz, *Quasi-entropies for Finite Quantum Systems*, Rep. Math. Phys., **23**, 57-65, 1986.
8. D. Petz, *Geometry of Canonical Correlation on the State Space of a Quantum System*, J. Math. Phys., **35**, 780-795, 1994.
9. D. Petz, *Monotone Metrics on Matrix Spaces*, Lin. Alg. Appl., **244**, 81-96, 1996.
10. D. Petz and C. Sudar, *Geometries of Quantum States*, J. Math. Phys., **37**, 2662-2673, 1996.

PART III: QUANTUM DYNAMICS AND QUANTUM CHAOS

Symmetry in Phase Space of a Chaotic System

Boyka Aneva

INRNE, Bulgarian Academy of Sciences, 1784 Sofia, Bulgaria

Abstract. Finite symmetry in phase space is used for a geometrical interpretation of chaos quantization conditions, which relate the eigenvalues of a Hamiltonian operator with the non-trivial zeros of the Riemann zeta function.

INTRODUCTION

An outstanding challenge for mathematical physics is the hypothesis relating the non-trivial zeros of the Riemann zeta function with the spectrum of a Hermitian operator in a Hilbert space. This work was inspired by the idea of Berry and Keating, discussed in [1,2] and in a subsequent paper [3], that the real solutions E_n of $\zeta(1/2+iE_n)=0$ are energy levels, eigenvalues of a quantum Hermitian operator associated with the one-dimensional classical hyperbolic Hamiltonian $H_{cl}(x,p)=xp$, where x and p are the conjugate coordinate and momentum. They suggest a quantization condition generating Riemann zeros, which however they 'see no way to interpret geometrically'.

We propose that such a condition can be consistently interpreted as a boundary quantization condition of a hyperbolic dynamical system within conformal geometry.

To quantize the system one considers the dilation operator in the x space

$$H=\frac{1}{2}(xp+px)=-i\hbar\left(x\partial_x+\frac{1}{2}\right). \tag{1}$$

The solutions of the eigenvalue equation

$$H\psi_E(x)=E\psi_E(x) \tag{2}$$

are the eigenfunctions (C is an arbitrary constant)

$$\psi_E(\pm x)=C\theta(\pm x)e^{(iE/\hbar-1/2)log|x|}, \tag{3}$$

together with the corresponding momentum eigenfunctions

$$\phi_E(p)=\frac{C}{|p|^{1/2+iE/\hbar}}(h/\pi)^{iE/\hbar}\frac{\Gamma(1/4+iE/2\hbar)}{\Gamma(1/4-iE/2\hbar)}. \tag{4}$$

CP553, *Disordered and Complex Systems*, edited by P. Sollich, et al.

SYMMETRIES OF THE HYPERBOLIC HAMILTONIAN

Since the hyperbolic Hamiltonian is scale invariant under $x' = \lambda x, p' = \frac{1}{\lambda}p, \lambda > 0$, we shall extend the scale invariance to a larger group and discuss it from the point of view of conformal representation theory. Conformal symmetry in the one-dimensional x space contains translations $x' = x + a$, dilations $x' = \lambda x$, and special conformal transformations, which can be written down as a superposition of two conformal inversions R (i.e., Weyl reflections),

$$Rx = -\frac{1}{x} \tag{5}$$

and a translation $T_{-c}x = x - c$, i.e., $x' = RT_{-c}Rx$. The differential operators

$$P = -i\partial_x, \qquad D = -id - ix\partial_x, \qquad K = -2idx - ix^2\partial_x, \tag{6}$$

generate the Lie algebra of the real Moebius group $SO(2,1) = SL(2,R)/Z_2$. d is a complex number, a conformal (or scale) dimension that labels the irreducible representations (IR). We consider the unitary IR of the principal series (and its dual) characterized by the value $-d = -1/2 + i\rho$ (and $-d = -1/2 - i\rho$, respectively), with ρ real, and realized in the Hilbert space of functions of one variable (and a given parity) with the scalar product $(f_1, f_2) = \int_{-\infty}^{\infty} \bar{f}_1(x) f_2(x) dx$. There exists a pair of intertwining maps relating the representation $T^{\chi}, \chi = i\rho$ and the dual one $T^{\tilde{\chi}}, \tilde{\chi} = -i\rho$, defined by the convolution integral $(G_\chi f)(x_1) = \int_{-\infty}^{\infty} G_\chi(x_1 - x_2) f(x_2) dx_2$, with the kernel:

$$G_\chi(x) = \frac{N_\chi}{\sqrt{2\pi}} |x|^{-1+2i\rho}. \tag{7}$$

The Fourier transform $G_\chi(p) = \int_{-\infty}^{\infty} G_\chi(x) e^{-ipx} dx$ restricts the choice of the normalization constant N_χ, since $G_\chi(p) G_{\tilde{\chi}}(p) = 1$.

Turning back to the eigenvalue equation (2), we may interpret it as a condition for a scale invariance of a conformal wave function of dimension $d = 1/2 - iE/\hbar$. If we postulate invariance of the wave function $\psi(x)$ under translations, we get a *constant* eigenfunction. The wave function $\psi(x)$ is also invariant under conformal R-inversion:

$$\psi(x) = \frac{1}{x^{2d}} \psi\left(-\frac{1}{x}\right). \tag{8}$$

Formula (8) defines the coordinate function as an automorphic function of weight d in the space of the IR principal series representation, invariant with respect to the discrete Weyl subgroup of $SL(2,R)$, with a non-trivial element chosen to be the conformal R-inversion.

The symmetry properties of $\psi(x)$ suggest that the quantum wave function of the chaotic dynamical system (1) is a homogeneous function of degree $-d$ and an

R-inversion invariant automorphic function of weight d, where $-d$ labels the IR principal series representation T^{χ} and is hence equal to $-1/2 + iE/\hbar$, with E real. The complex-conjugate wave function $\bar{\psi}(x)$ corresponds to the dual representation $T^{\tilde{\chi}}$ with $-d = -1/2 - iE/\hbar$. More precisely, the wave function of the chaotic system is a distribution, the square root of the kernel (7) of the intertwining operator.

To show that the real values E are eigenvalues of a self-adjoint operator we consider the shifted Hamiltonian $1/2 - iH/\hbar = -x\partial_x$:

$$x\partial_x(1 - x\partial_x) = -x^2\partial_x^2. \tag{9}$$

On the RHS of (10) is the Laplace-Beltrami operator L on the real x line with eigenfunctions

$$\psi_{\pm E}\left(\frac{1}{|x|}\right) = Cx^{1/2 \pm iE/\hbar}, \qquad x > 0. \tag{10}$$

The position eigenfunctions $\psi(x)$ are even functions of x. As seen from the explicit expression (3) they are functions of the argument $|x|$ only and can be considered as functions defined on the positive real line $(0, \infty)$. This assumption identifies $\pm x$ and allows the use of a parametrisation such that the representation space can also be realized as the Hilbert space of functions $f(x)$, defined on the half line $x > 0$ and square-integrable with respect to the measure $x^{-2}dx$. It is known that the functions $\theta(x, d) = x^{1/2+i\rho} + c(1/2 + i\rho)x^{1/2-i\rho}$, where $c(d)c(1 - d) = 1$, with the boundary condition at a with an additional parameter κ,

$$\theta(a) = a\kappa\theta'(a), \tag{11}$$

form a complete set of eigenfunctionals of the self-adjoint operator L in the space of linear functionals on the unitary principal series IR Hilbert space of functions $f(x)$ defined on the positive line $a \leq x \leq \infty, a \geq 1$ and square-integrable with respect to the measure $x^{-2}dx$.

Since

$$x\partial_x(1 - x\partial_x)x^{1/2 \pm iE/\hbar} = \left(\frac{1}{4} + \frac{E^2}{\hbar^2}\right)x^{1/2 \pm iE/\hbar} \tag{12}$$

the $\psi_{\pm E}(1/x)$ are the eigenfunctions corresponding to the eigenvalues $1/4 + E^2/\hbar^2$ of the continuous spectrum of L. Hence $x^{1/2 \pm iE/\hbar}$ are eigenvectors of the operators $x\partial_x$ and $1 - x\partial_x$, corresponding to the eigenvalues $1/2 \pm iE/\hbar$ and $1/2 \mp iE/\hbar$ respectively. Thus if $1/4 + E^2/\hbar^2$ belongs to the spectrum of L then $1/2 \pm iE/\hbar$ belongs to the spectrum of the dilation operator $x\partial_x$ and vice versa, so that roughly speaking the dilation operator is the square root of the Laplace-Beltrami operator in the Hilbert space of functions with domain (a, ∞) and square integrable with respect to the measure $x^{-2}dx$.

It can be verified further that $\psi(x)$ are eigenvectors of L but corresponding to complex eigenvalues $1/4 - (1 \pm iE/\hbar)^2$. However, there is an operator which, when acting on the position eigenfunctions $\psi(x)$, yields the real eigenvalues $1/4 + E^2/\hbar^2$. It is the R-transformed Laplace-Beltrami operator denoted hereafter L_R:

$$L_R = -x\partial_x(1 + x\partial_x) = -x^2\partial_x^2 - 2x\partial_x. \tag{13}$$

Expressing L_R in terms of the Hamiltonian (1),

$$L_R = (1/2 - iH/\hbar)(1/2 + iH/\hbar), \tag{14}$$

we obtain that the $\psi(x)$ are eigenfunctions of L_R corresponding to the eigenvalues $1/4 + E^2/\hbar^2$, so that the operators $1/2 \pm iH/\hbar$ are formally the square roots of a positive operator. The eigenfunctions of L and L_R are related by the R transformation. The Laplace-Beltrami operator is densely defined and self-adjoint on the line $[1, \infty)$. The continuous R-transformation $x' = 1/x$ on the line $x > 0$, maps its eigenfunctions to the eigenfunctions of the R-transformed operator L_R, which is hence densely defined and self-adjoint on the interval $(0, 1]$. The two operators have to be equal on the common elements, namely the functions f defined for $x = 1$, which is the fixed point of the transformation $x' = 1/x$. The relation (8), which determines the R-invariant automorphic functions, serves as a boundary condition at the point $x = 1$ (or at any point $a \geq 1$, if $x' = a/x$), for coincidence of the inversed operator $L_R = (1 - iH/\hbar)(1 + iH/\hbar)$ with the self-adjoint Laplace-Beltrami operator and will be used in phase space as a quantization condition to generate a discrete spectrum.

The R-inversions in the x-space are easily enlarged to transformations in phase space, which are area preserving and leave invariant the operators $\pm H$. (The volume element in phase space is given by the area $xp = h$ of the Planck cell with sides l_x and l_p.) The transformations have the form:

$$\begin{aligned} T_1^{\pm}: \quad & x' = -\frac{h}{x} \qquad p' = \pm\frac{x^2 p}{h}, \\ T_2^{\pm}: \quad & x' = -\frac{h}{p} \qquad p' = \mp\frac{x p^2}{h}. \end{aligned} \tag{15}$$

The upper $\pm$-indices denote that the determinant of the transformation is ± 1. The $-T_1$ transform of x is the conformal R-inversion corresponding to the value $a = 2\pi$ in (11) (in units $\hbar = 1$). For $xp = h$ the transformations contain the canonical one $x' = -p, p' = x$, and the reflections $x' = \pm x, p' = \pm p$. Together with the transformations $-T_1^{\pm}$ and $-T_2^{\pm}$ (the identity is $-T_2^{+}$), they form the dihedral group $\mathcal{D}_4$ generated by reflections (S_1) about the x axis and (S_2) about the line l inclined at an angle $\pi/4$ to the positive x-axis. A fundamental region is the 2-simplex with vertices $(0,0), (\sqrt{h}, 0), (\sqrt{h}, \sqrt{h})$. The 2-simplices are joined two by two to form four 2-complexes (the Planck cells in the distinct quadrants). The transformations fall into two classes depending on whether they preserve (reverse) sign of volume element, i.e., preserve (reverse) orientation. To form a closed path we use the isometries that map one side of the fundamental region onto a side of an adjacent region to glue distinct regions together along their boundaries. By identifying all boundary points we impose as a boundary condition the requirement for invariance under the discrete subgroup of reflections. The conditions

$$\psi(x) = \frac{1}{x^{2d}}\psi(Rx), \qquad \psi(x) = \psi\left(\pm\frac{h}{p}\right), \qquad \phi(p) = \phi\left(\pm\frac{h}{x}\right) \tag{16}$$

relate values of ψ and ϕ at boundary points mapped into each other by an element s, generating the finite Weyl subgroup of the dihedral group.

A vertex i is assigned to each element $g_i, i = 1, 2, \ldots, 8$ of the group and a directed edge (i, j) connects two vertices iff $g_i = S_k g_j$ for a generator S_k of the group. A path extends from a given g to $S_{i_1}...S_{i_k}g$ and is closed if it can be presented in the form $S_{i_1}...S_{i_k} = 1$. This is a consequence of the relation $(S_iS_j)^{m(i,j)} = 1$, with $m(i,i) = 1$, for the Coxeter element [4] of the dihedral group, which is the product of two generating reflections, i.e., a rotation, $r = S_1S_2$, through $\pi/2$. A closed path $S_1S_2S_1S_2$ is always of the form rr, which is the word representation of a projective plane. Thus the relation for the Coxeter element forces identification of antipodal boundary points and introduces topology of a projective plane in phase space. At the boundary $\pm x$ and $\pm p$ are identified, the latter according to the Moebius topology equivalence relation $(-\sqrt{h}, p) \simeq (\sqrt{h}, -p)$. With the identification of boundary points a closed path begins from a given (initial) point $g_i = (x, p)$ and ends at a (final) antipodal one $g_f = (-x, -p)$. The closed paths γ_s fall into two homotopy classes depending on whether the final point is reached via a product of even or odd number of reflections S_2, i.e., whether orientation is preserved (det $= 1$) or reversed (det $= -1$) along the path. Hence the two homotopy classes form a representation of the fundamental group of the projective plane $\pi_1(\mathcal{RP}^2)$, which is the non-trivial homology group Z_2. The boundary conditions have to be compatible with the topology of the projective plane. Due to the twist the original 2-complexes (the Planck cells in the original quadrants) are mixed so that eq.(16) becomes

$$\begin{aligned} \psi(x)|_{xp=h} &= \frac{1}{x^{2d}}\psi\left(-\frac{1}{x}\right)|_{(-x)(-p)=h} \\ &= \frac{1}{x^{2d}}\psi\left(\frac{1}{x}\right)|_{x(-p)=h} \end{aligned} \tag{17}$$

and serves at the same time as a quantization condition generating a discrete spectrum. With $x = h/\pm p$, the expression (4) for the momentum eigenfunctions, and the functional equation for the Riemann zeta function we obtain from (17) the result:

$$\begin{aligned} x^{1/2}\zeta(1/2 - iE)\psi(x) = \pm p^{1/2}\zeta(1/2 + iE/\hbar)\phi(p), \\ x(\pm p) = h. \end{aligned} \tag{18}$$

The latter formulas, with $\pm x$ and $\pm p$ identified, are consistent only if either $\psi(x) = 0$ (and hence $\phi(p) = 0$), or

$$\zeta(1/2 + iE/\hbar) = \zeta(1/2 - iE/\hbar) = 0, \tag{19}$$

which provides a regularization procedure for the semiclassical approximation and generates a discrete real spectrum of the dilation Hamiltonian operator (1). Geometrically the conditions reflect the fact that the fundamental group of the projective plane is non-trivial. The defining homomorphism $\sigma : \pi_1(\mathcal{RP}^2, x) \to \pm 1 \simeq Z_2$ assigns to each path γ_s a number $\sigma(\gamma_s) = \pm 1$, according to whether orientation is preserved or reversed by transport around it in such a way, that $\sigma(\gamma_s) \neq 1$, for $\gamma_s \neq 1$. This is consistent with the group structure of the double point dihedral group with the factorization property $\mathcal{D}_4 = \mathcal{D}_2 \times Z_2$ (or equivalently as a semidirect product $Z_4 \cdot Z_2$) [5]. Hence the irreducible representations have a Z_2-grading too, each element of the pair $\sigma(\gamma_s)\psi(x)$ to be interpreted as associated with an orientation preserving or reversing homotopy class.

The relation with the $-$ sign on the RHS of (18) is exactly the quantization condition suggested by Berry and Keating in [3].

SUMMARY

We have implemented conformal symmetry to model a chaotic system. Conformal invariance in x space gives rise to the finite dihedral symmetry in phase space which is used to impose quantization conditions generating the Riemann zeros as a discrete spectrum of the quantum hyperbolic Hamiltonian. The Riemann zeros are the eigenvalues for which the latter is the square root of the energy operator of the conformally invariant metric.

ACKNOWLEDGMENTS

I would like to thank Ray Streater and the Organizing Commetee for the kind invitation to participate in the conference and for their hospitality at King's College.

REFERENCES

1. M. Berry, Riemann's zeta function: a model for quantum chaos?, in *Quantum Chaos*, eds. T.H. Seligman and H. Nishioka, *Lecture Notes in Physics* **263**, pp. 1-17 (Heidelberg, Springer, 1986).
2. M. Berry, in *Supersymmetry and Trace Formulae*, eds. I.V. Lerner, J.P. Keating, pp. 355-367 (NY, Plenum, 1999).
3. M. Berry and J. Keating, The Riemann zeros and eigenvalue asymptotics, *SIAM Review* **41**, 236-266 (1999).
4. J.E. Humphreys, *Reflection groups and Coxeter groups*, Cambridge, University Press, 1992.
5. A. Bovier, M. Luling and D. Wyler, *J. Math. Phys.* **22**, 1536 (1981).

A Topological Method for Studying Dynamical Systems in Classical Mechanics

V. L. Cartas

Dept. of Physics, "Dunarea de Jos" University of Galati, Romania

Abstract. We present a topological method for analysing dynamical systems, starting with a smooth (or, alternatively, real analytic) manifold M representing configuration space and its cotangent bundle T^*M with given symplectic structure [1]. We assume k independent first integrals in involution, $F = (f_1, \ldots, f_k) : T^*M \to \mathbf{R}^k$ to be given, one of which, f, represents the Hamiltonian. For each $c \in \mathbf{R}^k$, the set $\mathcal{I}_c := F^{-1}(c)$ is an invariant manifold for the vector field X_f [2]. Information about the dynamics can be obtained by studying the changes in topological type of these manifolds as c varies. We illustrate this with two examples, the compound pendulum and the plane Kepler problem; we see that a bifurcation in the nature of the dynamics occurs at points where the topological degree of the invariant manifold jumps.

INTRODUCTION

In a large number of problems arising in physics, it is possible to formulate the question in terms of the properties of the phase space, identified as the cotangent bundle of the configuration space M. Here, we assume that M is a differential manifold of class C^∞, or a real analytic manifold. The dimension n of M is the number of degrees of freedom. The local coordinates on M are the generalised coordinates q, and the coordinates on the fibres of T^*M are the generalised momenta. All the information about the dynamics of the system is contained in the Hamiltonian f which is a real C^∞ function on T^*M. The symplectic geometric structure of T^*M allows us to associate with f a vector field X_f, whose Poisson bracket with an observable (smooth function on T^*M) gives us the equation for the time-evolution of that observable [1].

THE TOPOLOGICAL PROGRAMME

We consider a manifold M of dimension n, the number of degrees of freedom, and a Hamiltonian $f \in C^\infty(T^*M)$; the latter is a Lie algebra under Poisson bracket.

CP553, *Disordered and Complex Systems*, edited by P. Sollich, et al.

We assume that k independent first integrals in involution with f are known, say

$$F = (f = f_1, \ldots, f_k) : T^*M \to \mathbf{R}^k, \qquad k \leq 2n - 1.$$

We study the submanifolds $\mathcal{I}_c := F^{-1}(c)$ as c runs over $\mathbf{R}^k$. These are invariant manifolds for the dynamics, generated by

$$X_f = \sum_{i=1}^{n} \left(\frac{\partial f}{\partial p_i} \frac{\partial}{\partial q^i} - \frac{\partial f}{\partial q^i} \frac{\partial}{\partial p_i} \right).$$

Here, q^i are generalised coordinates on M and p_i are the dual momentum coordinates on the fibres of T^*M. Our general programme has three parts.

I) To characterise the topological type of the invariant manifolds $\mathcal{I}_c$ for $c \in \mathbf{R}^k$.

II) To determine the points $c \in \mathbf{R}^k$ where the topological type jumps.

III) To study the dynamics generated by X_f on each $\mathcal{I}_c$.

To be more exact in II), we introduce for F the definition of the bifurcation points $K = K(F) \subseteq \mathbf{R}^k$ as the smallest subset of $\mathbf{R}^k$ for which the map

$$F|F^{-1}\left(\mathbf{R}^k - K\right) : F^{-1}\left(\mathbf{R}^k - K\right) \to \mathbf{R}^k - K$$

is locally trivial (in the differential sense); so for every $c_0 \in \mathbf{R}^k - K$ there exists a 'hood $\mathcal{U}_0$ of c_0 in $\mathbf{R}^k - K$ such that $F^{-1}(\mathcal{U}_0)$ and $\mathcal{U}_0 \times \mathcal{I}_{c_0}$ are diffeomorphic, the diffeomorphism being such that it maps $\mathcal{I}_{c_0}$ onto $\{c_0\} \times \mathcal{I}_{c_0}$. For every component of $\mathbf{R}^k - K$, F is a differential fibre [3–5]. Thus, II) asks us to find the set of bifurcation points $K(F)$; we seek the points $c_0 \in \mathbf{R}^k$ such that in the 'hood of c_0 we can find points c whose corresponding $\mathcal{I}_c$ have different topological types. The topological programme in general tells us very little about III).

SOME ELEMENTARY EXAMPLES OF THE PROGRAMME

In one degree of freedom the topological structure is determined by the structure of f. Consider the compound pendulum. Here, $M = S^1$, the circle, and q is the angular coordinate, defined modulo 2π. The Hamiltonian is

$$f(p, q) = p^2/2 - \omega^2 \cos q \tag{1}$$

where ω is a real parameter. The phase space is the cylinder $T^*M = S^1 \times \mathbf{R}$, and

$$X_f = p\frac{\partial}{\partial q} + \omega^2 \sin q \frac{\partial}{\partial p}. \tag{2}$$

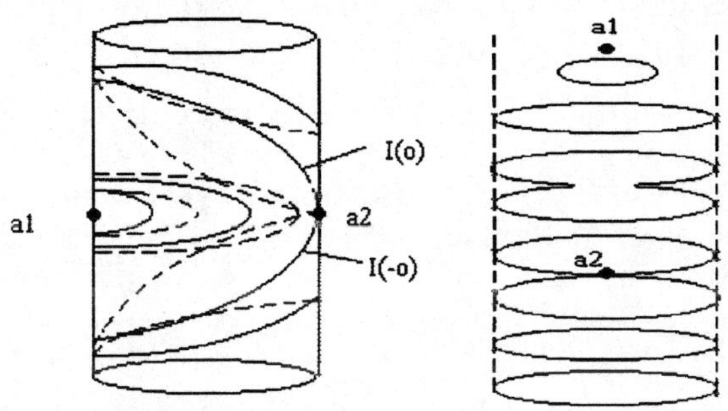

FIGURE 1. (a - left) The topological type of $\mathcal{I}_c$ changes in the vicinity of K. (b - right) From another point of view we can see that the bifurcation points arise at the points where the curve's topological degree jumps.

There are no other first integrals. There are two critical points to f: $a_1 = (0,0)$, a minimum, and $a_2 = (\pi, 0)$, a saddle point. Since $f(0,0) = -\omega^2$ and $f(\pi,0) = \omega^2$, we see that $K \subseteq \mathbf{R}$ is the two-point set $K = \{-\omega^2\} \cup \{\omega^2\}$. From Morse theory we deduce that the maps $f|f^{-1}(-\omega^2, \omega^2)$ and $f|f^{-1}(\omega^2, \infty)$ are differentiable fibres. The level structure of f, such as the orbits of the flow of X_f, is summarised thus: for $-\infty < c < -\omega^2$, $\mathcal{I}_c = \emptyset$; for $\omega^2 < c < \infty$, $\mathcal{I}_c = S^1 \sqcup S^1$, the disjoint union. There is a periodic current on each of the components. For $-\omega^2 < c < \omega^2$, $\mathcal{I}_c = S^1$ and again the current on $\mathcal{I}_c$ is periodic. The level sets are

$$\mathcal{I}_{-\omega^2} = \{a_1\}; \qquad\qquad \mathcal{I}_{\omega^2} = S^1 \vee S^1,$$

the union being joined at the point a_2. This is a bouquet of circles, an eight-shaped curve. The level set $\mathcal{I}_{-\omega^2}$ separates the empty class of orbits from the small oscillations. The level set $\mathcal{I}_{\omega^2}$ is also a separatrix; in its vicinity there are two types of motion: the small oscillations in $\mathcal{I}_c$ for $-\omega^2 < c < \omega^2$, and the full turns of the pendulum ($\mathcal{I}_c$ with $c > \omega^2$). Obviously, the topological type of $\mathcal{I}_c$ changes in the 'hood of K (fig. 1a). One way to see this change is exhibited in fig. 1b.

In the case of two degrees of freedom ($n = 2$) the characterisation of the motion is more difficult; so we consider a solvable case, the plane Kepler problem. The configuration manifold is $M = \mathbf{R}^2 - \{0\}$ and phase-space is a vector bundle: $T^*M = M \times \mathbf{R}^2$. We use polar coordinates $(r, \theta, p_r, p_\theta)$ on T^*M; θ is defined modulo 2π. The Hamiltonian f of this system is

$$f(r, \theta, p_r, p_\theta) = \frac{1}{2}\left(p_r^2 + p_\theta^2 r^{-2}\right) + V(r, \theta) \tag{3}$$

where $V : M \to \mathbf{R}$ is given by $V(r, \theta, p_r, p_\theta) = V_0(r) = -r^{-1}$, the Newtonian potential. It is well known that the particle moves in a plane and that angular momentum is conserved. Consider the function J on T^*M given by $J(r, \theta, p_r, p_\theta) = p_\theta$. The Poisson bracket, $\{f, J\}$ is:

$$\{f, J\} = \frac{\partial f}{\partial r}\frac{\partial J}{\partial p_r} + \frac{\partial f}{\partial \theta}\frac{\partial J}{\partial p_\theta} - \frac{\partial f}{\partial p_r}\frac{\partial J}{\partial r} - \frac{\partial f}{\partial p_\theta}\frac{\partial J}{\partial \theta} = 0. \tag{4}$$

Therefore J and f are in involution, and it is easy to verify that df and $dJ = dp_\theta$ are linearly independent at any point of T^*M. Put $F = (f, J) : T^*M \to \mathbf{R}^2$ and $c = (c', c'') \in \mathbf{R}^2$. Then the sets $\mathcal{I}_c = F^{-1}(c)$ are generically two dimensional invariant manifolds for the Hamiltonian vector field

$$X_f = p_r\frac{\partial}{\partial r} + r^{-2}p_\theta\frac{\partial}{\partial \theta} + r^{-3}\left(p_\theta^2 - r^{-2}\right)\frac{\partial}{\partial p_r}, \tag{5}$$

but also for $X_J = \frac{\partial}{\partial \theta}$, because f is a first integral for the dynamics generated by J. In this case the invariant manifold $\mathcal{I}_c$ is parallelizable: in $\mathcal{I}_c$ there are two vector fields $X_f|\mathcal{I}_c$ and $X_J|\mathcal{I}_c$ which are linearly independent at every point. Therefore, any connected component of $\mathcal{I}_c$ must be a torus $T^2 = S^1 \times S^1$ or a cylinder $S^1 \times \mathbf{R}$, because these are the only two parallelizable manifolds of dimension 2 of interest to us: both contain S^1 as a factor. In order to decide whether $\mathcal{I}_c$ is compact or not, we need to analyse the equations which define it:

$$\frac{1}{2}\left(p_r^2 + p_\theta^2 r^{-2}\right) + V_0(r) = c'; \qquad p_\theta = c''. \tag{6}$$

Therefore $p_r = \pm(2(c' - V_{c''}))^{1/2}$, in which $V_{c''}$ is the function $V_{c''}(r) = V_0(r) + c''^2/(2r^2)$, the "true" potential of the problem [6]. If we consider $L_c := \{(r, p_r) \in \mathbf{R}_+ \times \operatorname{Ran} p_r\}$ we easily obtain $\mathcal{I}_c \cong S^1 \times L_c$, and so the characteristic of $\mathcal{I}_c$ is given by the characteristic of L_c. In order to determine the latter, it is useful to consider the graph of $V_{c''}$ (fig. 2a). The cases are as follows.

$c'' = 0$

In this case, $V_{c''} = V_0$, and we have the cases:

- for $c' < 0$, $\mathcal{I}_c \cong S^1 \times \mathbf{R}$.
- for $c' \geq 0$, $\mathcal{I}_c \cong S^0 \times S^1 \times \mathbf{R}$.

$c'' \neq 0$

In this case, $V_{c''}$ has only one critical point, a minimum $r = c''^2$ and $V_{c''}(c''^2) = -1/(2c''^2) = v$, say. Then we have the cases:

- for $-\infty < c' < v$, $L_c = \emptyset$, so $\mathcal{I}_c = \emptyset$.
- for $c' = v$, $L_c = \{(c''^2, 0)\}$, so $\mathcal{I}_c \cong S^1$.
- for $v < c' < 0$, $L_c \cong S^1$, so $\mathcal{I}_c \cong S^1 \times S^1 \cong T^2$.
- for $0 < c' < \infty$, $L_c \cong \mathbf{R}$ and $\mathcal{I}_c \cong S^1 \times \mathbf{R}$.

It can be proved that the bifurcation set is

$$K = \left\{c = (c', c'') \in \mathbf{R}^2 : c' = -(2c'')^{-2}\right\} \cup \left\{c \in \mathbf{R}^2 : c' = 0\right\} \cup \left\{c \in \mathbf{R}^2 : c'' = 0\right\}. \tag{7}$$

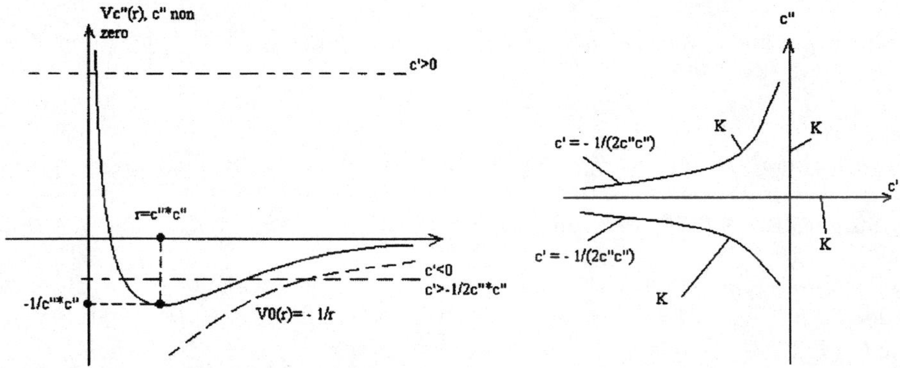

FIGURE 2. (a - left) The graph of $V_{c''}$. (b - right) The qualitative features of the bifurcation set.

The qualitative features of this set are shown in fig. 2b. It can be noticed that in the 'hood of K the topological type of $\mathcal{I}_c$ jumps. We can prove that the map $F|F^{-1}(\mathbf{R}^2 - K)$, which takes values in $\mathcal{R}^2 - K$, is locally trivial [7]. It is also possible to get a characterisation of the current induced by X_f in $\mathcal{I}_c$; for K it is similar to a translation current. Moreover, for $c' < 0$ the current in $\mathcal{I}_c \cong T^2$ is periodic, and therefore each orbit is compact.

CONCLUSIONS

From the above two examples, it is easy to see the effectiveness of the topological method in dynamics and to judge how this type of analysis might be useful for concrete problems. Some general observations might be made; the critical values of the first integrals occur in the bifurcation set. In fact, because the reaching of a critical value modifies the topological type of a function's level set (Morse theory [8,9]) we cannot have local triviality in the 'hood of a critical value. Thus, $K' \subseteq K$, where $K' = K'(F) = F(\tau(F))$ is the dependency set (the critical point set). Here, $F = (f_1, \ldots, f_k) : T^*M \to \mathbf{R}^k$ is a set of independent first integrals, with $f = (f_1, \tau(f))$.

In the second example, we obtained a part of K when we passed over the points with $c' = 0$, which provided the asymptote of the graph of $V_{c''}$; the existence of this asymptote is due to the fact that $M = \mathbf{R}^2 - \{(0,0)\}$ is not compact. Often, we may expect that the fact that T^*M or even M is not compact may contribute points to the bifurcation set K.

REFERENCES

1. R. Abraham and J. Marsden, *Foundations of Mechanics*, W. A. Benjamin, Inc., New York, Amsterdam, 1987.
2. I. Auslend, I Green and F. Hahn, *Flows on Homogeneous Spaces*, Ann. Math. Studies, **53**, 1983.
3. G. D. Birkhoff, *Dynamical Systems*, Amer. Math. Soc. Colloq. Publ., **9**, New York, 1927.
4. D. Burghelea and T. Hangan, *Introducere in topologia diferentiala*, Ed. St., Bucaresti, 1985.
5. R. Easton, *Regularization of Vector Fields by Surgery*, Jour. Diff. Equations, **15**, 1982.
6. A, Iacob, *Invariant Manifolds in the Motion of a Rigid Body*, Rev. Rom. Math. Purres Appl. **16**, 1982.
7. M. Ikeda and T. Fujitani, *On Linear First Integrals of Natural Ststems in Classical Mechanics*, Mathematica Japonicae, **12**, 2, 1970.
8. L. Landau and E. Lipschitz, *Mecanica*, Ed. Technica, Bucaresti, 1991.
9. J. Milnor, *Morse Theory*, Ann. Math. Studies, Princeton, 1961.

Chaos and Gram's Matrix

Mieke De Cock[1]

Instituut voor Theoretische Fysica
Katholieke Universiteit Leuven
Celestijnenlaan 200D
B-3001 Leuven, Belgium

Abstract. We propose to analyse the statistical properties of long sequences of vectors using the spectrum of the associated Gram matrix. Such sequences arise, e.g., by stroboscopic observation of a Hamiltonian evolution or by repeated action of a kicked quantum dynamics on an initial condition. We are interested in dynamical systems with a classical limit and we argue that, when the number of time steps, suitably scaled with respect to $\hbar$, increases, the limiting eigenvalue distribution of the Gram matrix reflects the possible quantum chaoticity of the original system as it tends to its classical limit.

Unlike in classical mechanics, where the root of chaotic phenomena lies in exponentially fast separation of phase space trajectories, the notion of quantum chaos is not reducible to a definite paradigmatic behaviour, but it refers to a variety of properties of quantum systems with classical limit [1].

Truly quantum dynamical systems with compact phase space are intrinsically finite dimensional by virtue of the uncertainty principle. The non-commutativity is measured by the quantization parameter $\hbar$ and the classical limit is reached for $\hbar \to 0$. As each state occupies the same volume $\hbar$, the dimension of the Hilbert space of states is $\frac{1}{\hbar}$. A quantum evolution in discrete time, also called kicked evolution, is then described by a unitary Floquet operator u. The time evolution between two consecutive kicks is in the Schrödinger picture given by $\varphi \mapsto u\varphi$.

The finite dimensionality implies that the spectrum of the evolution operator is discrete, leaving no room for decay of correlation functions, contrary to their classical counterparts. This reflects the non-commutativity of the $t \to \infty$ and the $\hbar \to 0$ limits and makes quantum chaos a matter of studying the dynamics at the right time scale.

One of the main goals is trying to understand the implications of the possible chaoticity of the classical system on the dynamical behaviour of its quantized version.

---

[1] Onderzoeker FWO, e-mail: mieke.decock@fys.kuleuven.ac.be

CP553, *Disordered and Complex Systems,* edited by P. Sollich, et al.

We face the problem of studying time sequences $\Phi = (\varphi, u\varphi, u^2\varphi, \ldots)$ generated by a Floquet operator u as it acts repeatedly on an initial condition φ and we propose to use the spectrum of Gram's matrix to analyse the statistical properties of such sequences.

In general, consider a sequence of K normalised vectors in an N-dimensional Hilbert space $\mathcal{H}_N$: $\Phi = (\varphi_1, \varphi_2, \ldots, \varphi_K)$. The Gram matrix associated with this sequence is

$$\Gamma^\Phi = [\langle \varphi_i, \varphi_j \rangle]_{i,j=1,\ldots,K}.$$

Γ^Φ is positive semidefinite and its rank equals the dimension of the space spanned by the φ_j. In particular, $\det \Gamma^\Phi \neq 0$ if and only if Φ is linearly independent. The spectrum $\Sigma(\Gamma^\Phi)$ of Γ^Φ is independent of the order of the φ_j in Φ and of a multiplication of the φ_j with a common phase. We shall argue that Gram matrices provide us with a statistical tool: the quantum counterpart of determining relative frequencies of letters in long words.

Let us first consider the classical case, i.e., we consider a classical word of length K, $\mathbf{i} = (i(1), i(2), \ldots, i(K))$, where the letters are chosen from an alphabet of N letters, $\mathbf{A} = (1, 2, \ldots, N)$. If we identify i with e_i for an orthonormal basis $\{e_1, e_2, \ldots, e_N\}$, $\mathbf{i}$ is in one-to-one correspondence with $\mathbf{e} = (e_{i_1}, e_{i_2}, \ldots, e_{i_K})$. By grouping the e_{i_l} with equal index j, the Gram matrix becomes block diagonal with blocks

$$E(j) = \begin{pmatrix} 1 & 1 & \ldots & 1 \\ \vdots & \vdots & \ddots & \vdots \\ 1 & 1 & \ldots & 1 \end{pmatrix},$$

where the dimension of $E(j)$ is precisely the multiplicity $m(j)$ of j in $\mathbf{i}$. As the spectrum of $E(j)$ consists of the non-degenerate eigenvalue $m(j)$ and the $(m(j) - 1)$-fold degenerate eigenvalue 0, we find that the spectrum of the Gram matrix determines precisely how many different numbers are visited in the sequence $\mathbf{i}$ with their multiplicity.

If the sequence shows some short periodicity, it will return to its initial state in a time period much shorter than the length of the alphabet, giving rise to many zeros and a few large eigenvalues in the spectrum. If, on the other hand, it has a long period, we expect more different points to be visited, which is reflected in a spectrum with many small eigenvalues.

For the general non-commutative case, the same interpretation remains valid but non-zero eigenvalues are generically non-degenerate and the spectrum is no longer limited to integers, i.e., the same vector can be visited a fractional number of times.

In general, we are interested in the limiting eigenvalue distribution of the Gram matrix when its dimension, the number of time steps, tends to infinity, appropriately scaled with respect to the quantum parameter. Therefore, we consider the limit of the empirical eigenvalue distribution

$$\rho_K(dx) = \frac{1}{K} \sum_{j=1}^{K} \delta(x - \gamma_j)\, dx, \tag{1}$$

where the γ_j are the eigenvalues of the matrix. The limiting distribution should reflect the possible quantum chaoticity of the original dynamics as it tends to its classical limit.

If successive states do not overlap, the corresponding Gram matrix equals the identity. This will typically happen for sufficiently localised initial conditions and sufficiently few time steps. Given an initial condition, we may look for the shortest time scale on which the Gram matrix becomes non-trivial, i.e., different from the identity. This means that we look for information about the approximate periodicity of the orbit. For integrable systems, Gram's matrix will become non-trivial already after a few time steps, whereas in the chaotic case, we expect the system to visit a sizeable portion of the available states.

We do not claim we can settle this question, but we shall present some examples to support this expectation.

We begin by considering a classical example, i.e., consider a discrete set of N elements $\mathbf{A} = (1, 2, \ldots, N)$ and words of length K as before. In the context of classical dynamical systems, such words appear in the following way. By covering the phase space by a partition of N patches of volume $\frac{1}{N}$, the dynamics (which we assume volume preserving and discrete in time) translates approximately into a bijection of the patches. So, after this coarse-graining procedure, we obtain for each N a one-to-one transformation of the set $\mathbf{A}$ that determines the evolution during one tick of the clock.

Suppose that, instead of looking at a sequence generated by a time evolution, we look at randomly generated words, where the letters are chosen w.r.t. the uniform probability measure on $\mathbf{A}$. Intuitively, we expect this to mimick somehow the most irregular behaviour. As before, we identify each letter with a unit vector of an orthonormal set and we consider the corresponding Gram matrix. Of course, the spectrum of such a Gram matrix is random, but it turns out that in the limit $K \to \infty$, $N \to \infty$, $\frac{K}{N} \to \tau$, the eigenvalue distribution converges to a deterministic one, more precisely stated in the following

Theorem: *The empirical eigenvalue distribution of the ensemble of classical random Gram matrices at density τ converges in probability to the discrete distribution $\rho = \{\rho(k) | k \in \mathbf{N}\}$ where*

$$\rho(j) = \frac{\tau^{j-1} e^{-\tau}}{j!} \; j \geq 1, \qquad \rho(0) = 1 - \frac{1}{\tau}(1 - e^{-\tau}). \tag{2}$$

Let us now consider some non-commutative examples. Quantum dynamical systems with chaotic classical limit are in general not amenable to an analytic treatment and one has to resort to numerical simulations. We shall present a stochastic model which can be treated in a rigourous way, but where the price to be paid is that the preservation of orthogonality of frames is lost. Furthermore, we present some numerical examples for more realistic quantum systems such as the baker map or the kicked top.

Our random vector model is constructed by picking randomly and independently K normalised vectors in an N-dimensional Hilbert space. With a random sequence $\Phi = (\varphi_1, \ldots, \varphi_K)$ is associated a random Gram matrix Γ^Φ of dimension K. The spectrum of Γ^Φ is of course random but again, it converges to a definite limit:

Theorem: *The empirical eigenvalue distribution of the ensemble of quantum random Gram matrices at density τ converges, in the limit $K = \tau N \to \infty$, weakly with probability one to the Marchenko-Pastur distribution* [2–4].

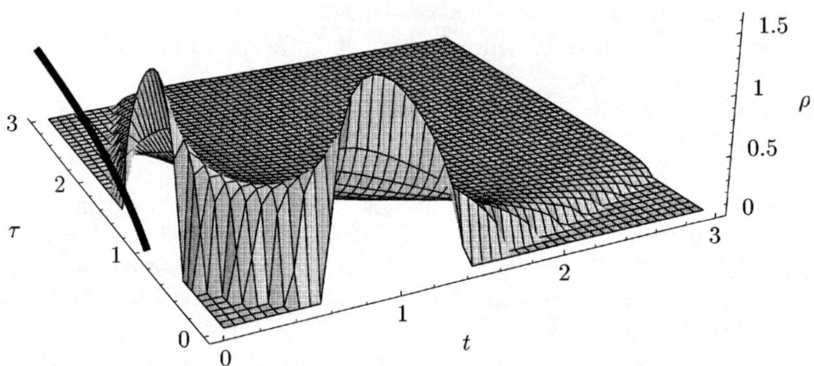

FIGURE 1. The Marchenko-Pastur distribution.

We now turn to some more realistic quantum systems, such as the quantum baker map [5,6] or the quantized version of the kicked top [7]. In the baker map, an N-dimensional Hilbert space (N even) is considered. Position and momentum operators are defined as translation operators acting on each other's eigenstates and where some periodicity condition is imposed on these states. Thus the normalised position eigenstates are $|n\rangle$, $n = 0, \ldots, N-1$ and the normalised momentum eigenstates are $|m\rangle$, $m = 0, \ldots, N-1$. The transformation functions $\langle n|m\rangle$ are defined as

$$(F_N)_{kn} = \frac{1}{N} e^{-i(2\pi/N)(k+1/2)(n+1/2)} \qquad k, n = 0, \ldots, N-1.$$

The unitary operator representing the dynamics of the quantum baker transformation is then given by

$$U = F_N^{-1} \begin{pmatrix} F_{N/2} & 0 \\ 0 & F_{N/2} \end{pmatrix}.$$

Taking the classical limit, which corresponds here to taking $N \to \infty$, results again in the classical baker's transformation. As the classical baker map is fully chaotic, we are not surprised that the spectrum of the associated Gram matrix fits quite well with the Marchenko-Pastur distribution at corresponding τ, as we see in Figure 2.

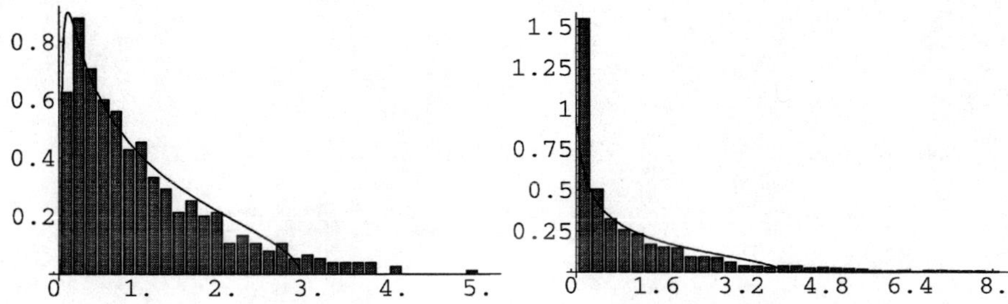

FIGURE 2. N=1000, (left) K=500 (right) K=1000

A last example is given by the quantized version of the kicked top. We consider a system characterised by an angular momentum vector $\mathbf{J} = (J_x, J_y, J_z)$, $[J_i, J_j] = i\varepsilon_{ijk}J_k$, $\mathbf{J}\cdot\mathbf{J} = j(j+1)$. The Hilbert space describing the system is $(2j+1)$-dimensional. The dynamics of $\mathbf{J}$ is described by the unitary Floquet operator

$$U = e^{-ipJ_y}e^{-i(k/2j)J_z^2}.$$

The last term describes an impulsive rotation around the z-axis by an angle proportional to J_z, while the first term accounts for a precession around the y-axis.

The classical limit is performed by defining rescaled quantities $\mathbf{J}/j$ and taking the limit $j \to \infty$. The quantum equations then turn into the classical ones. Depending on the values of the parameters p and k, the classical motion shows regular or chaotic behaviour. This makes this model an ideal testing ground for our claim concerning the spectrum of the Gram matrix: for values of the parameters where the classical system is regular, we expect a totally different picture for the spectrum than we do for the classical chaotic region. This is indeed what we see when we study the system numerically. In the pictures below, Figures 3 and 4, we start from a well-localised initial condition and we let it evolve under the unitary dynamics. We compute the corresponding Gram matrix and its spectrum.

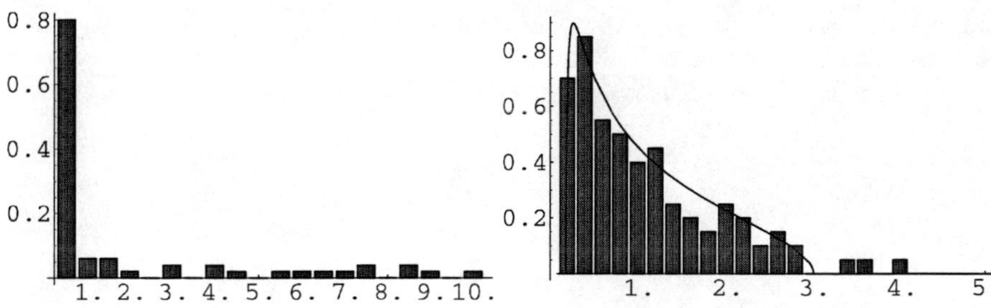

FIGURE 3. j=100, K=100, (left) k=1 (right) k=6.5, p=1.5

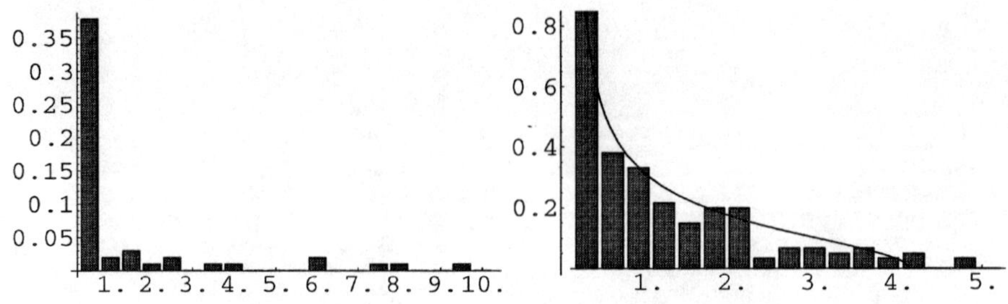

FIGURE 4. j=100, K=201, (left) k=1 (right) k=6.5, p=1.5

In these figures, we see clearly a difference in the eigenvalue distribution between the classical parameter domain of regular behaviour ($k = 1, p = 1.5$) and the classically chaotic region ($k = 6.5, p = 1.5$). The classical regularity is reflected in a huge degeneracy of the eigenvalue zero. The chaotic case on the other hand, resembles again the Marchenko-Pastur distribution at the corresponding τ (solid line). The exact form of the eigenvalue distribution remains an open question.

ACKNOWLEDGEMENTS

It is a pleasure to thank Mark Fannes and Pascal Spincemaille for very helpful discussions. I acknowledge financial support from FWO-project G.0239.96.

REFERENCES

1. Casati, G., and Chirikov, B., *Quantum Chaos*, Cambridge University Press, Cambridge, 1995.
2. Marchenko, V., and Pastur, L., *Math. USSR Sbornik* **1**, 457–483 (1967).
3. De Cock, M., Fannes, M., and Spincemaille, P., *J. Phys. A:Math. Gen.* **32**, 6547–6571 (1999).
4. De Cock, M., Fannes, M., and Spincemaille, P., *Europhys. Lett.* **49**, 403–409 (2000).
5. Balasz, N.L., and Voros, A., *Ann. Phys.* **190**, 1 (1989).
6. Saraceno, M., *Ann. Phys.* **199**, 37 (1990).
7. Haake, F., Kus, M., and Scharf, R., *Z. Phys. B* **65**, 381 (1987).

Spectra of Random Matrices Close to Unitary and Scattering Theory for Discrete-Time Systems

Yan V. Fyodorov

Department of Mathematical Sciences, Brunel University, Uxbridge, UB8 3PH, UK

Abstract. We analyse the statistical properties of complex eigenvalues of random matrices $\hat{A}$ which are close to unitary. Such matrices appear naturally when considering quantized chaotic maps within a general theory of open linear stationary systems with discrete time. The deviation from unitarity is characterized by the rank M and eigenvalues $T_i, i = 1, \ldots, M$ of the matrix $\hat{T} = \hat{\mathbf{1}} - \hat{A}^\dagger \hat{A}$. For the case $M = 1$ the problem is solved completely by deriving the joint probability density of the eigenvalues and calculating all n-point correlation functions. For the general case the correlation function of secular determinants is presented.

The theory of wave scattering can be looked at as an integral part of the general theory of linear dynamic open systems in terms of the input-output approach, see papers [1–3] and references therein. To make explicit the relation of the latter approach to the theory of quantum chaotic scattering [4,5] we recall that an open linear system is characterized by three Hilbert spaces: the space E_0 of internal states $\Psi \in E_0$ and two spaces $E_\pm$ of incoming $(-)$ and outgoing $(+)$ signals or waves also called input and output spaces, consisting of vectors $\phi_\pm \in E_\pm$. Acting in these three spaces are four operators, or matrices: a) the so-called fundamental operator $\hat{A}$ which maps any vector from the internal space E_0 onto some vector from the same space E_0, b) two operators $\hat{W}_{1,2}$, with $\hat{W}_1$ mapping incoming states onto an internal state and $\hat{W}_2$ mapping internal states onto outgoing states and c) an operator $\hat{S}_0$ acting from E_- to E_+.

We will be interested in describing the dynamics $\Psi(t)$ of an internal state with time t, provided we know the state at initial instant $t = 0$ and the system is subject to a given input signal $\phi_-(t)$. We assume the operators to be time-independent, restricting our considerations to stationary, or time-invariant, systems.

Let us begin with the case of continuous-time description. The requirements of linearity, causality and stationarity lead to a system of two dynamical equations:

$$i\frac{d}{dt}\Psi = \hat{A}\Psi(t) + \hat{W}_1\phi_-(t) \quad , \quad \phi_+(t) = \hat{S}_0\phi_-(t) + i\hat{W}_2\Psi(t) \tag{1}$$

CP553, *Disordered and Complex Systems*, edited by P. Sollich, et al.

The interpretation of these equations depends on the nature of the state vector Ψ and of the vectors $\phi_\pm$ and is different in different applications. In quantum mechanics one relates the scalar product $\Psi^\dagger\Psi$ to the probability of finding a particle inside the "inner" region at time t, whereas $\phi_\pm^\dagger\phi_\pm$ stands for the number of particles arriving at (resp. leaving from) the inner domain per unit time. The condition of particle conservation then reads $\frac{d}{dt}\Psi^\dagger\Psi = \phi_-^\dagger\phi_- - \phi_+^\dagger\phi_+$ and is compatible with the dynamics of Eq.(1) only provided the operators satisfy the following relations: $\hat{A}^\dagger - \hat{A} = i\hat{W}\hat{W}^\dagger$, $\hat{S}_0^\dagger\hat{S}_0 = \mathbf{\hat{1}}$ and $\hat{W}^\dagger \equiv \hat{W}_2 = -\hat{S}_0\hat{W}_1^\dagger$. In particular, this means that $\hat{A}$ can be written as $A = \hat{H} - \frac{i}{2}\hat{W}\hat{W}^\dagger$, with a Hermitian $\hat{H} = \hat{H}^\dagger$.

The meaning of $\hat{H}$ is transparent: it governs the evolution $i\frac{d}{dt}\Psi = \hat{H}\Psi(t)$ of an inner state Ψ when the coupling $\hat{W}$ between the inner space and input/output spaces is absent. As such, it is just the Hamiltonian describing the closed inner region. The fundamental operator $\hat{A}$ then has a natural interpretation as the effective non-selfadjoint Hamiltionian describing the decay of the probability from the inner region at zero input signal, $\phi_-(t) = 0$ for $t \geq 0$. If, however, the input signal is given in the Fourier-domain by $\phi_-(\omega)$, the output signal is related to it by:

$$\phi_+(\omega) = \left[\hat{S}(\omega)\hat{S}_0\right]\phi_-(\omega) \quad , \quad \hat{S}(\omega) = \mathbf{\hat{1}} - i\hat{W}^\dagger\frac{1}{\omega\mathbf{\hat{1}} - \hat{A}}\hat{W} \tag{2}$$

where we assumed $\Psi(t=0) = 0$. The unitary matrix $\hat{S}(\omega)$ is known in the mathematical literature as the characteristic matrix-function of the non-Hermitian operator $\hat{A}$. In the present context it is just the scattering matrix whose unitarity is guaranteed by the particle conservation law.

The contact with the theory of chaotic scattering is now apparent: the expression Eq.(2) has frequently been used in the physical literature [4,5] as a starting point for extracting universal properties of the scattering matrix for a quantum chaotic system within the so-called random matrix approach. The main idea underlying this approach is to replace the actual Hamiltonian $\hat{H}$ by a large random matrix and to calculate the ensuing statistics of the scattering matrix. The physical arguments in favor of such a replacement can be found in the cited literature. In particular, the statistical properties of complex eigenvalues of the operator $\hat{A}$ as well as related quantities have recently been studied in much detail [5–8]. Those eigenvalues are poles of the scattering matrix and have the physical interpretation of *resonances.*

In the theory presented above, the time t was a continuous parameter. On the other hand, a very useful instrument in the analysis of classical Hamiltonian systems with chaotic dynamics are the so-called area-preserving chaotic maps [9]. They appear naturally either as a mapping of the Poincaré section onto itself, or as result of a stroboscopic description of Hamiltonians which are periodic function of time. Their quantum mechanical analogues are unitary operators which act on Hilbert spaces of finite large dimension N. They are often referred to as evolution, scattering or Floquet operators, depending on the physical context where they are used. Their eigenvalues consist of N points on the unit circle (eigenphases). Numerical studies of various classically chaotic systems suggest that the eigenphases conform statistically quite accurately to the results obtained for unitary random matrices (Dyson circular ensembles).

Let us now imagine that a system represented by a chaotic map ("inner world") is embedded in a larger physical system ("outer world") in such a way that it describes particles which can come inside the region of chaotic motion and leave it after some time. Models of such type appeared, for example, in [10] where a kind of scattering theory for "open quantum maps" was developed based on a variant of the Lipmann-Schwinger equation.

On the other hand, in general systems theory [2,3], dynamical systems with discrete time are considered as frequently as those with continous time. For linear systems a "stroboscopic" dynamics is just a linear map $(\phi_-(n);\Psi(n)) \to (\phi_+(n);\Psi(n+1))$ which can generally be written as:

$$\begin{pmatrix} \Psi(n+1) \\ \phi_+(n) \end{pmatrix} = \hat{V}\begin{pmatrix} \Psi(n) \\ \phi_-(n) \end{pmatrix} \quad , \quad \hat{V} = \begin{pmatrix} \hat{A} & \hat{W}_1 \\ \hat{W}_2 & \hat{S}_0 \end{pmatrix} \tag{3}$$

Again, we would like to consider a conservative system, and the discrete-time analogue of the conservation law is: $\Psi^\dagger(n+1)\Psi(n+1) - \Psi^\dagger(n)\Psi(n) = \phi_-^\dagger(n)\phi_-(n) - \phi_+^\dagger(n)\phi_+(n)$ which amounts to unitarity of the matrix $\hat{V}$ in Eq.(3). It is easy to satisfy oneself that by choosing an appropriate basis in each of the corresponding spaces (input, output and internal) one can always bring such a unitary $\hat{V}$ into the following form:

$$\hat{V} = \begin{pmatrix} \hat{u}\sqrt{1-\hat{\tau}\hat{\tau}^\dagger} & u\hat{\tau} \\ \hat{\tau}^\dagger & -\sqrt{1-\hat{\tau}^\dagger\hat{\tau}} \end{pmatrix} \tag{4}$$

where the matrix $\hat{u}$ is unitary and $\hat{\tau}$ is a rectangular $N \times M$ diagonal matrix with the entries $\tau_{ij} = \delta_{ij}\tau_i,\ 1 \le i \le N\,,\ 1 \le j \le M, \quad 0 \le \tau_i \le 1$.

Such a form suggests a clear interpretation of the constituents of the model. Indeed, for $\hat{\tau} = 0$ the internal dynamics of the system amounts to $\Psi(n+1) = \hat{u}\Psi(n)$. We therefore identify $\hat{u}$ with a unitary evolution operator of the "closed" inner state domain decoupled from both the input and output spaces. Correspondingly, $\hat{\tau} \neq 0$ just provides a coupling that makes the system open and converts the fundamental operator $\hat{A} = \hat{u}\sqrt{1-\hat{\tau}\hat{\tau}^\dagger}$ to a contraction: $1 - \hat{A}^\dagger\hat{A} = \tau\tau^\dagger \ge 0$. As a result, the equation $\Psi(n+1) = \hat{A}\Psi(n)$ describes an irreversible decay of any initial state $\Psi(0)$ for zero input $\phi_-(n) = 0$, whereas for a nonzero input and $\Psi(0) = 0$ the Fourier-transforms $\phi_\pm(\omega) = \sum_{n=0}^{\infty} e^{in\omega}\phi_\pm(n)$ are related by a unitary scattering matrix $\hat{S}(\omega)$ given by:

$$\hat{S}(\omega) = -\sqrt{1-\hat{\tau}^\dagger\hat{\tau}} + \hat{\tau}^\dagger\frac{1}{e^{-i\omega}-\hat{A}}\hat{u}\hat{\tau} \tag{5}$$

Assuming further that the motion outside the inner region is regular, we should be able to describe generic features of open quantized chaotic maps choosing the matrix

$\hat{u}$ to be a member of a Dyson circular ensemble. Then one finds: $\hat{\tau}^{\dagger}\hat{\tau} = 1 - \left|\overline{\hat{S}(\omega)}\right|^2$, where the bar stands for the averaging of $\hat{S}(\omega)$ in Eq.(5) over $\hat{u}$. Comparing this result with [4,5] we see that the M eigenvalues $0 \leq T_i \leq 1$ of the $M \times M$ matrix $\hat{T} = \hat{\tau}^{\dagger}\hat{\tau}$ play the role of the so-called transmission coefficients and describe a particular way in which the chaotic region is coupled to the outer world.

We mention that some important particular examples of such systems can be found in recent papers [11]; these provided an important motivation for the present research. The particular contractive random matrices emerging there were, in fact, "truncations" of standard random unitary matrices. They were studied in much detail by the authors of a recent insightful paper [12]. The main goal of the present paper is to generalize the results of [12] to spectra of random contractions for a given deviation from unitarity.

The ensemble of general $N \times N$ random contractions $\hat{A} = \hat{u}\sqrt{1 - \hat{\tau}\hat{\tau}^{\dagger}}$ can be described by the following probability measure in the matrix space:

$$\mathcal{P}(\hat{A})d\hat{A} \propto \delta(\hat{A}^{\dagger}\hat{A} - \hat{G})dA \quad , \quad \hat{G} \equiv 1 - \hat{\tau}\hat{\tau}^{\dagger} \tag{6}$$

where $dA = \prod_{ij} dA_{ij}dA^{*}_{ij}$. It is natural to call the $N \times N$ matrix $\hat{\tau}\hat{\tau}^{\dagger} = \mathbf{1} - \hat{G} \geq 0$ the deviation matrix. It has M nonzero eigenvalues coinciding with the transmission coefficients T_i introduced above, the remaining $N - M$ eigenvalues being zero. The particular choice $T_i = 1$ corresponds to the case considered in [12].

Our first step is, following [6,12], to introduce the Schur decomposition $\hat{A} = \hat{U}(\hat{Z} + \hat{R})\hat{U}^{\dagger}$ of the matrix A in terms of a unitary $\hat{U}$, a diagonal matrix of the eigenvalues, $\hat{Z}$, and a lower triangular $\hat{R}$. One can satisfy oneself that the eigenvalues $z_1, \ldots, z_N$ are generically non-degenerate, provided all $T_i < 1$. Then, the measure written in terms of the new variables is given by $d\hat{A} = |\Delta(\{z\})|^2 d\hat{R}d\hat{Z}d\mu(U)$, where the first factor is just the Vandermonde determinant of the eigenvalues z_i and $d\mu(U)$ is the invariant measure on the unitary group. The joint probability density of complex eigenvalues is then obtained by integrating out $\hat{R}$ and $\hat{U}$ (some useful hints can be found in [6,12]). The integration over the unitary group poses a serious problem and the general case will be considered elsewhere [13]. However, when the rank M of the deviation matrix $\hat{\tau}\hat{\tau}^{\dagger}$ is unity, the integration can be performed by the methods of [6] and yields a very simple expression:

$$\mathcal{P}(\{z\}) \propto T^{1-N}|\Delta(\{z\})|^2 \delta\left(1 - T - |z_1|^2 \ldots |z_N|^2\right) \tag{7}$$

provided $0 \leq |z_l| \leq 1$ for all eigenvalues, and zero otherwise. Here $0 \leq T < 1$ is the only non-zero eigenvalue of the deviation matrix. Eq.(7) can be used to extract all n-point correlation functions:

$$R_n(z_1, \ldots, z_n) = \frac{N!}{(N-n)!} \int d^2 z_{n+1} \ldots d^2 z_N \mathcal{P}(\{z\}) \tag{8}$$

To achieve this, it is convenient to use the Mellin transform with respect to the variable $\zeta = 1 - T$, defined as $\tilde{R}_n(s; \{z\}_n) = \int_0^{\infty} d\zeta \zeta^{s-1} \left[(1-\zeta)^{N-1} R_n(\{z\}_n)\right]$. It is easy to see that such a transform brings $\mathcal{P}(\{z\})$ to a form suitable for exploitation of the orthogonal polynomial method [14]. The corresponding polynomials are $p_k(z) = \sqrt{(k+s)}z^n$ and are orthonormal with respect to the weight $f(z) = |z|^{2(s-1)}$ inside the unit circle $|z| \leq 1$. Following the standard route we find:

$$\tilde{R}_n(s;\{z\}_n) \propto \frac{\det\left[K(z_i,z_j)\right]|_{(i,j)=1,\ldots,n}}{s(s+1)\ldots(s+N-1)} \tag{9}$$

where the kernel is

$$K(z_1,z_2) = (f(z_1)f(z_2))^{1/2}\sum_{k=0}^{N-1} p_k(z_1^*)p_k(z_2) = |x|^{s-1}\left(s\phi(x) + x\frac{d}{dx}\phi(x)\right)|_{x=z_1^* z_2}$$

and $\phi(x) = (x^N - 1)/(x-1)$. Thus, the expression Eq.(9) can be rewritten as:

$$\tilde{R}_n(s;\{z\}_n) \propto \frac{\left(\prod_{l=1}^n |z_l|^2\right)^{s-1}}{s(s+1)\ldots(s+N-1)}\sum_{l=0}^{n} s^l q_l(\{z\}_n),$$

$$q_0(\{z\}_n) = \det\left[x\frac{d}{dx}\phi(x)|_{x=z_i^* z_j}\right]|_{i,j=1,\ldots,n} \quad \ldots \quad q_n(\{z\}_n) = \det\left[\phi(x)|_{x=z_i^* z_j}\right]|_{i,j=1,\ldots,n}$$

and can easily be Mellin-inverted to yield finally the original correlation functions in the following form:

$$\begin{aligned} R_n(\{z\}_n)|_{|z_n|\leq 1} &\propto \theta(T-1+a) \\ &\times T^{1-N}\sum_{l=0}^{n} q_l(\{z\}_n)\left(\frac{d}{da}a\right)^l\left[\frac{1}{a}\left(1-\frac{1-T}{a}\right)\right]^{N-1}|_{a=\prod_{i=1}^n |z_i|^2} \end{aligned} \tag{10}$$

where $\theta(x) = 1$ for $x \geq 0$ and zero otherwise. This equation is exact for arbitrary N. In the limit $N \gg n$ the leading order contribution to the correlation function, omitting the $\theta-$function factor, can be written as:

$$R_n(\{z\}_n) \propto T^{1-N}\frac{1}{a}\left(1-\frac{1-T}{a}\right)^{N-1}\det_{i,j=1,\ldots,n}\left(\frac{N(1-T)}{a-1+T}\phi(x) + x\frac{d}{dx}\phi(x)\right)|_{x=z_i^* z_j}$$

Further simplifications occur after taking into account that the eigenvalues z_i are expected to concentrate typically at distances of order $1/N$ from the unit circle. Then it is natural to introduce new variables y_i, ϕ_i through $z_i = (1-y_i/N)e^{i\phi_i}$ and to consider y_i to be of order unity when $N \to \infty$. As to the phases ϕ_i, we expect their typical separation to scale as $\phi_i - \phi_j = O(1/N)$. Now it is straightforward to perform the limit $N \to \infty$ explicitly and bring the above equation into the final form:

$$R_n(\{z\}_n) \propto e^{-g\sum_{i=1}^n y_i}\det\left[\int_{-1}^{1} d\lambda(\lambda+g)e^{-\frac{i}{2}\lambda\delta_{ij}}\right]_{i,j=1,n}$$

with $g = 2/T - 1$ and $\delta_{ij} = N(\phi_i - \phi_j) - i(y_i + y_j)$. The expression above coincides in every detail with that obtained in [6] for random matrices with rank-one deviation from Hermiticity provided one remembers that the mean linear density of phases ϕ_i along the unit circle is $\nu = 1/(2\pi)$. This completes the proof of universality for rank-one deviations. The general case is considered in [13].

For a general subunitary $\hat{A}$, one also succeeds in calculating a closely related object, namely, the correlation function of secular determinants:

$$I(z_1, z_2) = \left\langle \det\left(z_1 \mathbf{1} - \hat{A}\right) \det\left(z_2^* \mathbf{1} - \hat{A}^\dagger\right)\right\rangle_A \tag{11}$$

where the angular brackets stand for the averaging over the probability density in Eq.(6). For unitary $\hat{A}$, this correlation function has attracted much attention recently [9,15]. Relegating details to a more extended publication, we just present the final result in terms of the eigenvalues $p_i = 1 - T_i$ of the matrix $\hat{G}$:

$$I(z_1, z_2) = \sum_{k=0}^{N} (z_1 z_2^*)^k \begin{pmatrix} N \\ k \end{pmatrix}^{-1} \sum_{1 \le i_1 < i_2 < \ldots < i_k \le N} p_{i_1} p_{i_2} \cdots p_{i_k} \tag{12}$$

For random unitary matrices all $T_i = 0$ and the expression above reduces to $I(z_1, z_2) = \sum_{k=0}^{N} (z_1 z_2^*)^k$ in agreement with [15].

The author is obliged to B. Khoruzhenko and especially to H.-J. Sommers for stimulating discussions and suggestions. Financial support by SFB 237 "Unordnung und grosse Fluktuationen" as well as of the grant No. INTAS 97-1342 is acknowledged with thanks.

REFERENCES

1. M. S. Livsic *Operators, Oscillations, Waves: Open Systems*, Translations of Mathematical Monographs, v. 34 (AMS, Providence, RI, 1973)
2. D. Z. Arov *Sib. Math. Journ.* **20**, 149 (1979)
3. J. W. Helton *J. Funct. Anal.* **16**, 15 (1974)
4. C. H. Lewenkopf and H. A. Weidenmüller *Ann. Phys. NY* **212**, 53 (1991)
5. Y. V. Fyodorov and H.-J. Sommers *J. Math. Phys.* **38**, 1918 (1997)
6. Y. V. Fyodorov and B. A. Khoruzhenko *Phys. Rev. Lett.* **83**, 65 (1999)
7. H.-J. Sommers, Y. V. Fyodorov and M. Titov *J. Phys. A* **32** , L77 (1999)
8. K. Frahm et al *Europhys. Lett.*, **49**, 48 (2000); *Physica* A **278**, 469 (2000)
9. U. Smilansky in: *Supersymmetry and Trace Formulae: Chaos and Disorder*, edited by I. V. Lerner et al (Kluwer, NY, 1999), p. 173
10. F. Borgonovi, I. Guarneri and D. L. Shepelyansky *Phys. Rev. A* **43**, 4517 (1991)
11. M. Glück, A. R. Kolovski and H. J. Korsch *Phys. Rev. E* **60**, 247 (1999)
12. K. Zyczkowski and H.-J. Sommers *J. Phys. A* **33**, 2045 (2000)
13. Y. V. Fyodorov and H.-J. Sommers, e-preprint `nlin.CD/0005061` at `xxx.lanl.gov`
14. M. L. Mehta *Random Matrices* 2nd ed. (Academic Press, London, 1991)
15. F. Haake et al *J. Phys. A* **29**, 3641 (1996); S. Ketteman, D. Klakow and U. Smilansky *J. Phys. A* **30**, 3643 (1997)

Statistical Mechanics of Random Matrices: Application to Disordered Metals

Hiroshi Hasegawa[1]

Atomic Energy Research Institute, Nihon University, Kanda-Surugadai, Chiyoda-ku, Tokyo 101-0062, Japan

Abstract. An ensemble of $N \times N$ hermitians is constructed which satisfies *representation covariance* and *translation invariance* on the basis of maximum entropy and level dynamics, concluding that the resulting reduced distribution of N eigenvalues of a sample hermitian, $P(x_1, x_2, .., x_N)$, must be of the form $\prod_{j<k} c(x_j - x_k)$ in terms of a single function $c(x-y)$ of the difference of a pair of eigenvalues, x and y. It provides us with a concrete view of the level-statistics of quantum systems as the statistical mechanics of a 1-dimensional gas whose distribution function on $\{x_j\}$ may be represented as $P(\{x_j\}) = C_N \exp\left(-\beta \sum_{j<k} \phi(x_j - x_k)\right)$, where β is the symmetry parameter and

$$\phi(r) = \tfrac{1}{4}\log(1 + 2a^2\cos 2\theta/r^2 + a^4/r^4) \qquad a > 0; \quad 0 \leq \theta < \pi/2.$$

An additional restriction on the average level density provides its relation to the single-level potential $V(x)$ and the two-level correlation function $R(x-y)$, and furthermore, the possibility of *attractiveness* of the pair potential. The scheme allows one to incorporate the scaling theory of the metal-insulator transition by means of a *dimensionless conductance* $g(L)$ depending on *dimensionality*, which is outlined. As a result, the universal spectral statistics for 3D metals in the limit $L \to \infty$ are indicated.

INTRODUCTION

The idea that the quantum level statistics of a disordered system could be formulated as the statistical mechanics of a one-dimensional gas of interacting levels was the starting point of the formulation of random matrix theory by Dyson [1]. One curious question about this idea is the possibility of a phase transition occurring in such interacting systems, but at first sight the answer appears to be negative because of the one-dimensional nature of the problem and the strong repulsion acting between levels known as the Wigner repulsion. Here, we present a detailed *positive* answer to this question by way of example of the localization phenomena associated with a *metal-insulator transition.*

1) e-mail address: h-hase@mxj.mesh.ne.jp

CP553, *Disordered and Complex Systems*, edited by P. Sollich, et al.

INFORMATION THEORETICAL BASIS

On Balian's work in 1968

Balian [2] first introduced information theory into Random Matrix Theory. It was a strategy to generalize a known form of probability distribution: for example, the Wigner-Dyson form $\prod_{j<k}|x_j - x_k|^\beta$ could be modified so as to allow a level density in agreement with observations (see details below). The basic tool was the *maximum entropy principle*, which he designated as his *postulate* **B**. Importantly, he added a *postulate* **A** which he regarded as a more fundamental concept: this is the ansatz about a *metric* defined on matrix spaces, given by $ds^2 = \mathrm{Tr}dMdM^*$ for the distance between two infinitesimally distant matrices M and $M + dM$. Thus, first of all, we wish to answer the question about the most general Riemannian metric on $N \times N$ matrices for the entropy principle.

Possible Riemannian metrics on Matrix Spaces

We follow an axiomatic approach set up by Petz [3]. Let $\mathbf{K}_H$ denote a sesquilinear form defined on $\mathcal{M}_N$ (the set of all $N \times N$ matrices with complex coefficients), depending on a hermitian $H \in \mathcal{M}_N^s$ (the set of all $N \times N$ hermitians). Then,

(a) **symmetry** $\mathbf{K}_H(A^*, B^*) = \mathbf{K}_H(B, A)$ for every $A, B \in \mathcal{M}_N$

(b) **positive definiteness** $\mathbf{K}_H(A, A) \geq 0$ for all $A \in \mathcal{M}_N$ and $\mathbf{K}_H(A, A) = 0 \Rightarrow A = 0$. Later, we shall use a weakened (**b'**): positive definiteness for all $A \in \mathcal{M}_N^s$. These are the metric conditions on $\mathcal{M}_N$ and $\mathcal{M}_N^s$, respectively.

(c) **continuity of the map** $H \to \mathbf{K}_H$**:** a continuous map for every fixed A, B.

(d') **unitary covariance** $\mathbf{K}_{UHU^*}(UAU^*, UBU^*) = \mathbf{K}_H(A, B)$

(d'') **translational invariance** $\mathbf{K}_{H+\lambda I}(A, B) = \mathbf{K}_H(A, B)$

Conditions (d') and (d'') replace Petz's monotonicity condition (d) where H is a density matrix D (a special hermitian) by which the unitary covariance (d') also holds, enabling one to write $\mathbf{K}_H(A, A)$ in the H-diagonal representation. Accordingly, one can write

$$\mathbf{K}_H(A, A) = \sum_{j \neq k} c(x_j - x_k)^{-1}|A_{j,k}|^2 \qquad A \in \mathcal{M}_N, \tag{1a}$$

$$= \sum_{j<k} c(x_j - x_k)^{-1}|A_{j,k}|^2 \qquad A \in \mathcal{M}_N^s, \tag{1b}$$

where $c(r)$ is a real, continuous function and $c(r) > 0$. Upon integrating out the Gaussian distribution $\exp[-(1/2)\mathbf{K}_H(A, A)]$ over $A \in \mathcal{M}_N^s$ in Eq.(1b), one gets

$$P(\{x_j\}) = \mathrm{const} \prod_{j<k} c(x_j - x_k) \propto \exp[-\beta \sum_{j<k} \phi(x_j - x_k)]. \tag{2}$$

Maximum entropy principle under geometric and level-density constraints

For a Gaussian distribution: Maximizing $H[P] \equiv -\langle \log P \rangle_P$ with a fixed variance, i.e., Max $H[P] = H[P_G]$ with the constraints $\langle y^2 \rangle = \sigma^2$, $\langle y \rangle = 0$ gives

$$P(y) = P_G(y) = (2\pi\sigma^2)^{-1/2} e^{-y^2/2\sigma^2}. \tag{3}$$

At first sight Eq.(2) looks like the desired result, but we need a modification for *attractiveness* of the pair potential (see later).

For a prototype exponential family [4]: Let $C_1, C_2, .., C_n$ be a set of observables, and a result of (repeated) observations yields $\langle C_i \rangle_P = \eta_i$ $i = 1, 2, .., n$. A maximization of $H[P]$ under the above constraint yields the most unbiased distribution, called exponential family and given by

$$P_\theta = \exp\left[\sum_i \theta_i C_i - \psi(\{\theta_i\})\right] \quad \psi(\{\theta_i\}) = \log \int \exp\left[\sum_i \theta_i C_i\right] d\mu. \tag{4}$$

There exists a one-to-one correspondence between the parameter set $\{\eta_i\}$ and the Lagrange multiplier set $\{\theta_i\}$, and under the **potential condition** $\partial\theta_i/\partial\eta_j = \partial\theta_j/\partial\eta_i$, the covariance tensor of the measured values C_i is expressed as

$$\langle (C_i - \langle C_i \rangle)(C_j - \langle C_j \rangle) \rangle = \frac{\partial \eta_i}{\partial \theta_j} = \frac{\partial^2 \psi}{\partial \theta_i \partial \theta_j} \quad \text{(Fisher metric)}. \tag{5}$$

Beenakker's functional derivative [5] single level-density and two-level correlation function

We may apply the above exponential family framework to the observable level density $\rho(x) \equiv \sum_i \delta(x - x_i) = \mathrm{Tr}\, \delta(x - H)$ to write (the discrete index i corresponds to the continuous parameter x, i.e., C_i to $\rho(x)$, η_i to $\langle \rho(x) \rangle$, and θ_i to $V(x)$)

$$\begin{aligned} \frac{\delta \langle \rho(x) \rangle}{\delta V(x')} = \frac{\delta \langle \rho(x') \rangle}{\delta V(x)} &= -\beta(\langle \rho(x)\rho(x') \rangle - \langle \rho(x) \rangle \langle \rho(x') \rangle) \\ &\equiv -\beta R(x, x') \quad (R: \text{ Al'tshuler's correlation function}) \end{aligned} \tag{6}$$

$$\langle \rho(x) \rangle = -\beta \int_{-\infty}^{\infty} R(x, x') V(x') dx' \quad \text{(generalized Wigner semicircle law)}. \tag{7}$$

In Eq.(4), $d\mu = P(\{x_j\})\, dx_1 \dots dx_N$. Hence

$$P_\theta \to P_V; \qquad P_V(\{x_j\}) = C_N \exp\left[-\beta\left(\sum_{j<k} \phi(x_j - x_k) + \sum_i V(x_i)\right)\right]. \tag{8}$$

Remark. Al'tshuler's correlation function is related to Dyson's one through

$$R(x - x') = \delta(x - x') + X_2(x - x') - 1 = \delta(x - x') - Y_2(x - x') \tag{9}$$

in the unfolded scale $\langle \rho(x) \rangle = 1$.

Equilibrium state of Hamiltonian level dynamical system [6]

Level Dynamics has been used in a variety of approaches to quantum level statistics, but what we mean by this is exclusively its classical Hamiltonian framework, first proposed by Yukawa [7] for use as an equilibrium statistical mechanics, in particular, for an application to intermediate statistics. This is because two crucial constants of motion, the square of *angular momentum* and the *Hamiltonian* itself, yielded the essential ingredient for describing the statistics, namely, in the form $e^{-\beta\mathcal{H}-\gamma\mathcal{Q}}$. This is precisely Amari's exponential family with two Lagrangian multipliers $\beta(\neq$ previous one) and γ. Furthermore, the exponent of this form, $\beta\mathcal{H}+\gamma\mathcal{Q}$, represents a typical Riemannian metric with respect to all the components of angular momentum [6] provided both β and γ are real, positive parameters.

Detailed results of the intermediate statistics within the positive β and γ, are available [8,9], in particular, a smooth interpolation between Poisson and GUE (known from [10] to be exactly tractable). An interesting question that arose is a possibility other than the case $\beta, \gamma > 0$. Since $\beta(=\langle p^2\rangle^{-1}) > 0$ where p represents the linear momentum of a level, and is normalizable to 1, the question reduces to the possibility of a negative, or complex-valued γ. Let us first consider γ to be real but negative. The resulting quadratic form $\mathbf{K}_H(A, A)$ cannot be defined as a Riemannian metric everywhere in the manifold of $r : 0 < r \equiv x - y$, because then the manifold is separated into two regimes by a particular value of r, where $\mathbf{K}_H > 0$ in one of them and < 0 in the other, and at the boundary point, $\mathbf{K}_H = 0$. Its metricity can be recovered by taking the absolute value of $\mathbf{K}_H$ for any $r(> 0)$ except at that boundary value (see Fig. 1(i)).

In the explicit form of the variance function $c(r)$ in Eq.(2), derived for the model of level dynamics [8,10,11], we have $c(r) = 1/(1 + 1/\gamma r^2)$, i.e.,

$$\phi(r) = \log(1 + 1/\gamma r^2) \tag{10}$$

which shows that the critical behavior occurs at $r = r_c \equiv 1/\sqrt{-\gamma}$ for $\gamma < 0$ (see Fig. 1(i)).

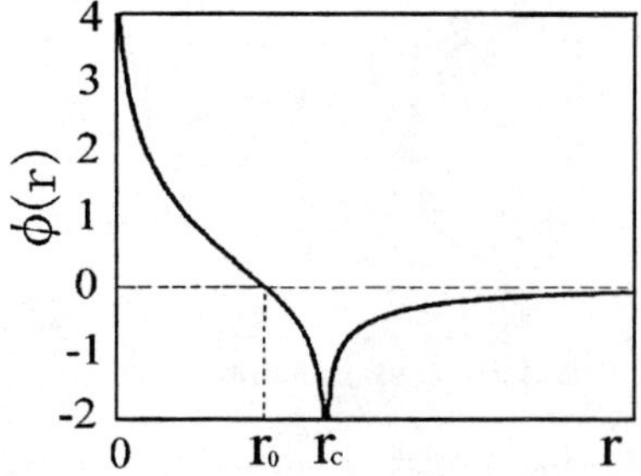

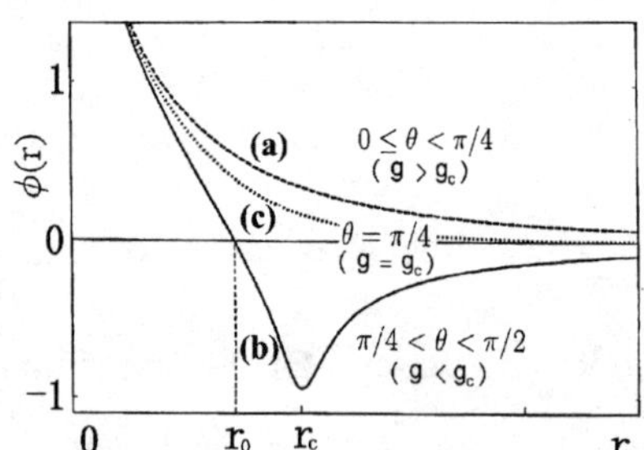

FIGURE 1. (i – Left) $\phi(r)$ from Eq. (10), for $\gamma < 0$. (ii – Right) Potentials from Eq. (11).

Complexification of Gaussian exponent

We envisage that the critical behavior shown in Fig. 1(i) reflects the metal-insulator transition in quantum level statistics with a sharp separation of the one-electron energy scale by the critical value r_c, giving rise to the metal phase ($0 < r < r_c$) and the insulator phase ($r > r_c$). However, the distribution $\prod_{j<k} |c(x_j - x_k)|$ for a real negative γ is a singular distribution which is intractable. The following procedure of complex parametrization in analogy with Breit-Wigner broadening will provide a physically meaningful treatment, in order to take the analysis further. In brief, whenever the c-function is replaced by a complex function, then each component of the product form may be replaced by its absolute magnitude.

Prescription 1. Given a metric $\mathbf{K}_H(A, A)$ subject to (a),(b),(c),(d′) and (d″) and represented by Eq.(1b). It is supposed that the real positive function $c(r)$ be replaced by a complex function denoted by $\hat{c}(r)$ which reduces to $c(r)$, when the complex factor in it is set to be real. Then, the complex form $\mathbf{K}_H(A, A)$ subject to the same axioms except (b) being replaced by (b′) and represented by Eq.(1a) may be used for reducing the Gaussian distribution, where $\hat{c}(-r) = \hat{c}(r)^*$, $c_{1a}(r) = \hat{c}(r)^{1/2}$ must hold, and conditioned by $Re[\hat{c}(r)^{1/2}] > 0$.

Accordingly, the resulting distribution $P(\{x_j\})$ for level dynamics is now given by

$$P(\{x_j\}) = C_N \prod_{j<k} |\hat{c}(x_j - x_k)| = C_N \exp\left[-\sum_{j<k} \frac{1}{2} \log\left(1 + 2\frac{a^2}{r^2}\cos 2\theta + \frac{a^4}{r^4}\right)_{r=x_j-x_k}\right] \tag{11}$$

where $\gamma^{-1} \equiv a^2 e^{-i2\theta}$, $a > 0$ and $0 \leq \theta < \pi/2$.

INCORPORATING THE SCALING THEORY

The scaling theory of the metal-insulator transition by Abrahams et al [12] states that the one-electron energy of a metal can be scaled by a linear system-size parameter L to give the quantity $g(L)$, called *dimensionless conductance*, which distinguishes the behavior of transition in d-dimensional metals, according to whether the β line crosses the abscissa log g or not, where β is defined by $\beta = \beta\{g(L)\} \equiv d(\log g)/d(\log L)$.

The case $d = 3$ (3D metal) is of a special interest, because it has a unique crossing point, independent of scale, which is denoted by g_c. Fig. 1(ii) shows the three different cases of ϕ-curves: (a) metal ($g > g_c$), (b) insulator ($g < g_c$) and (c) mobility-edge ($g = g_c$), subject to

Prescription 2. $ae^{-i\theta} \equiv g(L) - ig_c$, and $r_c \equiv g_c$ under scale (9), hence

$$\frac{g_c}{g(L)} = \tan\theta : \quad (a)\ 0 \leq \theta < \pi/4,\ (b)\ \pi/4 < \theta < \pi/2,\ (c)\ \theta = \pi/4. \tag{12}$$

It should be remarked that the complex parametrization in the above stems also from the Green function theory: $g_c + ig$ represents the *self energy* (Thouless, 1974), and on this basis detailed computations of spectral statistics can be performed.

UNIVERSAL SPECTRAL STATISTICS FOR 3D METAL

Prescription 2 above indicates that the critical conductance g_c be identified, indeed, with the critical energy r_c encountered in the singular point in Eq.(11) of the variance $c(r)$. Because of the said scale invariance of g_c, the real part of the self energy remains unchanged in the limit $L \to \infty$ where $g(L)$, the imaginary part, tends to vanish in the insulator phase. In the metal phase, $g(L) \to \infty$ and overwhelms g_c, corresponding to an ideally conducting metal. The *mobility edge state* exists in a thermodynamically meaningful sense in this case of $d = 3$. For dimension $d = 1, 2$, on the other hand, only $\theta = 0$ occurs.

Further details of our results will be presented elsewhere (to be submitted to Phys. Rev. E). Here, we mention our concrete result as regards the compressibility χ expressed in terms of g_c such that χ_c goes to 0 in the metal phase, to 1 in the insulator one, and $\chi = \chi_c$ always at the mobility-edge. This result could be obtained only for the $\beta = 2$ symmetry.

REFERENCES

1. F. J. Dyson, J. Math. Phys. **3**, 140 (1962).
2. R. Balian, Nuovo Cimento **B 57**, 183 (1968).
3. D. Petz, Linear Algebra and its Applications **244**, 81 (1996).
4. S.-I. Amari, Lecture Notes in Statistics, **28** (1985).
5. C.W.J. Beenakker, Phys. Rev. Lett. **70**, 1155 (1993); Rev. Mod. Phys. **69**, 731 (1997).
6. H. Hasegawa, Open Sys. & Information Dyn. (Kluver Academic Pub.) **4**, 359 (1997).
7. T. Yukawa, Phys. Rev. Lett. **54**, 1883 (1985); Phys. Lett. A **116**, 227 (1986).
8. H. Hasegawa, Open Sys. & Information Dyn. (Kluver Academic Pub.) **6**, 149 (1999).
9. H. Hasegawa and J.-Z. Ma, J. Math. Phys. **39**, 2564 (1998).
10. M. Gaudin, Nuclear Physics **85**, 545 (1966).
11. J.-L. Pichard and B. Shapiro, J. Phys.I (France) **4**, 623 (1994).
12. E. Abrahams, P.W. Anderson, D.C. Licciardello and T.V. Ramakrishnan, Phys. Rev. Lett. **42**, 673 (1979). See also H. Hasegawa and Y. Sakamoto, Prog. Theor. Phys. Suppl. **139**, 112 (2000).

Qualitative Properties of the Ground State in Nelson's Scalar Field Model: A Gibbs Measure-Based Approach

József Lőrinczi

Zentrum Mathematatik, Technische Universität München
Gabelsbergerstr. 49, 80290 München, Germany; lorinczimathematik.tu-muenchen.de

Abstract. We show how Gibbs measures can be associated with Nelson's massless scalar field model in its Euclidean representation. These Gibbs measures will be used for deriving qualitative properties of the Fock space ground state.

INTRODUCTION

Consider Nelson's massless model of a charged spinless point particle with position $q_t \in \mathbb{R}^d$ coupled to a scalar Bose field, given by the Hamiltonian

$$H_{\mathrm{N}} = \left(-\frac{1}{2}\Delta + V(x)\right) \otimes 1 + 1 \otimes \int \omega(k) a^*(k) a(k) dk \tag{1}$$
$$+ \int \frac{1}{\sqrt{2\omega(k)}} \frac{1}{(2\pi)^{d/2}} \left(\hat{\varrho}(k) e^{ikx} a(k) + \hat{\varrho}(-k) e^{-ikx} a^*(k)\right) dk.$$

H_{N} acts on $L^2(\mathbb{R}^d, dx) \otimes \mathcal{F}$, $\mathcal{F}$ being the symmetric Fock space over $L^2(\mathbb{R}^d, dx)$. The first term is a Schrödinger operator on the Hilbert space $L^2(\mathbb{R}^d, dx)$ describing the non-interacting particle under a confining potential V. The external potential $V : \mathbb{R}^d \to \mathbb{R}$ is a multiplication operator assumed here to be bounded from below, continuous and having the asymptotics $V(x) = C|x|^{2s} + o(|x|^{2s})$ for large $|x|$ with some constant $C > 0$ and exponent $s > 1$. This confining potential makes sure that the Schrödinger operator has a unique strictly positive continuous eigenfunction (ground state) ψ_0 lying at the bottom of its spectrum. The second term gives the free field energy with dispersion $\omega(k) = |k|$ and is written in terms of the Bose creation and annihilation operators $a^*(k), a(k)$ on Fock space satisfying the canonical commutation relations $[a(k), a^*(k')] = \delta(k - k')$, $[a(k), a(k')] = 0 = [a^*(k), a^*(k')]$. The last term in (1) accounts for the particle-field interaction. ϱ (its Fourier transform being denoted $\hat{\varrho}$) is the particle charge density that we take to be a radial, smooth, rapidly decreasing function whose integral is the full charge,

CP553, *Disordered and Complex Systems*, edited by P. Sollich, et al.

i.e. $\int \varrho(x)dx = e > 0$, which plays here the role of the coupling constant. For reasons of mathematical well-definedness ϱ has to be chosen such that a cutoff is introduced at both ends of the energy spectrum. The condition $\int_{\mathbb{R}^d} dk|\hat{\varrho}(k)|^2/|k|^3 < \infty$ implements an infrared cutoff. Under mild assumptions on ϱ it is easily seen by the Kato-Rellich theorem that H_{N} is self-adjoint on the domain of the self-adjoint operator $(-1/2\Delta + V(x)) \otimes 1 + 1 \otimes \int \omega(k)a^*(k)a(k)dk$, and it is also bounded from below.

An alternative to the Fock space representation of Nelson's model is a formulation in terms of Euclidean quantization. In this framework the particle and field are associated with specific stochastic processes. The Euclidean action for the particle is $S_{\mathrm{p}}(\{q_t\}) = \int \frac{1}{2}\dot{q}_t^2 dt$, hence the particle is associated with the $P(\phi)_1$-process q_t defined by the stochastic differential equation

$$dq_t = -\frac{\nabla\psi_0}{\psi_0}dt + dB_t \tag{2}$$

(dB_t denotes Brownian motion), and having the invariant measure $d\mathsf{N}^0 = \psi_0^2(x)dx$. The path measure of this process will be denoted by $\mathcal{N}^0$; for more details on this see [2] and references therein. The paths $Q = \{q_t\}$ are $\mathcal{N}^0$-almost surely continuous and q_t is a time-reversible process. The field in turn is described by a generalized Gaussian process $\xi(t,x)$ resulting from the Euclidean action $S_{\mathrm{f}}(\{\xi(t,x)\}) = \int_{\mathbb{R}}\int_{\mathbb{R}^d} \frac{1}{2}(\partial_t\xi(t,x)^2 + (\nabla\xi(t,x))^2)dxdt$. For any $f \in \mathcal{S}(\mathbb{R}^{d+1})$, i.e. the Schwartz space over $\mathbb{R}^d$, denote $\xi(f) = \int_{\mathbb{R}^{d+1}} f(t,x)\xi(t,x)dtdx$. The field $\{\xi(f), f \in \mathcal{S}(\mathbb{R}^{d+1})\}$ is described by the generalized Gaussian measure $\mathcal{G}$ on $\mathcal{S}(\mathbb{R}^{d+1})$ with zero mean and covariance

$$\mathrm{cov}_{\mathcal{G}}(\xi(f_1),\xi(f_2)) = ((-\partial_t^2 - \Delta)^{-1}f_1, f_2) = \int_{\mathbb{R}}\int_{\mathbb{R}^d} \frac{\hat{f}_1(k_0,k)\hat{f}_2^*(k_0,k)}{k_0^2 + k^2}dk_0dk. \tag{3}$$

The probability space is $\mathcal{S}'(\mathbb{R}^{d+1})$ with its associated Borel σ-field. It will be convenient to extend (3) to test functions of the form $f_t^{\varphi}(x',t') = \varphi(x')\delta(t'-t)$ with $\varphi \in \mathcal{S}(\mathbb{R}^d)$ and view $\xi_t = \xi(t,\cdot)$ as a stochastic process taking values in a suitable Hilbert space $\mathcal{B}_D$. Thus we will use the stationary Gaussian process $\xi_t(\varphi) = \xi(f_t^{\varphi})$, $\varphi \in \mathcal{S}(\mathbb{R}^d)$, $t \in \mathbb{R}$. Its stationary measure G defined on $\mathcal{S}'(\mathbb{R}^d)$ is again Gaussian with mean 0 and covariance $\mathrm{cov}_{\mathsf{G}}(\xi_0(\varphi_1),\xi_0(\varphi_2)) = 1/2\int_{\mathbb{R}^d} dk\hat{\varphi}_1(k)\hat{\varphi}_2^*(k)/k$. Moreover, $\xi_t(\varphi)$ is a time-reversible Markov process. The real Hilbert space $\mathcal{B}_D \subset \mathcal{S}'(\mathbb{R}^d)$ consists of functions with norm $||\xi||_{\mathcal{B}_D}^2 = \int_{\mathbb{R}^d\times\mathbb{R}^d} D(k_1,k_2)\hat{\xi}(k_1)\hat{\xi}^*(k_2)dk_1dk_2$ (where $\hat{\xi}^*(k) = \hat{\xi}(-k)$). Here D is the operator with kernel $D(k_1,k_2) = |k_1|_1\bar{D}(k_1,k_2)|k_2|_1$, where $\bar{D} = (-\Delta_k + |k|^2)^{-(d+1)}$ and $|k|_1 = |k|$ if $|k| < 1$ and $= 1$ otherwise. The process ξ_t is continuous with respect to this norm with probability 1.

The Euclidean interaction becomes $S_{\mathrm{int}}(\{q_t,\xi_t\}) = \int_{\mathbb{R}}\int_{\mathbb{R}^d} \varrho(x-q_t)\xi(t,x)dxdt \equiv \int \xi_t * \varrho(q_t)dt$. The joint particle-field process on the space $C(\mathbb{R}, \mathbb{R}^d \times \mathcal{B}_D)$ is described by the path measure $\mathcal{P}^0 = \mathcal{N}^0 \times \mathcal{G}$. The interacting system can be described by taking $\mathcal{P}^0$ as reference process and modifying it with a density given by

$$\frac{d\mathcal{P}_T}{d\mathcal{P}^0} = \frac{1}{Z_T} \exp\left(-\int_{-T}^{T} \xi_t * \varrho(q_t)dt\right) \tag{4}$$

Here

$$Z_T = \int e^{-\int_{-T}^{T} \xi_t * \varrho(q_t)dt} d\mathcal{P}^0 = \int e^{\int_{-T}^{T}\int_{-T}^{T} W(q_s - q_t, s-t)dsdt} d\mathcal{N}^0 \tag{5}$$

is the normalizing partition function with

$$W(x,t) = -\frac{1}{4}\int \frac{|\hat{\varrho}(k)|^2}{|k|} \cos kx \, e^{-|k||t|} dk \tag{6}$$

This is obtained by using that $\mathcal{G}$ is a Gaussian measure.

For any fixed path $Q = \{q_t : t \in \mathbb{R}\} \in C(\mathbb{R}, \mathbb{R}^d)$ consider the conditional distribution $\mathcal{P}_T^Q = \mathcal{P}_T(\,\cdot\,|\{q_t\} = Q)$ as a measure on $C(\mathbb{R}, \mathbb{R}^d \times \mathcal{B}_D)$. Since the action $\int_{-T}^{T} \xi_t * \varrho(q)dt$ is linear in ξ_t, $\mathcal{P}_T^Q$ is itself a Gaussian measure. We have then for $\mathcal{P}_T$ on $C([-T,T], \mathbb{R}^d \times \mathcal{B}_D)$ that $\mathcal{P}_T(E \times F) = \int_F \mathcal{P}_T^Q(E) d\mathcal{N}_T(Q)$, for all $T > 0$, where E and F are appropriate measurable sets. Here $\mathcal{N}_T$ is the probability measure on the finite-time particle path space $C([-T,T], \mathbb{R}^d)$ given by

$$\frac{d\mathcal{N}_T}{d\mathcal{N}^0} = \frac{1}{Z_T} \exp\left(-\int_{-T}^{T}\int_{-T}^{T} W(q_s - q_t, s-t)dsdt\right) \tag{7}$$

with the pair interaction potential W appearing in (6). Thus for every $T > 0$ the finite-time Gibbs measure $\mathcal{N}_T$ is the marginal of the joint particle-field path measure $\mathcal{P}_T$ obtained by integrating over the Gaussian variables and it is the stationary measure for Brownian motion under external potential V and pair interaction potential W appearing as an effective potential for the particle on averaging over the Bose field. Since $\mathcal{P}_T^Q$ is a Gaussian measure, for understanding the properties of the joint particle-field path measure we only need to understand the properties of $\mathcal{N}_T$. We denote the single time distribution of $\mathcal{N}_T$ by N_T.

GIBBS MEASURE

We will be interested below in whether the sequence of measures $\mathcal{P}_T$ has a limit as $T \to \infty$. We use the following notion of convergence. Let M be a metric space, and $C(\mathbb{R}, M)$ the space of continuous paths $\{X_t\}$ with values in M. For any interval $I_T = [-T,T] \subset \mathbb{R}$ let $\mathcal{C}_T \subset \mathcal{C}$ be a sub-σ field of the Borel σ-field $\mathcal{C}$ generated by the evaluations $\{X_t : t \in I_T\}$, i.e. $\mathcal{C}_T = \sigma(X_t, t \in I_T)$. We say that the sequence of probability measures $\{\mu_n\}$ on $C(\mathbb{R}, M)$ converges locally weakly to the probability measure μ if for any $0 \leq T < \infty$ the restrictions $\mu_n|_{\mathcal{C}_T}$ converge weakly to the measure $\mu_{\mathcal{C}_T}$. We have then the following result.

Theorem 1 (Existence of particle-field path measure) *For $d \geq 3$ and $0 < e \leq e^*$, with sufficiently small $e^* > 0$, there exists the local weak limit $\mathcal{P}_T \to \mathcal{P}$ as $T \to \infty$, and it is a probability measure of a stationary reversible Markov process on the path space $C(\mathbb{R}, \mathbb{R}^d \times \mathcal{B}_D)$. Moreover, the stationary distribution P of $\mathcal{P}$ can be obtained as the weak limit $\mathsf{P} = \lim_{T\to\infty} \mathsf{P}_T$, where P_T is the distribution of $\{q_0, \xi_0\}$ under $\mathcal{P}_T$.*

As pointed out above, this theorem results from the properties of the Gibbs measure $\mathcal{N} = \lim_{T\to\infty} \mathcal{N}_T$ defined by (7).

Theorem 2 (Existence of Gibbs measure) *Take an arbitrary increasing sequence $\{T_n\}$ of positive numbers such that $\lim_{n\to\infty} T_n = \infty$, and suppose $0 < e \leq e^*$ with some sufficiently small $e^* > 0$. Then the local weak limit $\lim_{n\to\infty} \mathcal{N}_{T_n} = \mathcal{N}$ exists and is a Gibbs probability measure; the limit does not depend on the particular choice of the sequence $\{T_n\}$. Moreover, the measure $\mathcal{N}$ is invariant with respect to time shifts and time reflections.*

Theorem 3 (Uniqueness of the Gibbs measure) *The Gibbs measure $\mathcal{N}$ is unique in DLR-sense whenever it exists.*

Theorem 4 (Typical path behaviour) *For every $0 < e \leq e^*$ there exists a constant $C > 0$ and a function $F : C(\mathbb{R}, \mathbb{R}^d) \to \mathbb{R}$ such that for large enough $|t|$*

$$|q_t| \leq (C \log |t| + 1)^{1/(s+1)} + F(\{q_t\}) \tag{8}$$

$\mathcal{N}$-almost surely.

Theorem 5 (Single-time distributions) *For every $0 < e \leq e^*$ we have that*

1. *the distributions S_T under $\mathcal{N}_T$ of positions q_τ at arbitrary moments of time $\tau \in \mathbb{R}$ are absolutely continuous with respect to $d\mathsf{N}^0 = \psi_0^2 dx$, i.e., there exist $c_1, c_2 \in \mathbb{R}$ such that $c_1 \leq d\mathsf{S}_T/d\mathsf{N}^0(q_\tau) \leq c_2$, $\forall q \in \mathbb{R}^d, T > 0$. Moreover, the limit $d\mathsf{S}/d\mathsf{N}^0(x) = \lim_{T\to\infty} d\mathsf{S}_T/d\mathsf{N}^0(x)$ exists pointwise;*

2. *the conditional distributions $\mathcal{N}_T(\,\cdot\,|q_\tau = q)$ converge locally weakly to $\mathcal{N}(\,\cdot\,|q_\tau = q)$, $\forall q \in \mathbb{R}^d$.*

Theorem 6 (Degree of mixing) *For every $0 < e \leq e^*$ and any pair of bounded functions F, G on $\mathbb{R}^d$*

$$\mathbb{E}_{\mathcal{N}}[F(q_s)G(q_t)] - \mathbb{E}_{\mathcal{N}}[F(q_s)]\mathbb{E}_{\mathcal{N}}[G(q_t)] \leq C \frac{\sup|F| \sup|G|}{|s-t|^\gamma + 1} \tag{9}$$

with some $\gamma > 1$ and a suitable constant prefactor $C > 0$.

For proofs and details we refer to [4]. Except for the uniqueness result the proofs are based on cluster expansion. Note that the exponent γ appearing in Theorem 6 is expected to be equal to the rate of decay of the potential, i.e. $d-1$. The statements listed above are actually true for a much larger class of pair interaction potentials than the W appearing in Nelson's model (see [4]) and therefore the results of this section are also applicable to other types of problems.

GROUND STATE PROPERTIES

Under the infrared (henceforth IR) cutoff condition $\int dk|\hat{\varrho}(k)|^2/|k|^3 < \infty$ and the conditions we have imposed on the external potential V, a unique strictly positive ground state Ψ of H_{N} exists in Fock space for all coupling constants [6]. However, Ψ actually exists for also more general V, possibly subject to restrictions on the coupling strength [1,6].

The main point of this section is that particular ground state expectations (i.e., scalar products in $L^2(\mathbb{R}^d, dx) \times \mathcal{F}$) can be obtained as Gibbs averages of suitable functions related to the observables involved. Moreover, deriving the qualitative properties of the ground state, i.e., obtaining meaningful bounds on these expectations, becomes possible by estimating these Gibbs averages. Here are some of these properties, obtained in [5,3].

Theorem 7 (Boson distribution) *Suppose the IR condition holds and the Gibbs measure $\mathcal{N}$ exists. Denote by $\Psi_n = \Psi_n(k_1, ..., k_n, q)$, $n \geq 1$, the basis elements of the Fock space, and write $p_n = \int |\Psi_n(k_1, ..., k_n, q)|^2 dk_1...dk_n dq$ for the probability distribution of n bosons occurring in the ground state. Put $h(\beta) = \sum_{n=0}^{\infty} p_n e^{-\beta n} = (\Psi, e^{-\beta N}\Psi)$, where $\beta \geq 0$ and N is the boson number operator. Then*

1. *With W as under (6) we have*

$$h(\beta) = \mathbb{E}_{\mathcal{N}}\left[\exp\left((e^{-\beta} - 1)\int_{-\infty}^{0} dt \int_{0}^{\infty} ds\, W(q_t - q_s, t - s)\right)\right] \tag{10}$$

   Moreover, $h(\beta)$ has an analytic continuation into the whole complex plane and formula (10) extends everywhere. For large n we have asymptotically $p_n \sim b^n/n!$.

2. *The average number of bosons in the ground state is*

$$(\Psi, N\Psi) = -\frac{dh}{d\beta}|_{\beta=0} = \mathbb{E}_{\mathcal{N}}\left[\int_{-\infty}^{0} dt \int_{0}^{\infty} ds\, W(q_t - q_s, t - s)\right] \tag{11}$$

   Moreover,

$$(\Psi, a^*(k)a(k)\Psi) \leq \frac{|\hat{\varrho}(k)|^2}{|k|^3} \tag{12}$$

Theorem 8 (Average field and field fluctuations) *Suppose the IR condition holds. Then the average field in the ground state is*

$$(\Psi, \xi(\varphi)\Psi) = \mathbb{E}_{\mathcal{N}}\left[\int_{-\infty}^{\infty} dt \int_{\mathbb{R}^d \times \mathbb{R}} \frac{\hat{\varrho}(k)}{\sqrt{|k|}}\hat{\varphi}(k')\cos(k - k')q_t\, e^{-|k||t|} dk dk'\right] \tag{13}$$

for every $\varphi \in \mathcal{S}(\mathbb{R}^d)$. If, moreover, φ is concentrated on a point $x \in \mathbb{R}^d$, then for large $|x|$ we have asymptotically $(\Psi, \xi(\varphi)\Psi) \sim 1/|x|$. Furthermore, field fluctuations increase by turning on the particle-field interaction, i.e. $\mathrm{var}_{\mathcal{P}^0}(\cdot, \xi) \geq \mathrm{var}_{\mathcal{P}}(\cdot, \xi)$.

Theorem 9 (Particle localization) *Suppose the ground state* Ψ *exists and denote* $\lambda(q) = ||\Psi(\cdot, q)||_{\mathcal{B}_D} = \left(\int \Psi(\xi, q)^2 d\mathcal{G}(\xi)\right)^{1/2}$. *There exist some constants* $C, \alpha > 0$ *such that*

$$\lambda(q) \leq Ce^{-\alpha|q|^{s+1}}, \quad \forall q \in \mathbb{R}^d. \tag{14}$$

If $0 < e \leq e^*$, *then there exist some constants* $c_2 \geq c_1 > 0$ *such that* $c_1\psi_0^2 \leq \lambda \leq c_2\psi_0^2$.

To conclude we mention that in dimension 3 the IR cutoff is the necessary and sufficient condition that in Fock space a ground state of H_{N} exists. This implies that if one allows infrared divergence, Fock space quantization and Euclidean quantization will not be equivalent for $d = 3$ (while for $d \geq 4$ the two Hamiltonians are equivalent); for further details we refer to [5]. However, by using the Gibbs-averages representation of ground state expectations we can show that the qualitative properties presented above stay valid even when the IR condition is removed [3].

ACKNOWLEDGMENTS

J.L. is supported by Deutsche Forschungsgemeinschaft within Schwerpunktprogramm "Interagierende stochastische Systeme von hoher Komplexität".

REFERENCES

1. Bach, V., Fröhlich, J. and Sigal, I.M., Quantum electrodynamics of confined nonrelativistic particles, *Adv. Math.* **137**, 299-395, 1998
2. Betz, V. and Lőrinczi, J., A Gibbsian description of $P(\phi)_1$-processes, 1999, submitted for publication
3. Betz, V., Hiroshima, F., Lőrinczi, J., Minlos, R.A. and Spohn, H., Ground state properties of Nelson's scalar model, TU München preprint, 2000
4. Lőrinczi, J. and Minlos, R.A., Gibbs measures for Brownian paths under the effect of an external and a small pair potential, 2000, submitted for publication
5. Lőrinczi, J., Minlos, R.A. and Spohn, H., The infrared behaviour in Nelson's model of a quantum particle coupled to a massless scalar field, TU-München preprint, 2000
6. Spohn, H., Ground state of a quantum particle coupled to a scalar Bose field, *Lett. Math. Phys.* **44**, 9-16, 1998

On the Application of Quantum L_p-Spaces

Adam W. Majewski[1]

IFTiA, University of Gdansk, 80-952 Gdansk, Poland

Abstract. We review some new developments in the theory of noncommutative spaces and indicate how these techniques may be used for constructions of Markov semigroups for certain classes of quantum systems. A special emphasis will be put on "quantum" features of considered maps.

INTRODUCTION

Let (X, μ) be a classical probability space. Here and subsequently, $L_p(X, \mu)$ will stand for the Banach space of p-integrable functions defined on X. Let $T_t : X \mapsto X$ be a one parameter semigroup of measurable transformations. Then, the triple (X, T_t, μ) defines a (classical) dynamical system, where the variable t signifies time. In order to formulate nearness and stability of time evolution, the space X is equipped with a topology compatible with the measure structure. The phase function f, say a continuous function on X, evolves according to the Koopman's operator $(V_t f)(x) \equiv f(T_t x)$ where $x \in X$ (cf. [8]). The Koopman's operators are isometries on the Banach spaces $L_p(X, \mu)$, $p \geq 1$, and unitary operators when restricted to the Hilbert space L_2 and transformations T_t are automorphisms. Let us mention that the relation of the point dynamics T_t with the Koopman's operators V_t was fully clarified by establishing the converse of Koopman's result (for details see [1,10,4]). We recall also that dynamical maps can be defined directly in terms of L_p-spaces, i.e., without any reference to the point dynamics. Then, using functional analysis one can study the time evolution as a family of transformations on $L_p(X, \mu)$. In particular, for the case $p = 2$, one can use the structure of Hilbert space and its basic tool, spectral analysis. Such an approach is extremely useful in the rigorous analysis of various properties of time evolution, especially those related to ergodicity.

The objectives of my talk are to review some new developments in the "quantization" of the framework for the theory of dynamical systems outlined above, and

[1] Supported by KBN grant PB/0273/PO3/99/16.

CP553, *Disordered and Complex Systems*, edited by P. Sollich, et al.

to describe certain constructions of quantum Markov semigroups. Moreover we will argue that the great advantage of our approach lies in the fact that the quantum maps considered are not only natural generalizations of their classical counterparts but also exhibit "genuine" quantum features.

QUANTUM DYNAMICAL MAPS

Quantum systems

To generalize the scheme just described to quantum theories, we associate with each (open) bounded region $\mathcal{O}$ in the space ($\mathbf{R}^\nu$ or $\mathbf{Z}^\nu$) the C^*-algebra $\mathcal{A}(\mathcal{O})$ representing physical operations performable within $\mathcal{O}$. As a result we get the net of algebras, i.e., the correspondence

$$\mathcal{O} \longmapsto \mathcal{A}(\mathcal{O}). \tag{1}$$

Moreover, $\mathcal{A} = \overline{\cup \mathcal{A}(\mathcal{O})}$, where the bar denotes an appropriate strong closure, represents all quasi-local observables of the quantum extended system. It is natural to view $\mathcal{A}$ as a noncommutative analogue of the space of bounded continuous functions.

As a next step, we wish to define the quantum counterpart of a certain distinguished (equilibrium) measure. This can be done as follows. Let $\{\mathcal{O}_n\}$ be a subnet consisting of an increasing sequence of open finite subsets of $\mathbf{R}^\nu$ or $\mathbf{Z}^\nu$ such that for every open finite subset $\mathcal{O}$ there is some finite positive integer $n(\mathcal{O})$ with the property: $\mathcal{O} \subset \mathcal{O}_k$ for all $k \geq n(\mathcal{O})$. Further, let us denote by H_n the Hamiltonian associated with the region $\mathcal{O}_n$. The choice of H_n amounts to specifying the class of interactions $\Phi(\mathcal{O}_n)$ for the region $\mathcal{O}_n$.

To define an equilibrium (Gibbs) state ω_β for an extended system one usually proceeds to the following limit:

$$\omega_\beta = \lim_n \omega_n \tag{2}$$

where $\omega_n(\cdot) = Z_n^{-1} Tr\{e^{-\beta H_n} \cdot\} \equiv Tr\{\varrho^{(n)} \cdot\}$, with β denoting the inverse temperature and Z_n the corresponding partition function. In a similar way, the time evolution α_t can be determined. The above program evidently involves some convergence questions and works perfectly under some additional conditions, e.g., for some class of interactions on a quantum spin lattice. Consequently, we arrive at the quantum probability space $(\mathcal{A}, \omega_\beta)$.

Quantum L_p-spaces

Recently we have proved that it is possible to associate with $(\mathcal{A}, \omega_\beta)$ a quantum counterpart of an interpolating family of L_p-spaces (see [12–14]). It should be pointed out that even in an equilibrium representation of $\mathcal{A}$ there are algebras associated with regions $\mathcal{O}$ such that a (faithful) trace is not defined. Hence, our construction of quantum L_p-spaces reflecting the general Haagerup theory of non-commutative integration [6], was based on the notion of thermodynamic limit. In particular, we have proved [14]:

Theorem 1 *With every Gibbs state ω_β on $\mathcal{A}$ we can associate an interpolating family of spaces $\{L_p(\omega_\beta, s)\}_{p\in[0,\infty),s\in[0,1]}$ in such a way that for any local observable $f \in \mathcal{A}(\mathcal{O})$ we have*

$$||f||^p_{L_p(\omega_\beta,s)} = \lim_n ||f||^p_{L_p(\omega^n,s)} \tag{3}$$

where the local $||\cdot||^p_{L_p(\omega^n,s)}$-norm is given by

$$||f||^p_{L_p(\omega^n,s)} = Tr|(\varrho^{(n)})^{\frac{s}{p}} \cdot f \cdot (\varrho^{(n)})^{\frac{1-s}{p}}|^p. \tag{4}$$

with $\varrho^{(n)}$ being the density matrix of ω restricted to $\mathcal{A}(\mathcal{O}_n)$.

Let us make a comment on this theorem. We got a two-parameter family of quantum L_p-spaces. The significance of the parameter p is the same as in the commutative case while the parameter s reflects, in general, non-symmetry of the imbedding of $\mathcal{A}$ into $L_p(\mathcal{A}, \omega_\beta)$. Finally, let us add that the utility of the interpolation scheme with respect to the additional parameter s was displayed in proofs of various estimates associated with quantum maps (cf. [14]).

Quantum Markov semigroups

In this subsection we indicate how L_p-space techniques may be used to produce a constructive and rigorous approach to quantum Markov evolution of large systems. By a constructive approach we mean one in which existence of the evolution of extended system is not postulated, as in a pure semigroup approach, but constructed on the basis of the local character of evolution in the bounded regions $\mathcal{O}_n$, i.e., the evolution should be constructed as the thermodynamic limit of the corresponding finite volume dynamics with an appropriate control of the convergence. In other words, we adopt the view that the dynamics of a extended quantum system has to be derived from intrinsic local properties of the system. Let us recall at this point that efforts in the construction and study of quantum Markov semigroups are as old as the equilibrium description of quantum systems. Thus to make progress it seems natural to use the new approach: quantum Liouville space technique based on quantum L_p-spaces. To illustrate this new approach we introduce a family of Markov generators and semigroups for a quantum spin system on a lattice, which are similar to a block spin flip stochastic dynamics of classical spin systems. We start with a definition of finite volume dynamics. For $\mathcal{O} \subset \Lambda$, finite subsets of the lattice, let $E_{\mathcal{O},\Lambda} : \mathcal{A} \longrightarrow \mathcal{A}$ be a map defined as follows

$$E_{\mathcal{O},\Lambda}(f) = Tr_{\mathcal{O}}\left(\gamma^*_{\mathcal{O},\Lambda} f \gamma_{\mathcal{O},\Lambda}\right) \tag{5}$$

with

$$\gamma_{\mathcal{O},\Lambda} \equiv \rho_\Lambda^{\frac{1}{2}} \left(Tr_{\mathcal{O}}\rho_\Lambda\right)^{-\frac{1}{2}} \tag{6}$$

where $Tr_{\mathcal{O}}$ is the partial trace over the Hilbert space associated with the region $\mathcal{O}$ while ρ_Λ is the density matrix of a finite volume Gibbs state ω_Λ.

Let $\mathcal{L}_{\mathcal{O},\Lambda}$ be an operator on $\mathcal{A}$ defined by

$$\mathcal{L}_{\mathcal{O},\Lambda} f \equiv E_{\mathcal{O},\Lambda}(f) - f \tag{7}$$

One can verify (see [12]) that $\mathcal{L}_{\mathcal{O},\Lambda}$ is well defined Markov generator. Therefore the corresponding semigroup $\mathcal{P}_t^\Lambda$ can be defined as $\mathcal{P}_t^\Lambda = e^{t\mathcal{L}^\Lambda}$. It has the following properties (see [12]):

Theorem 2 *(i) Positivity preserving: for any $f \in \mathcal{A}^+$ one has $\mathcal{P}_t^\Lambda f \geq 0$, (ii) Unit preserving, i.e., $\mathcal{P}_t^\Lambda \mathbf{1} = \mathbf{1}$, (iii) L_2-symmetry, i.e. $\langle \mathcal{P}_t^\Lambda(f), g\rangle_{\omega_\Lambda} = \langle f, \mathcal{P}_t^\Lambda(g)\rangle_{\omega_\Lambda}$, (iv) $||\mathcal{P}_t^\Lambda||_{L_2(\omega_\Lambda)\to L_2(\omega_\Lambda)} \leq 1$ and $\langle \mathcal{L}^{X,\Lambda}(f), f\rangle_{\omega_\Lambda} \leq 0$, (v) Invariance: for any $f \in \mathcal{A}$ $\omega_\Lambda\left(\mathcal{P}_t^\Lambda(f)\right) = \omega_\Lambda(f)$ where $\langle f, g\rangle_{\omega_\Lambda}$ stands for the inner product of $L_2(\mathcal{A}, \omega_\beta)$.*

It is clear that $\mathcal{P}_t^\Lambda$ is a well defined Markov semigroup. Such semigroups of block-spin flip type can be generalized to a thermodynamic limit of local Gibbs states on an infinite lattice, i.e., there exists the limit of $\gamma_{\mathcal{O},\Lambda}$ as $\Lambda \to \mathbf{Z}^\nu$ (see [14] for the proof). The main difficulty in carrying out this construction is that $\lim_{\Lambda\to\mathbf{Z}^\nu} \gamma_{\mathcal{O},\Lambda} = \gamma_0$ is an element of the von Neumann algebra $\mathcal{M} \equiv \pi_\omega(\mathcal{A})''$ obtained in the GNS construction with ω being a Gibbs state on the entire lattice. Nevertheless, in that case we can define $\mathcal{L}_{\mathcal{O}} f = E_{\mathcal{O}}(f) - f$ with $E_{\mathcal{O}}(f) = Tr_{\mathcal{O}}\{\gamma_0^* f \gamma_0\}$ where $f \in M$. More precisely (see [14])

Theorem 3 *Suppose that $|\beta| \leq \beta_0$, with β_0 a critical temperature (so that we are in the high temperature region) or, in the case of a one-dimensional lattice, that the interaction Φ is of finite range. Then $E_{\mathcal{O}}$ is a well defined positivity and unit preserving map on M into M which extends to a symmetric contraction on the quantum $L_2(M, \omega_\beta)$-space. In particular, $\mathcal{L}_{\mathcal{O}} f = E_{\mathcal{O}}(f) - f$ gives a well defined Markov generator of flip-spin type dynamics on infinite quantum system on a lattice.*

Here we touch only on a few aspects of the approach presented above. Some recent results are: other jump-type dynamics can be defined, e.g., the quantum counterpart of Kawasaki dynamics (see [15]). Secondly, diffusive type dynamics can also be defined and studied (see for example [2,3,5,13]). Consequently, a large variety of dynamical maps can be given within the constructive approach to the description of time evolution.

QUANTUM CORRELATIONS AND CONCLUDING REMARKS

Among the most emblematic concepts in quantum mechanics there is the idea of entanglement (cf. [16,17]). Let us recall that this concept enters into the description of quantum correlations between subsystems (cf. [11]), so in particular it plays a crucial role in quantum mechanical considerations. This explains why a better understanding of this concept is so important.

From a formal point of view, a state of a composite quantum system is called *inseparable (or entangled)* if it cannot be represented as a tensor product of states of its subsystems. On the contrary, a density matrix describes a *separable* state if it can be expressed as a convex combination of tensor products of its subsystem states. These definitions stem from the mathematical fact that, in general, the convex hull of product states is not a (weakly) dense subset in the state space of the tensor product of two (von Neumann) C^* algebras provided that both algebras are non-commutative (cf. [7]).

To examine this concept in the framework of our approach let us consider the simplest non-trivial case: we assume that $\Lambda = \mathcal{O}_1 \cup \mathcal{O}_2$ (with $\mathcal{O}_1 \cap \mathcal{O}_2 = \emptyset$) and apply quantum spin block flip dynamics. Let the state ω be of the form $\omega(\cdot) = Tr\varrho(\cdot)$ where $\varrho = \sum_i \lambda_i \varrho_i^I \otimes \varrho_i^{II}$, with $\lambda_i \geq 0$, $\sum_i \lambda_i = 1$, and ϱ_i^I (ϱ_i^{II}) defines a state on $\mathcal{A}(\mathcal{O}_1)$ ($\mathcal{A}(\mathcal{O}_2)$ respectively). Assume $f = F \otimes \mathbf{1}^{II}$, $g = \mathbf{1}^I \otimes G$, where $\mathbf{1}^I$ is the unit in $\mathcal{A}(\mathcal{O}_1)$, etc. Then the two-point correlation functions $\langle f \cdot g \rangle_\varrho$ factorizes and the system exhibits the classical correlations only. Now let us take into account a quantum dynamics, say of block spin flip type. Then, in general, $\langle \mathcal{L}_{\mathcal{O},\Lambda}(f) \cdot g \rangle$ does not factorize—so this type of dynamics produces additional quantum correlations. Let us add that the same can be observed for other discussed dynamics, e.g., for quantum exchange type dynamics. In summary: *the presented quantum maps can enhance quantum correlations.*

We wish to end this note with a remark that the quantum L_p-space technique, being so well suited to the description of quantum maps, still admits such "classical" constructions as the Koopman's one (see [9]). Namely, let us assume that $T : \mathcal{A} \to \mathcal{A}$ is a linear map. Denote by ι_p the imbbeding of $\mathcal{A}$ into $L_p(\mathcal{A}, \omega)$. Define

$$T^{(p)}(\iota_p(a) = \iota_p(Ta) \quad a \in \mathcal{A} \tag{8}$$

We say that T is p-integrable (with respect to ω) if the induced operator $T^{(p)}$ is L_p-bounded, in which case we denote its unique extension to $L_p(\mathcal{A}, \omega)$ by the same letter. In particular, if $J : \mathcal{A} \to \mathcal{A}$ is a Jordan automorphism (i.e., a linear bijective map such that $J(a^*) = J(a)^*$ and $J(a^2) = J(a)^2$) satisfying $\omega \circ J \equiv \omega$, then $J^{(p)}$ is an integrable map; one can even show that $J^{(p)}$ is an L_p-isometry.

Even the converse of the quantum Banach-Lamperti theorem can be discussed in this framework, namely, each L_p-isometry can be determined in terms of a Jordan map (cf. [18,19]). In particular, it seems that the quantum version of the Banach-Lamperti theorem reanimates some aspects of the Misra-Prigogine-Courbage theory of irreducible phenomena. Finally, let us add that in the quantum case the point dynamics is represented by a map on C^*-algebra.

Consequently, one can conclude that the basic ingredients of the theory of classical systems, which was discussed in the introduction, can be "quantized" and successfully applied to the construction of genuine quantum maps.

REFERENCES

1. Banach S., *Théorie des Opérations Lineaires*, Monografje Matematyczne, Warszawa 1932.
2. Cipriani F., *Dirichlet Forms and Markovian Semigroups on Standard Forms of von Neumann Algebras*, Thesis Trieste 1992.
3. Davies E.B. and Lindsay M., *Math. Zeit.* **210**, 379-411 (1992).
4. Goodrich K., Gustafson K., and B. Misra, *Physica* **102A**, 379-388 (1980).
5. Goldstein S., Lindsay J. M., *Math. Zeit.* **219**, 591-608 (1995).
6. Haagerup U., L_p-spaces associated with an arbitrary von Neumann algebra in *Algèbres d'opérateurs et leurs applications en physique mathématique*, Colloques internationaux du CNRS, No. 274, Marseille 20-24 juin 1977, 175-184. Éditions du CNRS, Paris 1979.
7. Kadison R. V. and Ringrose J. R., *Fundamentals of the Theory of Operator Algebras I and II*, Pure and Applied Mathematics, vol. 100, Academic Press, New York 1983 and 1986.
8. Koopman B., *Proc. Nat. Acad. Sci. USA* **17**, 315-318 (1931).
9. Labuschagne L. E., *Expo. Math* **17**, 429 -468 (1999)
10. Lamperti J., *Pacific J. Math.* **8**, 459-466 (1958).
11. Majewski A. W., *Open Systems & Information Dynamics*, **6**, 79 (1999).
12. Majewski A.W. and Zegarlinski B., Ł$_p$ Spaces, *Lett. Math. Phys.* **36**, 337-349 (1995)
13. Majewski A.W. and Zegarlinski B., Quantum Stochastic Dynamics I: Spin Systems on a Lattice, *Math. Phys. Elect. J* **1**, (1995) Paper 2.
14. Majewski A.W. and Zegarlinski B., *Rev. Math. Phys.* **8**, 689-713 (1996).
15. Majewski A.W. and Zegarlinski B., *Markov Processes Relat. Fields* **2**, 87-116 (1996).
16. Peres A., *Quantum Theory: Concepts and Methods*, Kluwer, Dordrecht, 1993.
17. Peres A., *Phys. Rev. Lett* **77**, 1413 (1996).
18. Watanabe K., *J. Operator Theory* **28**, 267-279 (1992).
19. Yeadon F. J., *Math. Proc. Camb. Phil. Soc.* **90**, 41-50 (1981).

The $1/r^2$ Integrable System: A Universal Hamiltonian for "Complex" Level Dynamics

Pragya Shukla

Department of Physics, Indian Institute of Technology, Kharagpur, India

Abstract. We review recent studies indicating that the $1/r^2$ model of interacting particles in 1-dimension is a universal generator of the eigenvalue dynamics of complex systems. The Hermitian nature of the associated operators, e.g., Hamiltonians, is suggested to be at the root of the universality. This knowledge can be very helpful in a better understanding of physical properties which are based on the perturbation sensitivity of the eigenvalues.

In this study, we reveal the ordered structure behind the chaotic world of complex systems and use it as a tool to understand the energy level behaviour. Recent work has indicated that the distribution of the energy levels, in energy-ranges where the quantum dynamics is delocalized with an ergodic classical limit, behaves like that of the particles of the Calogero Hamiltonian H_c [1]. The response of the distribution of the levels to a chaotic perturbation is also governed by an evolution equation which can be mapped to the time-dependent Schrödinger equation for the states of H_c [1]. Here the strength of the perturbation in the former plays the role of time in the latter. The latter is a generator of ordered motion of the particles interacting with each other via an inverse square interaction [2]. The similarity therefore suggests that the levels of a complex system respond, at least statistically, in an ordered and correlated way to an external perturbation. This tendency being independent of the nature of complexity results in universal level-statistics [3].

The ordered statistical evolution of 'ergodic' levels motivates us to seek a similar statistical tendency for the levels of a classically non-integrable system for which the corresponding quantum dynamics is localized; here the non-integrability need not be in a fully ergodic regime. The idea for such behaviour has its roots in the following knowledge. The response of the individual levels to changing non-integrability satisfies equations which are similar to the classical equations of motion of particles governed by an integrable Hamiltonian H_c of Calogero type [4]. Thus if individual levels can undergo an ordered motion in the parametric space (non-integrability), this feature should manifest itself in their statistical behaviour, too.

CP553, *Disordered and Complex Systems*, edited by P. Sollich, et al.

Our attempts in this direction show that the distribution of the levels due to a change in the nature (or degree) of the complexity can again be mapped to the non-stationary states of H_c. Here the time in the latter is now mapped to a function of the degree or nature of the complexity, in other words non-integrability, of the system (instead of perturbation strength as in the fully ergodic case). The mapping, along with the detailed analysis of the non-stationary states, can thus provide us with a new technique for the statistical analysis of complex level spectra.

Recently RM ensembles have been used to model physical systems with complicated interactions [7,8]. The logic in support of the model is that the missing information about the interactions can be mimicked by randomizing the associated generators of motion, that is, by taking their matrix-representations as random matrices. However, as the specific details of the complexity of an operator should be reflected in the associated RM model, the distribution of the matrix elements can be of various types. For example, for a Hamiltonian with non-integrable classical limit (least predictability of the long-term dynamics), the distribution can be chosen as Gaussian (the least information ensemble), with distribution parameters to be determined by the associated quantum dynamics. The corresponding RM model will thus belong to a generalized Gaussian ensemble (GGE) with matrix elements distribution given by $P(H) \propto e^{-f_1(H)\cdot f_2(H)}$ (with f_1 and f_2 arbitrary functions and H a typical matrix). The standard Gaussian ensembles (SGE) with matrix elements distribution given by $P(H) \propto e^{-\mathrm{Tr}H^2}$ are special cases of GGEs and many of their properties are already known. The various features of GGEs have, however, remained unknown so far. Similarly, non-Gaussian ensembles which can serve as good models for complex systems with various conditions on the associated quantum dynamics, are also needed. In fact the choice of a suitable random matrix model is very sensitive to the nature of its complexity. The statistical spectral analysis requires therefore a thorough probing of a wide range of random matrix ensembles, which is not easy. It is desirable, therefore, to identify a common mathematical structure among all the ensembles and analyze it to gain information about their properties. Our successful search in this direction leads to the Calogero-Moser Hamiltonian H_{cm} [9]. This is because both the eigenvalues of the ensembles, and a general state of H_{cm}, evolve in a similar way for arbitrary initial conditions. The varying nature of the complexity is reflected in different forms of the evolution parameter in each case. A complete investigation of H_{cm} can therefore help us in the spectral analysis of complex systems.

The evolution of the eigenvalues $\lambda_i(x)$ of a non-integrable Hamiltonian $H(x)$ as a function of x can be expressed as a set of first order differential equations in the continuously changing eigenbasis of $H(x)$ perturbed by $H'(x) \equiv \partial H(x)/\partial x$ [4–6]:

$$\frac{\mathrm{d}\lambda_i}{\mathrm{d}x} = H'_{ii}, \quad \frac{\mathrm{d}H'_{ii}}{\mathrm{d}x} = -\sum_{j\neq i} \frac{2 f_{ij} f_{ji}}{(\lambda_i - \lambda_j)^3} \tag{1}$$

$$\frac{\mathrm{d}f_{ij}}{\mathrm{d}x} = \sum_{k\neq i,j} f_{ik} f_{kj} \left(\frac{1}{\lambda_i - \lambda_k} + \frac{1}{\lambda_j - \lambda_k} \right) \tag{2}$$

where $f_{ij} = (\lambda_i - \lambda_j)H'_{ij}(x)$. These equations describe an incompressible flow of the variables $y_i \equiv (\lambda_i, H'_{ii}, f_{ij})$ in terms of the parameter x: As is obvious, the equation for each of the three kinds of velocity variables $\dot{y}_i = \frac{\mathrm{d}y_i}{\mathrm{d}x}$, with $y_i \equiv \lambda_i, H'_{ii}, f_{ij}$, does not contain y_i and therefore $\partial \dot{y}_i / \partial y_i = 0$ and $\mathrm{div}\, \dot{y} = 0$. A Hamiltonian flow is typical of such divergenceless flows and, indeed, the first two equations of motion (1-2) are of canonical form with a Hamiltonian $\hat{H} = \frac{1}{2}\sum_i p_i^2 + \frac{1}{2}\sum_{i,j;i \neq j} \frac{f_{ij}^2}{(\lambda_i - \lambda_j)^2}$ where the canonical momentum conjugate to λ_i is defined by $p_i = H'_{ii}$. (In order to distinguish between the Hamiltonians associated with the system and the eigenvalues, the latter will always be denoted $\hat{H}$). The Hamiltonian $\hat{H}$ is known as a generalized Calogero-Moser Hamiltonian and is integrable in nature. The eqs.(1-2) can therefore be viewed as a classical integrable system, with N degrees of freedom corresponding to the eigenvalues λ_i, and $N + N(N-1)/2$ degrees corresponding to the diagonal and off-diagonal elements of H', obeying appropriate Poisson bracket relations. Note that eqs.(1-2) are valid in general for any non-integrable system which implies that the eigenvalues of the corresponding quantum system evolve in an ordered way although the classical dynamics may become increasingly chaotic.

In principle, the trajectories for each eigenvalue in the parametric space x can be obtained by integrating their equations of motion (eqs.(1-2)); however, a statistical approach is more useful when dealing with systems having a large number of eigenvalues. Let us consider the statistical evolution of the eigenvalues of a system due to changing nature or degree of complexity. Under maximum entropy conditions, the effect of changing complexity on the eigenvalue statistics of the Hamiltonian can be modelled by the effect of the change in the distribution parameters of a GGE on their eigenvalues [9]. The latter analysis requires prior information about the effect of a small change in the matrix elements on the eigenvalues and eigenvectors; the related study is given in [9]. These results can then be used to study the evolution of the distribution $P(\mu, Y)$ of eigenvalues. The $P(\mu, Y)$ is governed by a partial differential equation which, after certain parametric redefinitions, turns out to be formally the same as the Fokker-Planck (FP) equation for the Brownian motion of particles interacting with each other via two-dimensional Coulomb interaction [8].

The steps can briefly be described as follows. Let us consider an ensemble of real symmetric matrices H, of dimension N, with matrix elements H_{kl} distributed as Gaussians with arbitrary variances and mean values. The distribution $\rho(H)$ of matrix H can therefore be taken as $\rho(H, y, b) = C \exp(-\sum_{k \leq l} \alpha_{kl}(H_{kl} - b_{kl})^2)$ with C as the normalization constant, y as the set of the coefficients $y_{kl} = \alpha_{kl} g_{kl} = \frac{g_{kl}}{2 < H_{kl}^2 >}$ and b as the set of all b_{kl}. Note that such a choice leads to a non-random Hamiltonian ($H_{kl} = b_{kl}$) in the limit $\alpha_{kl} \to \infty$ and therefore can model various real physical situations such as switching of disorder in a non-random Hamiltonian, e.g., metal-insulator transitions.

Let $P(\mu, y, b)$ be the probability of finding eigenvalues λ_i of H between μ_i and $\mu_i + \mathrm{d}\mu_i$ at a given y and b: $P(\mu, y, b) = \int \prod_{i=1}^{N} \delta(\mu_i - \lambda_i)\ \rho(H, y, b)\ \mathrm{d}H$ Our aim is to find a function Y of the coefficients α_{kl} and b_{kl} such that the evolution of $P(\mu, Y)$ in terms of Y satisfies a F-P equation similar to that of Dyson's Brownian motion model (Wigner-Dyson gas) [8]. For this purpose, we consider an evolution of P given by the sum S where $S \equiv 2\sum_{k\le l}(\gamma - y_{kl})\, y_{kl}\frac{\partial P}{\partial y_{kl}} - \gamma\sum_{k\le l} b_{kl}\frac{\partial P}{\partial b_{kl}}$ and γ is an arbitrary parameter. In this case, we use integration by parts, along with the eigenvalue equation $HU = U\Lambda$ with Λ as the matrix of eigenvalues λ_n and U as the unitary eigenvector matrix. Then P can be shown to satisfy following equation,

$$\frac{\partial P}{\partial Y} = \gamma\sum_n \frac{\partial}{\partial \mu_n}(\mu_n P) + \sum_n \frac{\partial}{\partial \mu_n}\left[\frac{\partial}{\partial \mu_n} + \sum_{m\neq n}\frac{\beta}{\mu_m - \mu_n}\right] P \tag{3}$$

with $\beta = 1$. A similar equation can also be obtained for the complex Hermitian Hamiltonians ($\beta = 2$) by considering an ensemble of matrices $H \equiv \{H_{kl;1} + iH_{kl;2}\}$ with distribution $\rho(H) \propto \exp(-\sum_{k\le l}\sum_{s=1}^{2}\alpha_{kl;s}(H_{kl;s} - b_{kl;s})^2)$ The left hand side of eq.(3) is same as the sum S which has been rewritten as $\frac{\partial P}{\partial Y}$ with Y given by the condition that

$$\frac{\partial P}{\partial Y} = \sum_{i=1}^{M}\frac{\partial P}{\partial X_i} \equiv 2\sum_{k\le l} y_{kl}(\gamma - y_{kl})\frac{\partial P}{\partial y_{kl}} - \gamma\sum_{k\le l} b_{kl}\frac{\partial P}{\partial b_{kl}}.$$

where $M \equiv N(N+1)$, $X_i = -\frac{1}{2}\ln(1 - \gamma y_{kl}^{-1}) + c_i$ for $i \le N(N+1)/2$ and $X_i = -\gamma^{-1}\ln b_{kl} + c_i$ for $i > N(N+1)/2$, and the c_i are arbitrary constants.

A parameter Y satisfying the above condition can be found by considering a transformation from the M-dimensional parameter space $\{X_i\}$ to another M-dimensional space $\{Y_1, \ldots, Y_M\}$ such that $Y_1 \equiv Y$ behaves as a center of mass variable in X-space; $Y_1 = \frac{1}{M}\sum_{i=1}^{M} X_i$. The Y_i, $i = 2, \ldots, M$ can be taken as the relative distances between the various X_i (for example, $Y_2 = X_1 - X_2, Y_3 = X_1 - X_3$ etc.) or more generally as $Y_i = \sum_j a_{ij}X_j$ with the condition that $\sum_{ij} a_{ij} = \delta_{1i}$ [9].

This can be clarified by a simple example: let us consider the case $N = 2$ with three parameters y_{11}, y_{12} and y_{22} in the ensemble $\rho(H) \propto e^{-\sum_{i,j=1;i\le j}^{2} H_{ij}^2}$. In this case,

$$X_1 = -\frac{1}{2}\ln(1 - \gamma y_{11}^{-1}) \qquad X_2 = -\frac{1}{2}\ln(1 - \gamma y_{12}^{-1}) \qquad X_3 = -\frac{1}{2}\ln(1 - \gamma y_{22}^{-1}),$$

and

$$S \equiv 2\sum_{k\le l}(\gamma - y_{kl})\, y_{kl}\frac{\partial P}{\partial y_{kl}} \equiv \sum_{i=1}^{3}\frac{\partial P}{\partial X_i}.$$

Now consider three variables Y_1, Y_2, Y_3 such that $X_1 = Y_1 + Y_2 + Y_3$, $X_2 = Y_1 - Y_2$ and $X_3 = Y_1 - Y_3$. It can easily be checked that if the sum S is written in terms of Y_1, Y_2 and Y_3, one gets $S = \frac{\partial P(Y_1,Y_2,Y_3)}{\partial Y_1}|_{Y_2,Y_3=\text{constant}}$. Thus the evolution curve of (y_{11}, y_{12}, y_{13}) gets mapped to a straight line parallel to the Y_1 axis. As Y_2, Y_3 remain constant during the evolution described by eq.(3), the choice of initial ensemble will fix the constant values of Y_2 and Y_3. However if the form of y_{11}, y_{12} is changed, for example, from power law to exponential, Y_2, Y_3 will take other constant values which implies now the transition will occur on a different straight line parallel to the Y_1 axis. Note that the different ensembles may correspond to the same value of Y_1 but different constant values of Y_2, Y_3. As the evolution of eigenvalues is governed by eq.(3) involving a variation of $Y_1 \equiv Y$ only, with the right hand side of eq.(3) remaining the same for different constant values of Y_2, Y_3, the different ensembles will have the same form for $P(\mu, Y_1, Y_2, Y_3)$ and thereby various correlations if their Y_1 values are equal. Thus in general the evolution of P, given by eq.(3), corresponds to the evolution due to the centre-of-mass motion in X-space with the relative distances held constant. Here the parameter $Y = \frac{1}{2N^2}\sum_{k\leq l}\left[\frac{1}{2}\ln\frac{y_{kl}}{|y_{kl}-\gamma|} - \frac{1}{\gamma}\ln|b_{kl}|\right] + \text{const.}$, a function of the relative values of the coefficients y_{kl} and b_{kl}, is a measure of the degree and nature of the complexity of a system and therefore it can be referred to as the "complexity parameter".

The steady state of eq.(3), $P(\mu, \infty) \equiv P_\infty = |\Delta(\mu)|^\beta e^{-\frac{\gamma}{2}\sum_k \mu_k^2}$, corresponds to $Y - Y_0 \to \infty$ (with Y_0 as the complexity parameter of the initial ensemble) This indicates that, in the steady state limit, the system tends to belong to one of the SGEs. The eq.(3) can, therefore, describe a transition from a given initial ensemble (with $Y = Y_0$) to either a GOE or GUE with $Y - Y_0$ as the transition parameter. The nonequilibrium states of this transition, given by non-zero finite values of $Y - Y_0$, are various Gaussian ensembles corresponding to varying values of the α_{kl} and b_{kl}, thus modelling different complex systems. The eq.(3) for $P(\mu, Y)$ can also be used to obtain the nth order density correlator $R_n(\mu_1, \ldots, \mu_n; Y)$, defined by $R_n = \frac{N!}{(N-n)!}\int P(\mu, Y)\mathrm{d}\mu_{n+1}\ldots\mathrm{d}\mu_N$. In the thermodynamic limit $N \to \infty$, the approach to equilibrium is abrupt and a rescaling of Y by the mean spacing $D(D = 1/R_1)$ is required to see a smooth transition, the new transition parameter being $\Lambda = \frac{Y-Y_0}{\gamma D^2}$.

The connection to the Calogero-Moser Hamiltonian can be shown as follows. It is easy to see that eq.(3) can be written as (with $\gamma = 1$, for simplicity)

$$\frac{\partial P}{\partial Y} = \sum_n \frac{\partial}{\partial \mu_n}|Q_N|^\beta \frac{\partial}{\partial \mu_n}\frac{P}{|Q_N|^\beta} \qquad \text{where} \qquad |Q_N|^\beta = |\Delta(\mu)|^\beta e^{-\frac{1}{2}\sum_k \mu_k^2}$$

with $\Delta(\mu) = \prod_{i<j}(\mu_i - \mu_j)$. The transformation $\Psi = P/|Q_N|^{\beta/2}$ can then be used to cast it in the suggestive form, $\frac{\partial\Psi}{\partial Y} = -\hat{H}\Psi$, where the 'Hamiltonian' $\hat{H}$ turns out to be the Calogero-Moser Hamiltonian:

$$\hat{H} = -\sum_i \frac{\partial^2}{\partial \mu_i^2} + \frac{1}{4}\sum_{i<j}\frac{\beta(\beta-2)}{(\mu_i - \mu_j)^2} - \sum_i V(\mu_i).$$

With the parabolic-confining potential and under the requirement (to take account of the singularity in H) that the solutions vanish as $|\mu_i - \mu_j|^{\beta/2}$ when μ_i and μ_j are close to each other, $\hat{H}$ has well-defined (completely symmetric) eigenstates and eigenvalues [12]. The "state" ψ or $P(\mu, Y)$ can therefore be expressed as a sum over eigenvalues and eigenfunctions which on integration over the initial ensemble $\rho(H_0)$ leads to the joint probability distribution $P(\mu, Y)$ and thereby static (at a single parameter value) density correlations R_n. The above correspondence can also be used to map the multi-parametric correlations of levels to multi-time correlations of the particle-positions.

The choice of the initial ensemble depends on the nature of the quantum dynamics associated with the physical system. For example, if the changing complexity results in a localization to delocalization transition (and vice versa), it is appropriate to choose $\rho(H_0)$ as that of the diagonal matrices ($\rho(H_0) \propto e^{-\sum_i [H_0]_{ii}^2}$) which corresponds to a Poisson distribution for the eigenvalues μ_0 and fully localized eigenstates. The equilibrium distribution ($Y - Y_0 \to \infty$), given by SGE, corresponds to complete delocalization of eigenstates. The localization $\to$ delocalization transition can therefore be mapped onto a time-evolutin of a general state ψ of CM Hamiltonian with "time" $Y - Y_0$ from an initial state $\psi_0 = P(\mu_0)/|Q(\mu_0)|^{\beta/2}$. Here the intermediate states of localization (represented by various RMEs depending on the values of y_{kl}'s and b_{kl}'s) correspond to the nonstationary states at various Y values. The eigenvalue correlations for various localization cases will therefore be similar to the particle-correlations in the nonstationary states. However the latter are known so far only for $\beta = 2$ although partial information about two point particle-correlations is available also for $\beta = 1, 4$ [1].

Fortunately the two-point correlation $\tilde{Y}_2(r;\Lambda)$ for the CM Hamiltonian $\beta = 2$ with initial state of particles given by a Poisson distribution has already been obtained [1] and, as discussed above, is also valid for the the complex systems which can be modelled by the generalized Gaussian ensembles of Hermitian matrices:

$$\tilde{Y}_2(r;\Lambda) - \tilde{Y}_2(r;\infty) = \frac{4}{\pi}\int_0^\infty \mathrm{d}x \int_{-1}^{1} \mathrm{d}z\, \cos(2\pi r x)\exp\left[-8\pi^2\Lambda x(1+x+2z\sqrt{x})\right] \times \left(\frac{\sqrt{(1-z^2)}(1+2z\sqrt{x})}{1+x+2z\sqrt{x}}\right).$$

Here $\Lambda = (Y - Y_0)/D^2$ and $\tilde{Y}_2(r,\infty) = \frac{\sin^2(\pi r)}{\pi^2 r^2}$ (the GUE limit). As can easily be checked, the above equation has the correct limiting behaviour, that is, $\tilde{Y}_2 = 1$ for $\Lambda \to 0$ (the Poisson case) and $\tilde{Y}_2 = \tilde{Y}_2(r;\infty)$ for $\Lambda \to \infty$. As is obvious, $\tilde{Y}_2$ depends on Y and therefore on the nature of localization which, for finite Λ-values (in the limit $N \to \infty$), results in various types of level-statistics intermediate between Poisson and GUE. For $r \ll \sqrt{\Lambda}$, $\tilde{Y}_2$ can be approximated as $\tilde{Y}_2 = \frac{6}{\pi^2\Lambda}\frac{\sin^2 \pi r}{\sinh^2 r\sqrt{6/\Lambda}}$ which is similar to the result given by the SUSY technique for a GGE with power law decay of variances [13]. As is obvious from the form of $\tilde{Y}_2$, it is very different from both Poisson as well as GUE for finite Λ-values. This indicates that the ensembles with distribution parameters giving rise to a finite Λ (in the limit $N \to \infty$) do not reach stationarity even for infinite size of their matrices, and since their properties are very different from those of the equilibrium ensembles, they can be referred to as "critical" [13].

The $\tilde{Y}_2$-behaviour can also be used to obtain the level compressibility χ [13]: $\chi(\Lambda) = 1 - \int_{-\infty}^{\infty} \mathrm{d}r \tilde{Y}_2(r;\Lambda)$. Thus $\chi(\Lambda)$, being a smooth function of Λ, would smoothly change between its values at $\Lambda = 0$ ($\chi = 1$) and at $\Lambda \to \infty$ ($\chi = 0$) and for any non-zero, finite Λ value $0 < \chi(\Lambda) < 1$. As a fractional value of χ is an indicator of the multifractality of the associated eigenvectors [13], any ensemble with finite Λ value will support multifractality of eigenvectors.

In this paper, we have reviewed the equivalence between the eigenvalue motion of systems with changing degree or nature of the complicated interactions and the dynamics of particles with $1/r^2$ interaction in one dimension. We find that the $1/r^2$ model of interacting particles in one dimension is a universal Hamiltonian governing the eigenvalue motion. Note that the main term in the RM model of complex systems, responsible for the correspondence with the CSM Hamiltonian, is due to the repulsion between eigenvalues. Its origin lies in the transformation from matrix space to eigenvalue space; since this is the same for all Hermitian ensembles (belonging to the same symmetry class), the correspondence with the CSM Hamiltonian should persist for almost all of them irrespective of the distribution of their matrix elements. Also note that a suitable rescaling of the system Hamiltonian $H(x)$ will lead to a confinement of the eigenvalues within a quadratic potential [6]. The motion of each eigenvalue can then be shown to be governed by a generator which is similar to $\hat{H}$ governing the evolution of the statistical state of the eigenvalues; the only difference lies in the coefficient of the $1/r^2$ interaction in the two cases. In the first case, the coefficient is a function $|f_{ij}|^2$ depending on the internal degree of freedom [6,5], and in the second case, a constant $\beta(\beta - 2)$ depending on the degrees of freedom for each matrix element of the system Hamiltonian. Thus intuition suggests a relation between the average value of $|f_{ij}|^2$ and β and can give us some information about the missing ensembles corresponding to other β-values.

REFERENCES

1. Pandey, A., *Chaos, Soliton and Fractals* **5** 1275 (1995).
2. Calogero, F., *J. Math. Phys.* **10**, 2191 (1969).
3. Shukla, P., *Phys. Rev. E* **59**, 5207 (1999).
4. Nakamura, K., and Lakshmanan M., *Phys. Rev. Lett.* **57**, 1661 (1986).
5. Hasegawa, H., and Ma J.-Z., *J. Math. Phys.* **39**, 2564 (1998).
6. Haake, F., *Quantum Signature of Chaos*, Springer, Berlin, 1991.
7. Guhr, T., Muller-Groeling A., and Weidenmuller H.A. *Phys. Rep.* **299** 189, (1998).
8. Mehta, M.L., *Random Matrices*, Boston, Academic Press, 1991.
9. Shukla, P., *Phys. Rev. E* bf 62, 2098 (2000).
10. Narayan, O., and Shastry, B.S., *Phys. Rev. Lett.* **71**, 2106 (1993).
11. Shastry, B.S., "The $1/r^2$ Integrable Hamiltonian: A Universal Hamiltonian for Quantum Chaotic Systems", in *Correlated effects in low dimensional systems—1993*, edited by N. Kawakami and A. Okiji, Proceedings of 16th Taniguchi International Symposium on the Theory of Condensed Matter, October 23-29, 1993, Shima, Japan.
12. Ha, Z.N.C., *Phys. Rev. Lett.* **73**, 1574 (1994) and *Nucl. Phys. B* **435**, 604 (1995).
13. Kravtsov, V.E., and Muttalib, K.A., *Phys. Rev. Lett.* **79**, 1913 (1997).

The Distribution of the Modes of Coupled Quartic Oscillators

Mitsuyoshi Tomiya* and Naotaka Yoshinaga†

*Department of Applied Physics, Faculty of Engineering, Seikei University, Kichijyoji-Kitamachi 3-3-1, Musashinoshi, Tokyo 180-8633, Japan.
†Department of Physics, Faculty of Science, Saitama University, Shimo-okubo 255, Urawashi, Saitama 338-0825, Japan.

Abstract. The mode fluctuation distribution (MFD) of quartic oscillators coupled by quartic perturbations is studied numerically. Changing a single coupling strength, the system can be continuously transferred from the integrable to the chaotic regime. It is demonstrated that the MFDs in two- and three-dimensional potential systems are a very sensitive measure of integrability, as in billiard systems. On the other hand, even in the region much before the chaotic volume saturates, the MFD already becomes almost a Gaussian distribution, as in the completely chaotic regime.

INTRODUCTION

Recently the mode fluctuation distribution (MFD) has been proposed as an alternative tool that detects the chaotic nature of systems [1,2]. It has been predicted that the MFD can reflect the chaoticity of the system more faithfully than the nearest neighbor spacing distribution (NNSD). It was also shown numerically that the MFD in billiard systems works fairly well [3,4]. Up to now, most studies in quantum chaology have been carried out in two-dimensional systems, which provide the most elementary and simplest examples of chaotic systems. To study more realistic systems, we should proceed to the investigation of three-dimensional systems. In the billiard systems the contribution of families of neutral periodic orbits (quasi stable orbits) generally dominates the fluctuation properties, especially in three dimensions or higher. Though the contribution of isolated unstable orbits is generic to the chaotic properties of the system, it becomes harder to observe because of the "sea" caused by those neutral orbits. Then Primack et al. ought to have successfully developed a skillful technique to eliminate the contribution of neutral orbits [5,6].

On the other hand, smooth potential systems essentially have no family of neutral orbits irrespective of its dimensions. It is expected that only the generic orbits (isolated unstable orbits) contribute to the fluctuation properties by nature. This

CP553, *Disordered and Complex Systems*, edited by P. Sollich, et al.

is an advantage of potential systems. Therefore we investigated a two-dimensional potential system of quartic oscillators [7]. In this talk the MFD in two and three dimensions is studied for a smooth coupled quartic oscillator potential system [8].

MODE FLUCTUATION DISTRIBUTION

A spectral staircase function $N(E) \equiv \sum_{n=1}^{\infty} \theta(E - E_n)$ and a spectral density $d(E) \equiv dN(E)/dE$ are defined by a discrete energy spectrum $\{E_n\}$ of a quantum system with a bounded potential. The staircase function can be separated into a mean smooth part $\langle N(E) \rangle$ and a mode fluctuation part $N_{fl}(E) : N(E) = \langle N(E) \rangle + N_{fl}(E)$. Note that the bracket $\langle \ldots \rangle$ denotes the average over an interval which is much larger than the mean energy level spacing $\langle d \rangle^{-1}$ and sufficiently smaller than the energy E under consideration.

As in the billiard systems [1,2,9] and also in coupled oscillators [7,8], our system is expected to have a saturated value $\Delta_\infty(E)$ of the spectral rigidity $\Delta_3(L, E)$ at a sufficiently large L. The saturated value turns out to be the second moment of the fluctuation part

$$\Delta_\infty(E) \Rightarrow \left\langle \frac{\langle d \rangle}{L} \int_{-L/2\langle d \rangle}^{L/2\langle d \rangle} N_{fl}(E + \varepsilon)^2 \, d\varepsilon \right\rangle \quad \text{at } L \gg L_{\max}, \tag{1}$$

where $L_{\max}$ corresponds to a scale $h \langle d \rangle / T_{\min}$, which is an energy scale $h/T_{\min}$ normalized by the mean energy level spacing $\langle d \rangle^{-1}$ and $T_{\min}$ is the period of the shortest classical closed orbit in the system. The MFD is the distribution of the fluctuation part $N_{fl}(E)$ normalized by the second moment, i.e., the variable

$$W(E) = \frac{N_{fl}(E)}{\sqrt{\Delta_\infty(E)}}. \tag{2}$$

Thus its average must be zero and its variance must be 1 by definition. It is hypothesized that its distribution becomes Gaussian when the system is chaotic.

COUPLED QUARTIC OSCILLATORS

In this talk, we consider systems of two- and three-dimensional quartic oscillators with a single coupling constant λ. The Hamiltonian we assume is

$$H = \frac{1}{2}\left(p_x^2 + p_y^2 + p_z^2\right) + V(x, y, z), \tag{3}$$

where the potential is given as

$$V(x, y, z) = 3x^4 + 2y^4 + z^4 - \lambda\left(x^2y^2 + y^2z^2 + z^2x^2\right). \tag{4}$$

Note that we set dimensionless units: $\hbar = m = 1$ $(h = 2\pi)$, where m is the mass of a classical point particle in the system. A non-zero value of λ distorts the three-dimensional quartic potential, but keeps the potential $V(x, y, z)$ homogenous. If we set $y = 0$ and $p_y = 0$, this becomes our two-dimensional system. Thus chaoticity does not depend on an energy range considered. For $\lambda = 0$, this system becomes separable, that is, a superposition of three integrable systems with quartic potentials. From the viewpoint of quantum mechanics, if $\lambda > 1.75877\ldots$ (3-dim) ($\lambda > 2\sqrt{3} \cong 3.464\ldots$ (2-dim)), the potential $V(x, y, z)$ no longer forms a bound quantum system. Then the search for a discrete spectrum of the system becomes meaningless. Therefore, we must keep $\lambda \leq 1.75877\ldots$ (3-dim) ($\lambda \leq 3.464\ldots$ (2-dim)) in order to keep a bound system.

Our Hamiltonian has the classical scaling property [7,8,10,11]. Owing to the homogenous potential $V(x, y, z)$, the system is invariant under the transformation

$$E \to a^4 E, \quad x \to ax, \quad y \to ay, \quad z \to az, \quad t \to t/a. \tag{5}$$

Under the scaling (5) the classical dynamical properties, e.g., the Poincaré surface of sections, the equipotential surfaces and the orbits in the phase space, etc., are geometrically similar. The closed orbit of length ℓ and period T is transformed as $\ell \to a\ell$ and $T \to T/a$, if $E \to a^4 E$. Thus it is helpful not only for classical dynamics but also for semiclassical analysis of quantum properties. However, the search for classical periodic orbits, even the second shortest orbit, is difficult.

In two dimensions, the Poincaré section is very instructive for seeing the structure of phase space classically [7]. On the other hand, in three dimensions, the Poincaré section is not as powerful a tool as it is in two dimensions [8]. The energy surface is five-dimensional in the six-dimensional phase space. In order to make a two-dimensional Poincaré section, three additional constraints are necessary in general and it is likely that some arbitrary orbits have zero measure on the Poincaré surface of section. This means that the majority of the orbits cannot hit the section. Therefore in order to investigate the classical dynamical features of our system, we calculate maximal Lyapunov exponents of classical orbits whose initial data are randomly generated under the condition of the same energy [12]. At $h \approx 0$, the distribution of the maximal Lyapunov exponents, of course, has only one bunch around zero Lyapunov exponent. It implies that for $\lambda < 1.0$ the system is simply hyperbolic. Around $0.4 < \lambda < 1.0$, the distribution shows a gradual transition toward a simply hyperbolic system. In the region $0.0 < \lambda < 0.4$, the distribution of Lyapunov exponents implies a complicated structure of the phase space [8].

We compute quantum energy levels by numerical diagonalization of the truncated matrix of the Hamiltonian (3) in the basis of three (two) independent harmonic oscillators [13]. We calculate 2500-5000 (3000-5000) eigenvalues from 23000- (8500-15000)-dimensional truncated matrices of the Hamiltonian (3). The convergence of the calculated eigenvalues has been confirmed by changing the frequencies of the three (two) harmonic oscillators. If we take larger λ, the convergence becomes worse and fewer eigenvalues from the ground state can be reliable. In three dimensions, the level density is much higher than that in two dimensions. Much higher precision is necessary to numerically evaluate up to about the same number of eigenvalues. At least six significant digits are required to keep the spectral statistics from becoming meaningless. We have to pay much attention to the convergence of the numerical diagonalization. Therefore we prepare a larger basis for the truncated matrix of the Hamiltonian (3) than in two dimensions. The Hamiltonian has a discrete symmetry and only eigenvalues corresponding to wave functions of the odd-odd-odd (odd-odd) parity are calculated. This means that what we calculate is the desymmetrized system: one eighth of the system, where a classical point particle is confined in the region $x \geq 0$, $y \geq 0$ and $z \geq 0$, and is reflected by walls at $x = 0$, $y = 0$ and $z = 0$.

RESULTS AND CONCLUTIONS

The Brody parameter is estimated numerically for each value of λ, using 2500-5000 (3000-5000) eigenvalues from the ground state [14]. For $\lambda > 1.0$ the value of the Brody parameter varies widely in a range of about 0.2. Considering the precision of our calculation in estimating the Brody parameter, however, it is unlikely due to some unknown physical complexity rather than real fluctuation. The convergence of eigenvalues in a numerical diagonalization of the truncated Hamiltonian becomes worse in a region of larger λ. The λ-dependence of the Brody parameter in three dimensions shows that the NNSD transforms smoothly from the Poisson distribution to the Wigner distribution [8]. There is no plateau which is the effect of the separation of the phase space by KAM tori as in two-dimensional systems, where they prevent the phase space from overall mixing [7]; in three-dimensional systems, the Arnold diffusion among KAM tori avoids the separation. This shows a clear contrast to the two-dimensional results and also the good correspondence between the properties of classical dynamics and its quantum counterpart [7,8].

In the exactly integrable case $\lambda = 0$, the saturated spectral rigidity is semiclassically estimated [15] in three (two) dimensions as

$$\Delta_\infty(E) = \frac{9}{2\pi^2} \sum_{m,n,k=1}^{\infty} \frac{m^2 n^2 k^2 / C_1^3 C_2^3 C_3^3}{\left(\frac{m^4}{C_1^3} + \frac{n^4}{C_2^3} + \frac{k^4}{C_3^3}\right)^{5/2}} \cdot E^{3/2} \cong 0.0030 E^{3/2} \quad \text{(3-dim)} \tag{6}$$

$$\Delta_\infty(E) = \frac{3}{2\pi^2} \sum_{m,n=1}^{\infty} \frac{m^2 n^2 / C_1^3 C_3^3}{\left(\frac{m^4}{C_1^3} + \frac{n^4}{C_3^3}\right)^{7/4}} \cdot E^{3/4} \cong 0.0257 E^{3/4} \quad \text{(2-dim)} \tag{7}$$

where $C_1 = 3^{1/3} \cdot \left(\frac{3\pi}{2K}\right)^{4/3}$, $C_2 = 2^{1/3} \cdot \left(\frac{3\pi}{2K}\right)^{4/3}$, $C_3 = \left(\frac{3\pi}{2K}\right)^{4/3}$, $K = F\left(\frac{\pi}{2}, \frac{1}{\sqrt{2}}\right) \cong 1.85407$ and with F standing for the elliptic function of the first kind. On the other hand, from numerically estimated energy levels we have $\Delta_\infty(E) = 0.0028E^{3/2} + 1.34$ $(0.0254E^{3/4} + 0.146)$ [7,8], which conforms to the semiclassical analysis (6) ((7)). This implies that the Berry-Tabor treatment [15] works fairly well even in the case

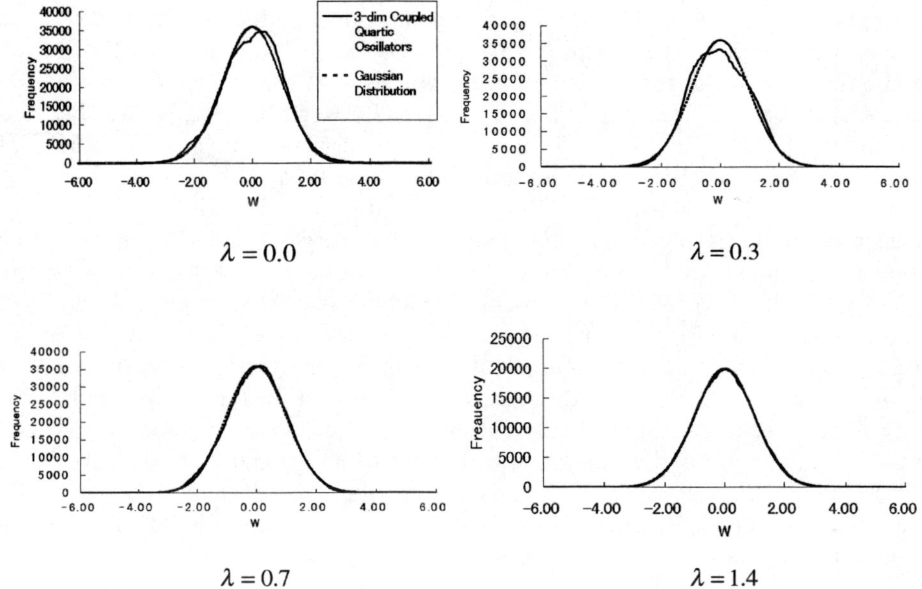

FIGURE 1. The mode fluctuation distributions of three-dimensional coupled quartic oscillators which are gven by the histogram of the variable $W(E)$.

of potential systems. Whereas if the system is sufficiently chaotic, $\lambda = 1.4$ (2.7), the saturated spectral rigidity is also semiclassically estimated as

$$\Delta_\infty(E) \cong \frac{1}{\pi^2} \ln\left(e \frac{2\pi \langle d \rangle}{T_{\min}}\right) - \frac{1}{8} \quad = 0.152 \ln E - 0.195 \quad \text{(3-dim)} \tag{8}$$

$$(= 0.0760 \ln E - 0.0616 \quad \text{(2-dim)}), \tag{9}$$

where we have used the energy dependence $T_{\min} \cong 1.41E^{-1/4}$ for the shortest period, which is evaluated by searching classical periodic orbits numerically, and the level density $\langle d \rangle \cong 0.0412E^{5/4}$ $(0.152E^{1/2})$. The latter is estimated from the calculated quantum energy levels, and e is the Euler number. From numerically obtained energy levels, the energy dependence of $\Delta_\infty(E)$ is well fitted as $\Delta_\infty(E) = 0.171 \ln E - 0.068$ at $\lambda = 1.4$ ($\Delta_\infty(E) = 0.0802 \ln E - 0.0454$ at $\lambda = 2.7$). Including our result of the integral limit, this means that the semiclassical treatment works as well in three-dimensional systems [8] as in two-dimensional system [7].

In Fig.1 we show the MFDs of three dimensional coupled quartic oscillators. At λ=0.0, the MFD shows a skewed distribution, which is a characteristic feature of integrable systems [4,7,8,16]. However, it changes rapidly as λ increases. At $\lambda = 0.3$ the distribution of numerical maximal Lyapunov exponents shows that the system is still fairly near integrable, although its MFD already has a distorted form. Then at $\lambda = 0.7$, the MFD is almost like the Gaussian distribution, but the distribution of the exponents means the system is still in between the integrable and the chaotic regimes. Finally at $\lambda = 1.4$, the NNSD becomes almost the Wigner distribution and the distribution of the exponents also implies that the system is chaotic. For the case of two dimensions, see [8]. The relations between the NNSD and the MFD are roughly the same in the two- and three-dimensional systems. The MFD is shown to be more sensitive to integrability than chaoticity [7,8].

In this talk we have studied the spectral fluctuation properties of two- and three-dimensional potential systems. Particularly the MFD of the three-dimensional system is examined for the first time [8]. We found that the MFD is less faithful to chaoticity than the NNSD in potential systems [5], which was also indicated in billiard systems [17]. We showed that the semiclassical treatment works also in three dimensions, at least when our system is either exactly integrable or sufficiently chaotic. We have found the sign of the Arnold diffusion in the NNSD. The Brody parameter increases gradually without a plateau as the coupling constant is changed [8]; it shows a clear contrast to the two-dimensional case. In conclusion, the MFD is much more sensitive to integrability than to chaoticity in our potential systems, as in billiard systems. Thus the MFD should be considered as a measure of integrability of a system rather than of chaoticity.

REFERENCES

1. R. Aurich, J. Bolte and F. Steiner, Phys. Rev. Lett. 73, 1356 (1994).
2. R. Aurich, F. Scheffler, and F. Steiner, Phys. Rev. E51, 4173 (1995).
3. M. Tomiya and N. Yoshinaga, J. Phys. Soc. Jpn, 66, 3318 (1997).
4. R. Aurich, A. Bäcker and F. Steiner, Int. J. Mod. Phys. B11, 805 (1997).
5. M. Sieber, H. Primack, U. Smilansky, I. Ussishkin and H. Schanz, J. Phys. A: Math. Gen. 28, 5041 (1995).
6. N. Yoshinaga, K. Matsumura and M. Tomiya, Prog. Theor. Phys. 98, 869 (1996).
7. M. Tomiya and N. Yoshinaga, Phys. Rev. E58, 8017 (1998).
8. M. Tomiya and N. Yoshinaga, to be published in J. Phys. Soc. Jpn (2000).
9. M. V. Berry, Proc. R. Soc. London A400, 229 (1985).
10. K. M. Atkins and G. S. Ezra, Phys. Rev. E51, 1822 (1995).
11. D. Ullmo, M. Grinberg and S. Tomsovic, Phys. Rev. E54, 136 (1996).
12. J. Bolte, B.Müller and Ascäfer, Ergodic Properties of Classical SU(2) Lattice Gauge Theory, hep-lat/9906037 ver.2.
13. S. Graffi, V. R. Manfredi and L. Salasnich, Mod. Phys. Lett. B9, 747 (1995).
14. M. Tomiya and N. Yoshinaga, J. Stat. Phys. 83, 215 (1996).
15. M. V. Berry and M. Tabor, Proc. R. Soc. Lond., A349, 101 (1976); J. Phys. A: Math. Gen. 10, 371 (1977).
16. P. M. Bleher, Z. Cheng, F. J. Dyson, and J. L. Lebowitz, Commum. Math. Phys. 154, 433 (1993).
17. H. Alt, A. Bäcker, C. Dembowski, H.-D. Gräf, R. Hofferbert, H. Rehfeld and A. Richter, Phys. Rev. E58, 1737 (1998).

PART IV:
REACTION-DIFFUSION EQUATIONS

Nonlinear Relation between Diffusion and Conductivity for Levy Flights

V. E. Arkhincheev
Buryat Science Center of the Russian Academy of Sciences,
670047 Ulan-Ude, Russia

Abstract. The relation between diffusion and conductivity is studied for a Levy flight diffusion case. It is shown that due to the "anomalous" character of Levy hops the particle mobility is a nonlinear function of the electric field in arbitrary weak fields. A crossover to usual diffusion occurs if the displacements at every step are finite.

INTRODUCTION

A feature of the Levy flight diffusion is that in each step a particle may move by an arbitrarily large distance, so that the root-mean-square displacement per unit time appears to be infinite [1]. Numerical simulation of diffusion via Levy hops shows that the points visited by a diffusing particle form spatially well-separated clusters. From more in-depth consideration one can see that that each cluster consists of a set of clusters, so that a structure of self-similar clusters appears [2].

The probability distribution function in the Fourier representation has the form:

$$P(k,t) \propto e^{-A|k|^{\mu}t} \tag{1}$$

where A and μ are positive magnitudes, $1 < \mu < 2$. Such stable distributions are called Levy distributions. A more detailed discussion of Levy hops is given in [3].

In the present communication the particle drift or the relation between diffusion and conductivity is studied when there is Levy diffusion in the system. For the case of usual classical diffusion and linear response (Ohm's law) this problem was considered by A. Einstein and the well known Einstein relation was obtained. However in the case of Levy hops a question about the existence of an Einstein relation arises. The problem is that the diffusion coefficient, defined in the usual way as $D = \lim\limits_{t\to\infty} \frac{x^2(t)}{t}$, diverges in a Levy flight diffusion case.

Consequently, there are two possibilities: Either the particle mobility tends to infinity, which is nonsense from a physical point of view, or the Einstein relation is broken.

CP553, *Disordered and Complex Systems*, edited by P. Sollich, et al.

Below it will be shown that instead of the Einstein relation a new nonlinear relation between mobility and diffusion coefficient appears.

EINSTEIN RELATION AND ITS GENERALIZATION

Let us recall the well-known Einstein arguments. Let there be in the system the diffusion $J_d = -D\nabla n$ and the field $J_f = \mu En$ currents. In equilibrium the diffusion current J_d is compensated by the field current J_f, and the distribution function must have Boltzmann's form:

$$J_d + J_f = 0, \; N_{eq} \propto e^{-U/kT} \tag{2}$$

where U is the potential energy, T is temperature, and k is Boltzmann's constant.

Before applying analogous arguments to Levy flights consider the assumptions used in deriving the Einstein relation. There are the three following assumptions:

i) the Boltzmann's statistics

ii) the expression for the diffusion current in the usual classical form

iii) the linear Ohm's law

Let us try to understand which of these assumptions need to be modified. Firstly, the assumption about Boltzmann's statistics is not essential, since its type is determined by the statistical properties of the system, and we will retain it. Secondly, the diffusion current has a different form and we write it in a general operator form:

$$J_d = -\hat{K}n = -iA\vec{k}|k|^{\mu-2}n \tag{3}$$

And finally we write the field current as $J_f = nV$, where V is the drift velocity.

By taking a definition for the derivative of the fractional order in the form of the set [4], one can get a general formula for the drift velocity:

$$V = e^{U/kT} \lim_{\epsilon \to 0} (\triangle^2 + \epsilon)^{(\mu-2)/4} \nabla \exp(-\frac{U}{kT}) \tag{4}$$

where $\triangle$ is the Laplace operator.

In a homogeneous electrical field $U = -qEr$ we recover that the drift velocity depends on the electric field in a nonlinear way:

$$V = A(\frac{q\vec{E}}{kT})^{\mu-1} \tag{5}$$

It should be emphasized that this nonlinearity occurs in arbitrarily weak fields and is a consequence of the unusual character of diffusion. The power of nonlinearity is described by the critical index of the Levy hop diffusion.

This is a preliminary result which we obtain below in an exact way.

RANDOM WALKS OF LEVY AND PARTICLE DRIFT IN THE ELECTRIC FIELD

Let us consider a one-dimensional discrete analog of a Levy flight [1]. Let the probability that a particle occupies the l-th site after n steps be $P_n(l)$ and let $f(l)$ be the probability distributions of hops over lengths. So the master equation for complex diffusion has the form:

$$P_{n+1}(l) = \sum_{m=-\infty}^{\infty} f(l-m)P_n(m) \tag{6}$$

To simulate a Levy flight the following function is used for $f(l)$:

$$f(l) = \sum_{n=0}^{\infty} a^{-n}(\delta_{l,-b^n} + \delta_{l,b^n}) \tag{7}$$

where $\delta_{n,m}$ is the Kronecker delta and a and b are the parameters of the Levy flight. Then after Fourier transformation the structure function for such a random walk is equal to:

$$\lambda = \sum_{n=0}^{\infty} a^{-n}\cos(kb^n) \tag{8}$$

One can establish the nonanalytic power-law behavior with k by means of a Mellin transformation, or with the help of formulas of set summation of Poisson. For details see [1].

Let us now introduce an anisotropy into the random walk on self-similar clusters. By virtue of the specific nature of Levy hops a particle can move in one hop over an arbitrary distance b^n. For this reason a small anisotropy $(1+\alpha)$, with $\alpha = qEs/kT$, when particles move on a small distance s, becomes exponentially large on large distances b^n. Since at each step a diffusing particle leaves a site, the sum of probabilities W_+ and W_- of motions parallel and antiparallel, respectively, to the field must be equal 1:

$$W_+ + W_- = 1$$

Hence we get the expressions for probabilities of motion parallel and antiparallel to the field:

$$W_\pm = \frac{(1\pm\alpha)^{b^n}}{(1+\alpha)^{b^n} + (1-\alpha)^{b^n}} \tag{9}$$

Therefore, the structure function $\lambda(k;E)$ in the case of diffusion via Levy hops in the electrical field equals:

$$2\lambda(k;E) = \sum_{n=0}^{\infty} a^{-n}[\cos(kb^n) + i\sin(kb^n)(W_+ - W_-)] \tag{10}$$

As for usual diffusion the second term contains the drift velocity for small $k \to 0$:

$$V = i\frac{\partial\lambda(k;E)}{\partial k}\,|_{k\to 0} = \sum_{n=0}^{\infty}(\frac{b}{a})^n * (W_+ + W_-) \approx \sum_{n=0}^{\infty}(\frac{b}{a})^n \tanh(\alpha b^n) \qquad (11)$$

where $\tanh(x)$ is the hyperbolic tangent.

Using the Poissson formula we obtain after some calculations the formula for the velocity:

$$V(E) = \alpha/2 + \alpha^{\mu-1}\left[\sum_{m=-\infty}^{\infty}\int_1^{\infty}\tanh(z)z^{-\gamma_m}dz + \int_o^{\alpha}\tanh(z)z^{-\gamma_m}dz\right] \qquad (12)$$

where the exponent is equal to:

$$\gamma_m = \mu + 2\pi i m/\ln b.$$

It is easy to see that for weak fields the second term in brackets is less than the first term. Thus, in arbitrarily weak electric fields one can get the nonlinear field dependence of velocity (5).

TRANSITION FROM ORDINARY DIFFUSION TO LEVY DIFFUSION

To understand this nonlinearity as a result of "anomalous" diffusion in addition to Levy hops we allow for ordinary classical diffusion. The simplest way to do this is to introduce a finite length ξ at each step. So we obtain random walks in which ordinary diffusion alternates with Levy hops.

$$f(l) = \sum_{n=0}^{\infty} a^{-n}[\delta_{l,-(b^n+\xi)} + \delta_{l,(b^n+\xi)}] \qquad (13)$$

Anisotropy is introduced into random walks by the method described above: we replace the hop length with the quantity $b^n + \xi$. Thus the structure function in an electric field and for finite hop length is:

$$\lambda(k,\xi,\alpha) = \sum_{n=0}^{\infty} a^{-n}[\cos(kb^n + k\xi) + i\sin(kb^n + \xi)(W_+ - W_-)] \qquad (14)$$

And after calculations by Poisson's method we obtain the following results: in arbitrarily weak fields the velocity is nonlinear in the field, eq. (5), and crosses over to linear behavior in strong fields:

$$V \simeq E\xi^{2-\mu}, \qquad qE\xi/kT \gg 1 \qquad (15)$$

Thus the particle velocity in an electric field has two asymptotic limits in accordance with two diffusing regimes: Levy hops and ordinary diffusion.

So in the considered diffusion problem the new length L_E, governed by the electric field, emerges:

$$L_E = \frac{kT}{qE} \tag{16}$$

The physical meaning of this new field length is the following: at these scales the ordered motion prevails over diffusion. This gives as an estimate for the particle velocity:

$$V = \frac{L_E}{t_E} \tag{17}$$

where t_E is the diffusion time for displacement L_E. For ordinary diffusion we obtain again the well-known Einstein relation and for Levy flights the formula (5).

ACKNOWLEDGMENTS

The author would like to thank the organizing comittee of the conference and Prof. R. Streater for their wonderful hospitality, and Dr. E. Barkai (MIT) for interesting discussions. This work was supported by the Russian Foundation for Basic Research (Grants No. 99-02-17355, No.00-01-10766).

REFERENCES

1. Hughes, B. D., Shlesinger, M. F., and Montroll, E. W., *Proc. Natl. Acad. Sci (USA)* **78**, 3287 (1981).
2. Mandelbrot, B., *Fractals: Form, Chance and Dimension*, Freeman, San-Francisco, 1977.
3. Zumofen, G., and Klafter, J., *Physica D* **69**, 436 (1993).
4. Arkhincheev, V. E., Baskin, E. M., and Batiyev, E. G., *J. Non.-Cryst. Solids* **90**, 21 (1987).
5. Arkhincheev, V. E., Nomoev, A. V., *Physica A* **269**, 293 (1999).

Steady States and Long Time Asymptotics of Solutions of Streater's Models

Piotr Biler

Mathematical Institute, University of Wrocław
pl. Grunwaldzki 2/4, 50-384 Wrocław, Poland

Abstract. We study Streater's models which describe the evolution of the density of a cloud of particles in an external potential and a temperature, coupled through a Poisson equation. We introduce an entropy and consider stationary solutions in a bounded domain, or in an unbounded domain with a confining external potential. The (relative) entropy controls the convergence of time-dependent solutions to the stationary ones.

INTRODUCTION

We consider the system of parabolic-elliptic equations for the density of charged particles $u \geq 0$, temperature $\theta > 0$ and the potential ϕ

$$\begin{aligned} u_t &= \nabla \cdot \left[\kappa \left(\nabla u + \frac{u}{\theta}(\epsilon \nabla \phi + \zeta \nabla \phi_0)\right)\right] \\ (u\theta)_t &= \nabla \cdot (\lambda \nabla \theta) + \nabla \cdot [\kappa(\theta \nabla u + \epsilon u \nabla \phi + \zeta u \nabla \phi_0)] \\ &\quad + (\epsilon \nabla \phi + \zeta \nabla \phi_0) \cdot \left[\kappa \left(\nabla u + \frac{u}{\theta}(\epsilon \nabla \phi + \zeta \nabla \phi_0)\right)\right] \\ -\Delta \phi &= u \end{aligned} \tag{1}$$

in a connected domain $\Omega \subset \mathbb{R}^d$, supplemented with the boundary conditions

$$\begin{aligned} \partial_n u + \frac{u}{\theta}(\epsilon \partial_n \phi + \zeta \partial_n \phi_0) &= 0 \quad \text{(no mass flux)} \\ \partial_n \theta &= 0 \quad \text{(no heat flux)}, \end{aligned} \tag{2}$$

where ∂_n denotes the normal external derivative on the boundary $\partial\Omega$ and $\kappa > 0$, $\lambda > 0$. For the potential ϕ we consider either the Dirichlet boundary condition $\phi = 0$ (perfect conductor) when Ω is bounded or the "free" condition $\phi = E_d * u$, where E_d is the fundamental solution of the Laplacian in $\mathbb{R}^d$. In order to cover the cases with or without given external potential ϕ_0 and Poisson coupling via ϕ, we shall assume that $\langle \epsilon, \zeta \rangle$ takes the values $\langle 0, 0 \rangle$, $\langle 0, 1 \rangle$, $\langle 1, 0 \rangle$ or $\langle 1, 1 \rangle$. We assume that ϕ_0 satisfies for some $T > 0$ the condition

CP553, *Disordered and Complex Systems*, edited by P. Sollich, et al.

$$e^{-\phi_0/T} \in L^1(\Omega) . \tag{3}$$

These models describe the dynamics of Brownian particles in the presence of an external potential and/or a self-consistent potential given by a Poisson-type coupling, as well as the accompanying thermodynamic processes. Recently, R. F. Streater derived such models in the case of the interaction of particles with an external potential only, cf. [1,2], as generalizations of the Smoluchowski equation. The paper [3] gave an extension of Streater's models to the interacting particles case. Streater's models extend classical Nernst–Planck–Debye–Hückel drift-diffusion systems for charged particles (cf., e.g., [4,5]) and those for gravitationally attracting particles (formally with $\epsilon = -1$, cf. [6,7]) that do not take into account the evolution of the temperature. Mathematical properties of solutions of these *isothermal models* are quite well understood now. In particular, finite time blow-up of solutions may occur for attracting particles, while solutions for electrically interacting particles are global in time and tend to steady states.

Here we will focus on the existence and uniqueness of stationary solutions for fixed charge or mass and energy, and on the convergence of solutions of the initial value problem to these stationary solutions. The existence of solutions to the Cauchy problem is only partially known, and a proof of a global existence result is open and seems very difficult, cf. comments in [3]. We will use the entropy functional, which yields a natural *a priori* estimate that can be used to control the behaviour of solutions for long time and the structure of the set of stationary solutions. The entropy was, of course, known before in the context of parabolic equations but has probably been underestimated in the study of simpler models for which other estimates controlling the relaxation were easy to build, cf. [8] and [9].

The results presented in this paper have been obtained in collaboration with Jean Dolbeault, Maria J. Esteban (both of Université de Paris-Dauphine), Grzegorz Karch (Wrocław), Peter A. Markowich (Vienna) and Tadeusz Nadzieja (Zielona Góra). A part of these results is taken from the recent paper [10]. The preparation of this paper was supported by the grant KBN 50/P03/2000/18, a KBN–ÖAD project and the Royal Society during the visit to King's College in London. The author wishes to thank Andrzej Raczyński for interesting remarks.

The boundary conditions (2) guarantee that total mass (or rather charge) $M = \int_\Omega u\,dx$ and energy $E = \int_\Omega u\left(\theta + \zeta\phi_0 + \frac{\epsilon}{2}\phi\right)\,dx$ are preserved, i.e. the first law of thermodynamics is satisfied. We define also the entropy by $W = \int_\Omega u \log\left(\frac{u}{\theta}\right)\,dx$. We can easily see that

If Ω is of class C^1 and if $\langle u, \theta, \phi\rangle$ is a smooth solution of (1)-(2), then the entropy W is decreasing

$$\frac{dW}{dt} = -\int_\Omega \lambda \frac{|\nabla\theta|^2}{\theta^2}\,dx - \int_\Omega \kappa u \left|\frac{\nabla u}{u} + \frac{1}{\theta}(\epsilon\nabla\phi + \zeta\nabla\phi_0)\right|^2\,dx \le 0\,, \tag{4}$$

so that the model (1) is compatible with the second law of thermodynamics.

Using the entropy production relation (4) we see that
Any smooth stationary solution $\langle u, \theta, \phi \rangle$ *of (1) with given* $M > 0$ *and* $E > 0$ *is such that* u *and* ϕ *are determined by the nonlinear Poisson–Boltzmann equation*

$$u := -\Delta\phi = M \frac{e^{-(\phi+\phi_0)/\theta}}{\int_\Omega e^{-(\phi+\phi_0)/\theta}\,dx}, \tag{5}$$

and θ *is a positive constant given by the energy relation* $E = M\theta + \int_\Omega u\left(\phi_0 + \frac{1}{2}\phi\right)dx$.

The case $\langle \epsilon, \zeta \rangle = \langle 0, 0 \rangle$ is the simplest one since in a bounded domain u and θ are constants, with $u \equiv M/|\Omega|$ and $\theta \equiv E/M$.

In the case $\langle \epsilon, \zeta \rangle = \langle 0, 1 \rangle$ the stationary solution is obviously given by $u = u_{\infty,\theta} \equiv Me^{-\phi_0/\theta}\left(\int_\Omega e^{-\phi_0/\theta}\,dx\right)^{-1}$, where θ is a constant determined by the condition $E = E(\theta) = M\theta + \int_\Omega \phi_0 u_{\infty,\theta}\,dx$. The main result here is:
The above problem has at most one solution. Moreover, if we define the numbers $E_\pm = \lim_{\pm(\theta - T_\mp)\searrow 0}\left(M\theta + \int_\Omega \phi_0 u_{\infty,\theta}\,dx\right)$, *with* $T_- = \inf\{\theta > 0 \ : \ E(\theta) > -\infty\}$ *and* $T_+ = \sup\{\theta > 0 \ : \ E(\theta) < \infty\}$, *then the solution exists if and only if* $E \in (E_-, E_+)$. *If* ϕ_0 *is bounded from below, then* $T_- = 0$, $E_- = \inf \phi_0$. *If* ϕ_0 *is bounded both from above and from below (which is possible under (3) only if* Ω *is bounded), then* $T_+ = E_+ = \infty$.

Assume that $\langle u, \theta, \phi \rangle$ is a stationary solution in the case of a bounded Ω and $\langle \epsilon, \zeta \rangle = \langle 1, 0 \rangle$. According to (4), we have $\frac{\nabla u}{u} + \frac{1}{\theta}\nabla\phi = 0$ a.e. with respect to the measure $u(x)\,dx$. Then $\Delta\theta = 0$ means that θ is a constant, and $\psi = \phi/\theta = C - \log u$ is a solution of the Poisson–Boltzmann equation (5), where θ is determined by the energy $E = M\theta + \frac{1}{2}\|\nabla\psi\|^2_{L^2(\Omega)}\,\theta^2$. For given M and E, ψ is the solution of

$$-\sigma(\|\nabla\psi\|_{L^2(\Omega)})\,\Delta\psi = \frac{e^{-\psi}}{\int_\Omega e^{-\psi}\,dx}, \quad \psi_{|\partial\Omega} = 0, \tag{6}$$

where $\sigma(t) = (-1 + \sqrt{1 + \chi t^2})/t^2$, $\chi = 2E/M^2$. We prove that
If Ω *is a bounded domain of class* C^1, *then (6) has a unique solution for any* $M > 0$ *and* $E > 0$.
We refer to [11] for a complete proof of a similar statement, based on the idea that any solution ψ of (6) is a critical point of the strictly convex functional $J[\psi] = F(\|\nabla\psi\|_{L^2(\Omega)}) + \log\left(\int_\Omega e^{-\psi_+}\,dx\right)$, where $F(t) = \sqrt{1 + \chi t^2} - \log\left(1 + \sqrt{1 + \chi t^2}\right)$.

In the case $\langle \epsilon, \zeta \rangle = \langle 1, 1 \rangle$ we are able to obtain the following existence result:
Let $M > 0$ *and* $E > E_{\phi_0,\Omega} := \inf\left\{\int_\Omega u\left(\phi_0 + \frac{1}{2}\phi\right)dx \ : \ u \in L^1_+(\Omega),\ \int_\Omega u\,dx = M\right\}$, *where* ϕ *is given by (1)-(2). Assume furthermore that* $d \geq 2$ *and that* ϕ_0 *is an external potential which is bounded from below in* Ω *and satisfies (3) for some sufficiently large* T. *Then, there exists at least one* $u \in L^\infty_+(\Omega)$ *such that the Poisson–Boltzmann equation (5) has a solution.*

The proof is based on a variational argument involving directly the entropy. Its critical level is defined as a max-min level instead of simply being the minimal level of the convex functional $W[u,\theta] = \int_\Omega u \log\left(\frac{u}{\theta}\right) dx$ well defined on the set $X = \left\{\langle u,\theta\rangle \in L^1_+(\Omega)\times L^\infty_+(\Omega) : \int_\Omega u\,dx = M, \quad \int_\Omega u\left(\theta + \phi_0 + \frac{1}{2}\phi\right)dx = E\right\}$. The question of the uniqueness remains largely open. In special cases however, it is possible to give some results, e.g., when $d = 1$, $\phi_0 = |x|$, $\Omega = \mathbb{R}$ or $\Omega = \mathbb{R}^d$, $d \geq 2$, $\phi_0 = |x|^\beta$.

The case of gravitational interaction is much more complicated, and even if the external potential is absent, the structure of the set of steady states (existence–uniqueness vs. nonexistence–multiplicity) is complicated and depends in a very sensitive manner on geometric properties of the domain. For examples of the problem with a singular external potential, see, e.g., [7]. We study in this subsection stationary solutions of (1) with the third (Poisson) equation replaced by $\Delta\phi = u$ which corresponds to the gravitational attraction of particles, i.e. the equation

$$u := \Delta\phi = M\frac{e^{-\phi/\theta}}{\int_\Omega e^{-\phi/\theta}\,dx} \tag{7}$$

with either the Dirichlet boundary condition or the (physically relevant) free condition. Scaling the potential $\phi = \theta\psi$ solving (7), the energy is $E = M\theta + \frac{1}{2}\theta^2\int_\Omega \psi\Delta\psi\,dx$. In other words, we need that

$$\left(\frac{E}{M^2}\right)m^2 = m + \frac{1}{2}\int_\Omega \psi\Delta\psi\,dx \equiv \mathcal{E}(m) \tag{8}$$

holds, where $m = M/\theta$ and $\psi = \psi_m$ solves the unscaled Poisson–Boltzmann equation

$$\Delta\psi = m\frac{e^{-\psi}}{\int_\Omega e^{-\psi}\,dx}. \tag{9}$$

Thus, the problem of finding a steady state satisfying (7) with a given energy E and mass $M > 0$ is equivalent to find a solution of the equation (8) in the range of admissible m's (the solutions of (9) depend on m continuously along the branches). As it is well known, see also examples below, (9) can have nontrivial solutions either for $m \in (0, m_\Omega)$ or $m \in (0, m_\Omega]$ with some $0 < m_\Omega \leq \infty$. Moreover, in the case of star-shaped domains $\Omega \subset \mathbb{R}^d$, $d \geq 2$, one has $m_\Omega < \infty$, while, e.g., for annuli $\Omega \subset \mathbb{R}^d \quad m_\Omega = \infty$.

Example. If $\Omega = B(0,1) = \{x \in \mathbb{R}^2 : |x| < 1\}$ we have for a fixed m only a unique radially symmetric solution of (9) given by $\frac{\partial}{\partial r}\psi = 4r\left(r^2 + 8\pi/m - 1\right)^{-1}$ with $r = |x|$ and $m \in (0, 8\pi)$. Since $\lim_{m\nearrow 8\pi}\mathcal{E}(m) = -\infty$, a steady state of (7) exists for each $E \in \mathbb{R}$ and each $M > 0$.

If $\Omega = B(0,1) \subset \mathbb{R}^d$, $d > 2$, the problem (9) is no longer integrable but it can be reduced to the study of a dynamical system (introduced by I. M. Gelfand) in the plane, cf. [7]. Using this idea, one checks that $\inf_{m\in(0,m_\Omega)} \mathcal{E}(m) > -\infty$. This means that (8) does not have solutions for $E/M^2 \ll -1$. Remark that if $3 \le d \le 9$, the solutions of (9) for given m are not, in general, unique.

Conjecture. *Suppose that a bounded domain $\Omega \subset \mathbb{R}^d$ is such that (9) has solutions for all $m > 0$, i.e. $m_\Omega = \infty$. Then for some $c_\Omega > 0$*

$$\mathcal{E}(m) \ge m - c_\Omega m^2. \tag{10}$$

If this conjecture is true, we will have no steady states of (7) for $E/M^2 \ll -1$, while for $E/M^2 > c_0$ with some $c_0 \in \mathbb{R}$ steady states do exist.

This conjecture is supported by the analysis of (9) in annuli $\Omega = \{x \in \mathbb{R}^d : a < |x| < A\}$, $0 < a < A < \infty$. Indeed, the conjectured estimate from below (10) holds true for *radially symmetric* solutions in annuli which exist for *all* $m > 0$. However, there are also nonradial solutions of (9) in certain cases.

Example. (cf. [6], p. 523) If $\Omega \subset \mathbb{R}^2$ is a "long" rectangle, i.e. $\Omega = (0,a) \times (0,b)$ with $b \gg a > 0$, then $m_\Omega < \infty$ but there exists a steady state maximizing the functional $I[\psi] = \log\left(\int_\Omega e^{-\psi}\,dx\right) + \frac{1}{2}\int_\Omega \psi\Delta\psi\,dx$ (considered in the class of potentials ψ corresponding to the densities $u \equiv \Delta\psi \ge 0$, $\int_\Omega u\,dx = m$ and satisfying boundary conditions) for $m = m_\Omega$, unlike it was in the case of a disc in the plane. This is a consequence of the relation $\sup_{m<m_\Omega} I[\psi] < \infty$ which permits us to prove that a maximizing sequence for I with $m \nearrow m_\Omega$ converges to a maximizer of I for $m = m_\Omega$. Therefore a solution of (9) exists for $m = m_\Omega$ and thus $\inf_{m\le m_\Omega} \mathcal{E}(m) > -\infty$. In this case, steady states of (7) do not exist for $E/M^2 \ll -1$.

We believe that the situations described in examples for balls, annuli and rectangles are (qualitatively) generic for all bounded domains Ω.

RELATIVE ENTROPY AND CONVERGENCE TO STATIONARY SOLUTIONS

Consider a solution $\langle u, \theta, \phi\rangle$ of (1) in a bounded domain Ω and denote by $\langle u_\infty, \theta_\infty, \phi_\infty\rangle$ a stationary solution with same charge and energy such that

$$\lim_{t\to\infty} W[u(t,.), \theta(t,.), \phi(t,.)] = W[u_\infty, \theta_\infty, \phi_\infty]\,.$$

The existence of such a solution follows from the entropy production equation (4); at least for a sequence $t_n \nearrow \infty$, the sequence $\langle u, \theta, \phi\rangle(t_n + .\,,.)$ converges to some $\langle u_\infty, \theta_\infty, \phi_\infty\rangle$. We may define the relative entropy by

$$\begin{aligned}\Sigma[u,\theta,\phi] &= W[u,\theta,\phi] - W[u_\infty,\theta_\infty,\phi_\infty] \qquad (11)\\ &= \int_\Omega s_1\left(\frac{u}{u_\infty}\right) u_\infty\,dx + \int_\Omega s_2\left(\frac{\theta}{\theta_\infty}\right) u\,dx + \frac{1}{2\theta_\infty}\int_\Omega |\nabla\phi - \nabla\phi_\infty|^2\,dx,\end{aligned}$$

where $s_1(t) = t\log t + 1 - t$ and $s_2(t) = t - 1 - \log t$ are two nonnegative strictly convex functions such that $s_1(1) = s_2(1) = 0$. Our main result is:

Let $\langle u, \theta, \phi \rangle$ be a solution of (1) in a bounded domain Ω. Then

$$\frac{1}{M}\|u-u_\infty\|^2_{L^1(\Omega)}+\frac{2}{\theta_\infty}\|\nabla\phi-\nabla\phi_\infty\|^2_{L^2(\Omega)} \\ +C[\theta,u](t)\,\|\theta-\theta_\infty\|^2_{L^1(\Omega,u(t,x)\mathbf{1}_{\{\theta<\ell\theta_\infty\}}dx)} \le 4\,\Sigma[u,\theta,\phi](t)\,, \tag{12}$$

for all $\ell \ge 1$, where $C[\theta, u] = (M\theta_\infty^2 \cdot \max\{1, \ell\})^{-1}$ so that each solution $\langle u, \theta, \phi \rangle$ of (1) converges as $t \to \infty$ to a unique stationary solution $\langle u_\infty, \theta_\infty, \phi_\infty \rangle$ with the same charge and energy.

The proof follows from (4). Indeed, the relative entropy $\Sigma[u, \theta, \phi](t)$ is bounded from below and

$$\frac{d}{dt}\Sigma[u,\theta,\phi](t) = -\int_\Omega \lambda\frac{|\nabla\theta|^2}{\theta^2}\,dx - \int_\Omega \kappa u\left|\frac{\nabla u}{u}+\frac{1}{\theta}(\epsilon\nabla\phi+\zeta\nabla\phi_0)\right|^2 dx \le 0\,,$$

and this inequality is strict unless $u \equiv u_\infty$: the only possible limit of $\frac{d}{dt}\Sigma[u, \theta, \phi](t)$ is zero. The bound (12) is proved using the relative entropy and the Csiszár–Kullback type inequality, see [12].

This answers the question raised in Section 4.2 of [3]. The result does not mean that the stationary problem has a unique solution for any given charge and energy, but that the limit $\langle u_\infty, \theta_\infty, \phi_\infty \rangle$ for a given solution is unique; it does not depend on the sequence $t_n \nearrow \infty$.

Note that for the isothermal models with linear and nonlinear diffusion considered in [5,13] we proved much better results including convergence rate as well as intermediate asymptotics described by solutions of the diffusion equation if $\zeta = 0$.

REFERENCES

1. Streater R. F., *J. Stat. Phys.* **88**, 447 (1997).
2. Streater R. F., *J. Math. Phys.* **38**, 4570 (1997).
3. Biler P., Krzywicki A., Nadzieja T., *Rep. Math. Phys.* **42**, 359 (1998).
4. Biler P., Hebisch W., Nadzieja T., *Nonlinear Analysis T. M. A.* **23**, 1189 (1994).
5. Biler P., Dolbeault J., to appear in *Annales Henri Poincaré* **1** (2000).
6. Caglioti E., Lions P.-L., Marchioro C., Pulvirenti M., *Comm. Math. Phys.* **143**, 501 (1992).
7. Biler P., Nadzieja T., *Adv. Diff. Eq.* **3**, 177 (1998).
8. Toscani G., *Quart. Appl. Math.* **57**, 521 (1999).
9. Arnold A., Markowich P. A., Toscani G., Unterreiter A., to appear in *Comm. Partial Diff. Eq.* **25** (2000).
10. Biler P., Dolbeault J., Esteban M., Karch G., to appear in *Adv. Diff. Eq.*
11. Desvillettes L., Dolbeault J., *Comm. Partial Diff. Eq.* **16**, 451 (1991).
12. del Pino M., Dolbeault J., *preprint* Ceremade no. 9905 (1999).
13. Biler P., Dolbeault J., Markowich P. A., to appear in *Transport Th. Stat. Phys.*

Scaling in Nonlinear Parabolic Equations: Applications to Debye System

Grzegorz Karch[1]

Instytut Matematyczny, Uniwersytet Wrocławski
pl. Grunwaldzki 2/4, 50-384 Wrocław, POLAND
e-mail: karch@math.uni.wroc.pl
http://www.math.uni.wroc.pl/~karch

Abstract. The Cauchy problem for parabolic equations with quadratic nonlinearity is studied. We investigate the existence of global-in-time solutions and their large-time behaviour assuming some scaling property of the equation as well as of the norm of the Banach space in which the solutions are constructed. We apply these abstract results to the study of the Debye system arising in the theory of electrolytes.

INTRODUCTION

Following [1], we would like to describe in this note a method of constructing global-in-time solutions of the Cauchy problem for the parabolic equation

$$u_t = \Delta u + \mathbf{B}(u,u) \tag{1}$$

supplemented with the initial condition

$$u(x,0) = u_0(x). \tag{2}$$

Here $u = u(x,t)$, $x \in I\!R^n$, and $t \in [0,T)$ for some $T \in (0,\infty]$. We assume that the nonlinear term $\mathbf{B}(\cdot,\cdot)$ is defined by a bilinear form acting on $u(x,t)$ with respect to the spatial variable only. This nonlinearity will also be assumed to satisfy a scaling property. To set it up, first given $f : I\!R^n \to I\!R$ we define the rescaled function

$$f_\lambda(x) = f(\lambda x)$$

for each $\lambda > 0$. We extend this definition to all $f \in \mathcal{S}'$ (tempered distributions) in the standard way.

1) The preparation of this paper was supported by the grant KBN 50/P03/2000/18 and the Royal Society during the author's visit to King's College London.

CP553, *Disordered and Complex Systems*, edited by P. Sollich, et al.

Definition 1 *The bilinear form* $\mathbf{B}(\cdot,\cdot)$ *is said to have the scaling order equal to* $b \in I\!R$ *if*

$$\mathbf{B}(f_\lambda, g_\lambda) = \lambda^b \big(\mathbf{B}(f,g)\big)_\lambda \tag{3}$$

for any $\lambda > 0$ *and all* $f, g \in \mathcal{S}'(I\!R^n)$, *for which the both sides of (3) make sense.*

Our main requirement is that *the bilinear form* $\mathbf{B}(\cdot,\cdot)$ *in (1) has the scaling order equal to* $b \in [0,2)$. There are several important equations of the form (1) satisfying assumption. In the next section, we mention some of them, however, the Debye system from the theory of electrolytes is our main example. Note here that one more condition will be imposed on $\mathbf{B}(\cdot,\cdot)$ below (cf. (7)).

Given a Banach space E a solution $u(t)$ of (1)-(2) is interpreted as an E-valued mapping defined on $[0,T)$. It is known that if Δ is the infinitesimal generator of a C_0-semigroup on E and if, e.g., $\mathbf{B}(u(\cdot),u(\cdot)) \in C([0,T);E)$, it follows that

$$u(t) = e^{t\Delta}u_0 + \int_0^t e^{(t-\tau)\Delta}\mathbf{B}(u(\tau),u(\tau))\,d\tau. \tag{4}$$

On the other hand, a solution of (4) does not necessarily satisfy (1)-(2). Hence, let us stress that, throughout the remainder of this paper, by solutions to (1)-(2) we always mean solutions to (4).

Given $u_0 \in \mathcal{S}'$, the heat semigroup has the following form $e^{t\Delta}u_0 = p_t * u_0$ where the Gauss-Weierstrass kernel is denoted by $p_t(x) = (4\pi t)^{-n/2}\exp(-|x|^2/(4t))$. In our case $e^{t\Delta} : E \to E$ will be a bounded operator for any $t \geq 0$. Moreover, for $u_0 \in E$, $e^{t\Delta}u_0 \in C((0,T);E)$, but $e^{t\Delta}u_0$ will tend to u_0 as $t \searrow 0$ in the sense of $\mathcal{S}'$ only. In this paper, we do not require that $e^{t\Delta}$ is a strongly continuous semigroup on E. This does not allow us to apply the scheme of the existence proof directly from e.g. [2]. Here, a solution is constructed in the space $\mathcal{F}([0,T];E)$ containing E-valued measurable functions which are bounded on $[0,T]$ in the norm of E and which attain the initial value $u(0)$ as $t \searrow 0$ in the sense of tempered distributions.

Let us briefly describe the main idea of this paper. Suppose we are able to construct local-in-time solutions to (1)-(2) in a Banach space E consisting of tempered distributions. Assume that $\mathbf{B}(\cdot,\cdot)$ satisfy the condition (3) for some $b \in [0,2)$. We are going to show that these two assumptions combined with a scaling property of $\|\cdot\|_E$ allow us to obtain global-in-time solutions to (1)-(2) for suitably small initial data. To get such results we introduce a new Banach space of distributions which, roughly speaking, is a homogeneous Besov-type space modeled on E. This approach allows us to get solutions for initial data less regular than those from E. In this abstract setting, we also study large-time behaviour of constructed solutions. We find a simple condition (in terms of decay properties of the heat semigroup) which guarantees that solutions to (1)-(2) have the same asymptotic behaviour as $t \to \infty$. These abstract results will be applied to the study of the Cauchy problem for the Debye system.

MODEL EXAMPLES

The semilinear heat equation with a quadratic nonlinearity $u_t = \Delta u + au^2$ and the diffusion-convection equation $u_t = \Delta u + (\bar{b}, \nabla(u^2))$, where $u : \mathbb{R}^n \times [0, T) \to \mathbb{R}$, are perhaps the simplest examples of (1) (here, $a \in \mathbb{R}$ and $\bar{b} \in \mathbb{R}^n$ are constant). Obviously, the nonlinear term has the scaling order equal to 0 in the case of the first equation and 1 in the case of the second one. The Navier-Stokes system for an incompressible fluid with no potential force in $\mathbb{R}^n$ is our next example. Here, the nonlinear term satisfies (3) with $b = 1$ (see [1], for details).

The following nonlocal parabolic equation

$$u_t = \Delta u + \nabla \cdot (u \nabla \varphi_u), \tag{5}$$

where the coefficient $\nabla \varphi_u$ is determined from u via the potential $\varphi_u = G_n * u$, G_n being the fundamental solution of the Laplacian in $\mathbb{R}^n$, appears in nonequilibrium statistical mechanics. Here $u : \mathbb{R}^n \times [0, T) \to \mathbb{R}$ is the density of particles in $\mathbb{R}^n$ interacting with themselves through the gravitational potential. The Cauchy problem for (5) was considered by Biler in [3], and we refer there for further references. The order of scaling of the nonlinear term in (5) equals 0. One can check this at once writing $\nabla \varphi$ as $\nabla G_n * u$ and recalling that ∇G_n is homogeneous of degree $-n + 1$.

The Debye system from the theory of electrolytes consists of two parabolic equations

$$u_{1t} = \Delta u_1 - \nabla \cdot (u_1 \nabla \varphi) \qquad u_{2t} = \Delta u_2 + \nabla \cdot (u_2 \nabla \varphi) \tag{6}$$

coupled through the electric potential φ satisfying $\varphi = \varphi_{u_1 - u_2} = G_n * (u_1 - u_2)$. This system describes the evolution of the densities u_1 and u_2 of ions in an electrolyte. We refer to [4] for a more detailed interpretation of the system (6), as well as for references to other papers. Moreover, the recent paper [5] contains results on the large time behaviour of solutions to (6) for integrable initial data. System (6) has the form of (1) with $u = (u_1, u_2)$. As in the previous example, the nonlinearity has the order of scaling equal to 0.

LOCAL EXISTENCE OF SOLUTIONS

This is a usual situation, when one deals with a semilinear evolution equation $u_t = \Delta u + J(u)$ in a given Banach space E that the nonlinear mapping is "singular" in E, but $e^{t\Delta} J$ is locally Lipschitz on E for each $t > 0$. There are several papers where this property was used to construct local-in-time solutions. The definition (given below) of an *adequate space* to the problem (1)-(2) is a variation of those ideas. However, in our case, we shall also use it to construct global-in-time solutions for suitably small initial data.

Definition 2 *The Banach space $(E, \|\cdot\|_E)$ is said to be adequate to the problem (1)-(2) if the following conditions are satisfied:*

i. $\mathcal{S} \subset E \subset \mathcal{S}'$ and the both inclusions are continuous;

ii. the norm $\|\cdot\|_E$ on E is translation invariant, i.e.

$$\text{for all} \quad f \in E \quad \text{and} \quad \xi \in \mathbb{R}^n, \quad \|f(\cdot + \xi)\|_E = \|f\|_E;$$

iii. for all $f, g \in E$, $\mathbf{B}(f,g) \in \mathcal{S}'$; moreover there exist $T_0 > 0$ and a positive function $\omega \in L^1(0, T_0)$ such that

$$\|e^{s\Delta}\mathbf{B}(f,g)\|_E \le \omega(s)\|f\|_E\|g\|_E \tag{7}$$

for any $f, g \in E$ and $s \in (0, T_0)$.

A standard reasoning based on the Banach contraction principle leads to the local existence of solution to (1)-(2) in an adequate space E.

Theorem 1 *Let a Banach space E be adequate to the problem (1)-(2). Given $u_0 \in E$ there exists $T = T(u_0) \in (0, T_0]$ and a unique solution of the problem (1)-(2) in the space $\mathcal{X}T = \mathcal{F}([0,T]; E)$. Moreover, if the maximal time T_* of existence u is finite and if $T_* < T_0$, then $\lim_{t \to T_*} \|u(t)\|_E = \infty$.*

In [1, Section 3], we prove that *the space $L^p(\mathbb{R}^n)$ is adequate to equation (5) and to the Debye system (6) provided $n/2 < p < n$.* This fact is obtained combining the Hölder inequality with the well-known estimates for the heat semigroup and its derivatives

$$\begin{aligned} \|e^{t\Delta} f\|_{L^p} &\le C t^{n(1/p - 1/q)/2} \|f\|_{L^q}, \\ \|\nabla e^{t\Delta} f\|_{L^p} &\le C t^{n(1/p-1/q)/2 - 1/2} \|f\|_{L^q}, \end{aligned} \tag{8}$$

which are valid for each $1 \le q \le p \le \infty$, every $t > 0$, $f \in L^p(\mathbb{R}^n)$, and is positive constants C.

The paper [1] contains several other examples of spaces (the Marcinkiewicz weak L^p space, the Lorentz spaces, and the Morrey spaces) which are adequate to Cauchy problems for some other nonlinear parabolic equations.

NEW CLASS OF BANACH SPACES

Let $(E, \|\cdot\|_E)$ be a Banach space. *The norm $\|\cdot\|_E$ is said to have scaling degree equal to ℓ, if $\|f_\lambda\|_E = \lambda^\ell \|f\|_E$ for each $f \in E$ such that $f_\lambda \in E$ and for all $\lambda > 0$.* It is evident that the usual norms of the space $L^p(\mathbb{R}^n)$ has the scaling degree equal to $-n/p$. In this work, Banach spaces endowed with norms having this property are denoted by E_p.

Now fixing a Banach space $E \subset \mathcal{S}'$ we introduce a new space of distributions denoted by BE^α, which is, loosely speaking, a homogeneous Besov space modeled on E.

Definition 3 *Let $\alpha \geq 0$. Given a Banach space E imbedded continuously in $\mathcal{S}'$, we define*

$$BE^{\alpha} = \{f \in \mathcal{S}' : \|f\|_{BE^{\alpha}} \equiv \sup_{t>0} t^{\alpha/2} \|e^{t\Delta} f\|_E < \infty\}.$$

Let $E = L^p(I\!R^n)$ for a moment. It is not difficult to prove (using (8)) that $L^q(I\!R^n) \subset BE^{\alpha}$ for each $1 \leq q \leq p \leq \infty$ and $\alpha = n(1/q - 1/p)$. In this case, the norm $\|\cdot\|_{BE^{\alpha}}$ is equivalent to the standard norm of the homogeneous Besov space $\dot{B}^{-\alpha}_{p,\infty}$ introduced via a dyadic decomposition.

Now our goal is to define a sufficiently large class of Banach spaces E, for which BE^{α} is complete.

Definition 4 *We denote by $\mathcal{K}$ a class of all Banach spaces $E \subset \mathcal{S}'$ such that for some $q \in [1, \infty]$, $e^{t\Delta} : E \to L^q$ is a bounded operator for every $t > 0$.*

It is proved in [1, Lemma 4.2] that the space $(BE^{\alpha}_p, \|\cdot\|_{BE^{\alpha}_p})$ is a Banach space provided $E_p \in \mathcal{K}$ and has a norm with the scaling degree equal to $-n/p$ for some $p \in [1, \infty]$.

GLOBAL-IN-TIME SOLUTIONS

Suppose that the Banach space E_p is adequate to the problem (1)-(2) and has a norm with the scaling degree equal to $-n/p$. Let $0 \leq \alpha \leq b + n/p$. There exists a constant $C > 0$, independent of t, f, g, such that

$$\|e^{t\Delta}\mathbf{B}(f,g)\|_{BE^{\alpha}_p} \leq C t^{(\alpha-b-n/p)/2} \|f\|_{E_p} \|g\|_{E_p}$$

for all $t > 0$ and $f, g \in E_p$. This auxiliary inequality plays the crucial rôle in the construction of solutions to (1)-(2) in the space BE^{α}_p. The scaling properties of $\|\cdot\|_{E_p}$ and of $\mathbf{B}(\cdot,\cdot)$ are essential in its proof.

In the following theorem, the global-in-time solutions to (1)-(2) in the space BE^{β}_p are constructed for suitably small initial data.

Theorem 2 *Fix $p > n/(2-b)$ satisfying $p(1-b) < n$. Denote $\beta = 2 - b - n/p$. Suppose that E_p is a Banach space with the following properties*

i. E_p is adequate to the problem (1)-(2);
ii. the norm $\|\cdot\|_{E_p}$ has scaling degree $-n/p$;
iii. $E_p \in \mathcal{K}$.

There exists $\varepsilon > 0$ such that for each $u_0 \in BE^{\beta}_p$ satisfying $\|u_0\|_{BE^{\beta}_p} < \varepsilon$ there is a solution of (1)-(2) for all $t \geq 0$ in the space

$$\mathcal{X} \equiv \mathcal{F}([0,\infty) : BE^{\beta}_p) \cap \{u : (0,\infty) \to E_p \ : \ \sup_{t>0} t^{\beta/2} \|u(t)\|_{E_p} < \infty\}.$$

This is the unique solution satisfying the condition $\sup_{t>0} t^{\beta/2}\|u(t)\|_{E_p} \leq 2\varepsilon$.

ASYMPTOTIC BEHAVIOUR OF SOLUTIONS

The following result is an immediate consequence of Theorem 2.

Theorem 3 *Let the assumptions from Theorem 2 hold. Assume that u and v are two solutions of (1)-(2) constructed in Theorem 2 corresponding to the initial data $u_0, v_0 \in BE_p^\beta$, respectively. The following conditions are equivalent:*

$$\lim_{t\to\infty} t^{\beta/2}\|e^{t\Delta}(u_0 - v_0)\|_{E_p} = 0. \tag{9}$$

and

$$\lim_{t\to\infty} t^{\beta/2}\|u(\cdot,t) - v(\cdot,t)\|_{E_p} = 0,$$

provided $\varepsilon > 0$ is sufficiently small.

Theorem 3 asserts that two solutions of (1)-(2) have the same large time behaviour provided their initial data differ from each other by a distribution f satisfying the condition $\lim_{t\to\infty} t^{\beta/2}\|e^{t\Delta}f\|_{E_p} = 0$.

Now, let us apply Theorem 3 to the study of existence and stability of self-similar solutions to (1)-(2). Other applications to the Burgers equation, diffusion-convection equations, and Navier-Stokes system can be found in [1].

Note first that if a function u solves (1) then for each $\lambda > 0$ the rescaled function $\Lambda_{2-b}u(x,t) \equiv \lambda^{2-b}u(\lambda x, \lambda^2 t)$ is also a solution of (1). The solutions satisfying the scaling invariance property $\Lambda_{2-b}u \equiv u$, for any $\lambda > 0$, are called *forward self-similar solutions*. They have the form

$$U(x,t) = t^{-(2-b)/2}U(xt^{-1/2}, 1)$$

for all $x \in \mathbb{R}^n$ and $t > 0$. Observe that if $u_0(x) = \lim_{t\to 0} t^{-(2-b)/2}U(xt^{-1/2}, 1)$ exists, then u_0 is necessarily homogeneous of degree $b-2$. On the other hand, a self-similar solution to (1)-(2) can be obtained directly from Theorem 2, taking u_0, homogeneous of degree $b-2$ and suitably small.

Now by Theorem 3, every self-similar solution $v(x,t) = t^{-(2-b)/2}V(xt^{-1/2})$ describes the large time behaviour of those general solutions whose initial data u_0 satisfy (9), where $v_0(x) = \lim_{t\to 0} t^{-(2-b)/2}V(xt^{-1/2})$ in the sense of distributions.

REFERENCES

1. Karch G., *J. Math. Anal. Appl.* **234**, 534 (1999).
2. Weissler F. B., *Indiana Univ. Math. J.* **29**, 79 (1980).
3. Biler P., *Studia Math.* **114**, 181 (1995).
4. Biler P., Hebisch W., Nadzieja T., *Nonlinear Analysis T.M.A.* **23**, 1189 (1994).
5. Biler P., Dolbeault J., *Annales Henri Poincaré* **1** (2000). To appear.

Estimates for Fundamental Solutions of Parabolic Equations with Application to a Nonlinear Problem

V. Kondratiev[†1], V. Liskevich[*] and Z. Sobol[*]

[†] *Department of Mathematics and Mechanics*
Moscow State University, Moscow, Russia
[*] *School of Mathematics*
University of Bristol, Bristol BS8 1TW

Abstract. Gaussian estimates for fundamental solutions for a class of degenerate parabolic equations are discussed. Applications to exterior problems for semi-linear elliptic equations are given.

INTRODUCTION

The aim of this note is to study some qualitative properties of second-order elliptic and parabolic equations in divergence form with measurable coefficients. The coefficients in the main part are allowed to have a point degeneracy and growth at infinity which makes the conditions different from the classical situation of uniformly elliptic equations [1]. In the first section we study linear equations and by means of a general result due to L. Saloff-Coste [8] we obtain two-sided Gaussian estimates for the fundamental solution. For this we provide a Poincaré inequality needed. We also discuss classes of potentials preserving Gaussian estimates for fundamental solutions. In the second part we study an exterior problem for a semi-linear second-order elliptic equation and obtain necessary and sufficient conditions for non-existence of positive solutions in terms of a polynomial type nonlinearity. The problems of this type attract much attention due to applications in geometry (see, e.g. [4]). We state the main result for the equation without lower order terms and discuss its generalisation to the equation containing zero-order perturbations. Because of restricted space we provide only hints to the main steps of the proofs. A full size paper with complete proofs (and a proper list of references) is in preparation [5].

[1)] Supported by EPSRC through Grant GR/L76891.

CP553, *Disordered and Complex Systems*, edited by P. Sollich, et al.

ESTIMATES FOR FUNDAMENTAL SOLUTIONS

In this section we discuss estimates for the fundamental solution to the parabolic equation in $\mathbf{R}^n$, $n \geq 3$,

$$\partial_t u(x,t) = \sum_{i,j=1}^{n} \partial_{x_i} a_{ij}(x)|x|^\alpha \partial_{x_j} u(x,t), \tag{1}$$

where $a = (a_{ij})$ is a uniformly elliptic matrix of measurable functions and $0 \leq \alpha < 2$. Of course, if $\alpha = 0$, the two-sided Gaussian estimates are well known [1]. It turns out that one can obtain corresponding estimates for the fundamental solution of (1) using a generalisation of Moser's approach (see [8]).

The solution to (1) is understood in the sense of weak solutions. Otherwise one can construct the self-adjoint operator, corresponding to the right-hand side of (1), by means of the form method. The corresponding semigroup (on L^2) will be the resolution operator to the Cauchy problem associated with (1). (The two methods lead to the same notion of solution.)

Let us consider the quadratic form in $L^2(\mathbf{R}^n)$:

$$E(u) = \sum_{i,j=1}^{n} \int_{\mathbf{R}^n} a_{ij}(x)|x|^\alpha \partial_{x_i} u \partial_{x_j} u dx, \quad u \in C_c^1(\mathbf{R}^n).$$

The form is closable since $|x|^{-\alpha} \in L^1_{loc}$. Its closure associates a non-negative self-adjoint operator A in L^2. The semigroup e^{-tA} is a semigroup of integral operators whose integral kernel is the fundamental solution to (1).

According to [8,2] in order to obtain two-sided estimates for the fundamental solution we need to introduce a new distance in $\mathbf{R}^n$.

Definition 1 $d(x,y) = \sup\{\phi(x) - \phi(y) \; : \phi \in W^{1,\infty}_{loc}(\mathbf{R}^n)\,, |\nabla\phi|^2|x|^\alpha \leq 1\}$

It is readily seen that $\mathbf{R}^n$ endowed with the distance d is a complete metric space. One can prove the following estimates for the volume of the metric ball $B^d(x,r)$.

Lemma 1 *There exists $c_1, c_2 > 0$ such that*

$$c_1(r^{\frac{2n}{2-\alpha}} + r^n|x|^{\frac{\alpha n}{2}}) \leq |B^d(x,r)| \leq c_2(r^{\frac{2n}{2-\alpha}} + r^n|x|^{\frac{\alpha n}{2}}).$$

In the next theorem we state the Gaussian two-sided estimate for the fundamental solution which is well known for the cases of uniformly elliptic operators, sub-elliptic operators of the Hörmander type and Laplace-Beltrami operators on some classes of smooth manifolds (see, e.g. [1,2,8]).

Theorem 1 *Let $p(t,x,y)$ be the fundamental solution to (1). Then there exist $C_1, C_2, C_3, C_4 > 0$ such that*

$$C_1 \frac{e^{-C_2 \frac{d(x,y)^2}{t}}}{|B^d(x,\sqrt{t})|} \leq p(t,x,y) \leq C_3 \frac{e^{-C_4 \frac{d(x,y)^2}{t}}}{|B^d(x,\sqrt{t})|} \tag{2}$$

for all $x, y \in \mathbf{R}^n$ and $t > 0$.

The proof is based on the Poincaré inequality in the form: there are $P > 0, \kappa \geq 1$ such that

$$\int_{B^d(x,R)} \left| u - |B^d(x,R)|^{-1} \int_{B^d(x,R)} u dz \right|^2 dz \leq PR^2 \int_{B^d(x,\kappa R)} |x|^\alpha |\nabla u|^2 dz$$

for all $R > 0, x \in \mathbf{R}^n, u \in C^1(\mathbf{R}^n)$, the doubling property and the equivalence of the original topology in $\mathbf{R}^n$ with the one induced by the distance d (see [8]).

Remark 1 *Estimate (2) is equivalent to the parabolic Harnack principle for the equation (1) (see [3,8]). In particular this implies that every non-negative solution to (1) is either strictly positive or identically zero.*

Corollary 1 *Let $\Gamma(x,y)$ be the fundamental solution to the elliptic equation $\sum_{i,j=1}^n \partial_{x_i} a_{ij}(x)|x|^\alpha \partial_{x_j} u(x) = 0$. Then there exist $c, C > 0$ such that*

$$c \int_{d(x,y)^2}^{\infty} |B^d(x,\sqrt{t})|^{-1} dt \leq \Gamma(x,y) \leq C \int_{d(x,y)^2}^{\infty} |B^d(x,\sqrt{t})|^{-1} dt.$$

In particular,

$$c|x|^{2-n-\alpha} \leq \Gamma(x,0) \leq C|x|^{2-n-\alpha}.$$

This follows from Theorem 1 and the equality $\Gamma(x,y) = \int_0^\infty p(t,x,y)dt.$

Perturbation by a potential. One can extend the result of Theorem 1 to the case of equations with singular lower order terms. Here we confine ourselves to zero order perturbations (called potentials). The proper class of potentials which preserves local in time two-sided Gaussian estimates on the fundamental solutions is the so-called Kato-class. In our setting it can be defined as follows: $V \in L^1_{loc}$ is said to belong the Kato-class if $\lim_{r \to 0} \sup_{x \in \mathbf{R}^n} \int_{d(x,y)<r} \Gamma(x,y)|V(y)|dy = 0$. In fact, it suffices to require the last limit to be less than one, in which case it is said that V belongs to the enlarged Kato-class. To illustrate typical singularities of the Kato-class potentials we note that the potential $\frac{\mathbf{1}_R}{|x|^2(\log \frac{1}{|x|})^{1+\epsilon}}$, $\epsilon > 0$, $R < 1$, is from the Kato-class for $\alpha = 0$. ($\mathbf{1}_R$ is the characteristic function of the ball of radius R centered at the origin.)

In order to have global in time Gaussian estimates for the fundamental solution, one needs a more restrictive condition on the potential, namely, $\sup_{x \in \mathbf{R}^n} \int_{\mathbf{R}^n} \Gamma(x,y)|V(y)|dy < 1$, which we will call the global Kato-class.

In the next section we will also use an assumption on the potential, controlling its behaviour at infinity, which is some times called Green-tight:

$\lim_{R\to\infty} \sup_{x\in\mathbf{R}^n} \int_{|y|>R} \Gamma(x,y)|V(y)|dy = 0$. One can see that if a potential V is from the Kato-class and Green-tight then $\sup_{x\in\mathbf{R}^n} \int_{\mathbf{R}^n} \Gamma(x,y)|V(y)|dy < \infty$.

If a potential V satisfies the global Kato-class condition then the assertion of Theorem 1 holds with $p(t,x,y)$ being the fundamental solution to the equation

$$\partial_t u(x,t) = \sum_{i,j=1}^{n} \partial_{x_i} a_{ij}(x)|x|^\alpha \partial_{x_j} u(x,t) + V(x)u(x,t).$$

We refer the reader to Zhang [9] for the proof whose method carries over in our setting (see also [6]).

It is of course natural to investigate the validity of the Gaussian estimates for the fundamental solutions of parabolic equations in presence of first order perturbations. Most known results for such equations are obtained under additional assumptions that the coefficients of the main part (a_{ij}) satisfy certain degree of smoothness (typically Hölder continuous). In the general case of measurable bounded (a_{ij}) for $\alpha = 0$ we refer to [7,6] where new classes of first order perturbations guaranteeing Gaussian estimates for fundamental solutions were introduced.

AN EXTERIOR PROBLEM FOR AN ELLIPTIC INEQUALITY

In this section we discuss an application of the main result of the previous section. We are concerned here with non-negative weak solutions to the inequality

$$\sum_{i,j=1}^{n} \partial_{x_i} a_{ij}(x)|x|^\alpha \partial_{x_j} u(x) + u(x)^q \le 0 \tag{3}$$

outside a compact set $K \subset \mathbf{R}^n$. It turns out that there exists a critical value of q separating the cases of existence and non-existence of positive solutions. The result reads as follows.

Theorem 2 *Let $n \ge 3$ and $0 \le \alpha < 2$. Then the following are equivalent:*

(i) For every compact set $K \subset \mathbf{R}^n$, every non-negative solution to (3) outside K is identically zero.

(ii) $q \le \dfrac{n}{n-2+\alpha}$.

Let us sketch the proof of non-existence of positive solutions. The next lemma is the key ingredient.

Lemma 2 *Let $0 \le W \in L^\infty_{loc}$. Let u be a non-negative weak solution to the inequality*

$$\sum_{i,j=1}^{n} \partial_{x_i} a_{ij}(x)|x|^\alpha \partial_{x_j} u(x) + W(x)u(x) \le 0$$

in a ball. Then there exists λ such that $W > \lambda$ implies $u \equiv 0$.

The proof of the lemma is based on standard quadratic estimates.

Lemma 3 *Let $u \ge 0$ be a non-trivial solution to (3). Then there exists $C > 0$ such that*

$$u(x) \ge C\Gamma(x, 0).$$

The proof of the lemma uses the weak maximum principle.

In order to prove the statement on non-existence of positive solutions in Theorem 2 one has to distinguish several cases.

Case 1. $1 \le q < \frac{n}{n+\alpha-2}$. Let R_0 be such that $K \subset B(0, R_0)$. Suppose that $u \ge 0$ is a nontrivial solution to (3) outside K. Then u solves (3) in any annulus $B(0, 2\rho R_0) \setminus B(0, \rho R_0)$ with $\rho \ge 1$. Scaling (3) to $B(0, 2R_0) \setminus B(0, R_0)$ and using Lemma 3 one gets a contradiction with Lemma 2.

Case 2. $q = \frac{n}{n+\alpha-2}$. To treat this case we consider the equation

$$\sum_{i,j=1}^{n} \partial_{x_i} a_{ij}(x)|x|^\alpha \partial_{x_j} v(x) + c|x|^{-2+\alpha} v(x) = 0. \tag{4}$$

One can show that for a sufficiently small c there exists a weak solution to (4) such that $v|_{\partial B(0,R_0)} = 1$ (can be made precise in a weak sense) and $v(x) \ge c|x|^{2-n-\alpha} \log |x|$ outside a ball. This is also based on Lemma 3. Comparison with this solution and a scaling argument lead to the desired conclusion by means of getting a contradiction with Lemma 2.

Case 3. $q \le 1$. The proof is based on the maximum principle and use of Lemmas 3 and 2.

In order to prove the existence of positive solutions for the supercritical case $q > \frac{n}{n+\alpha-2}$ one shows, by means of Schauder's fixed point theorem, that for any $0 < R_0 < R$, the problem

$$\sum_{i,j=1}^{n} \partial_{x_i} a_{ij}(x)|x|^\alpha \partial_{x_j} u_R(x) + u_R(x)^q = 0, \quad u_R|_{\partial B(0,R_0)} = c, \; u_R|_{\partial B(0,R)} = 0$$

has a non-trivial non-negative weak solution satisfying the inequality $u_R(x) \le b|x|^{2-n-\alpha}$ with b independent of R. For this again the comparison with fundamental solution of the corresponding linear equation is crucial. The sequence u_R converges (in $H^1_{loc}(\mathbf{R}^n \setminus B(0, R_0))$) to a non-trivial weak solution of the equation

$$\sum_{i,j=1}^{n} \partial_{x_i} a_{ij}(x)|x|^\alpha \partial_{x_j} u(x) + u(x)^q = 0$$

outside $B(0, R_0)$.

Equation with a potential. Next we consider the problem of existence of non-negative weak solutions to the inequality

$$\sum_{i,j=1}^{n} \partial_{x_i} a_{ij}(x)|x|^{\alpha} \partial_{x_j} u(x) + V(x)u(x) + u(x)^q \leq 0 \tag{5}$$

outside a compact set $K \subset \mathbf{R}^n$. We are interested in conditions (minimal possible) on the potential V under which a result analogous to Theorem 2 holds. We assume that V is from the Kato-class and is Green-tight with respect to the unperturbed equation (1) (see the previous section). Under these assumptions the assertion of Theorem 2 holds for equation (5).

The proof is similar to that of Theorem 2. Note however that the assumptions on the potential do not yield the global Kato-class condition which is needed for global in time Gaussian estimates. So in the supercritical case we choose R_0 in such a way that $\sup_x \int_{|y|>R_0} \Gamma(x,y)|V(y)|dy < 1$ and show the existence of a non-trivial solution to (5) outside the closure of the ball $B(0, R_0)$.

REFERENCES

1. Aronson D. *Non-negative solutions of linear parabolic equations*, Ann. Scuola Norm. Sup.Pisa **22** (1968), 607-694.
2. Davies E.B. *Heat Kernels and Spectral Theory,* Cambridge Tracts in Mathematics **92**, Cambridge University Press, 1989.
3. Fabes E.B. and Stroock D.W. *A new proof of Moser's parabolic Harnack inequality using the old idea of Nash*, Arch. Rat. Mech. **96** (1986), 327-338.
4. Kenig C. and Ni W.-M. *An exterior Dirichlet problem with applications to some nonlinear equations arising in geometry*, Amer. J. Math. **106** (1984), 689-702.
5. Konratiev V., Liskevich V. and Sobol Z., in preparation.
6. Liskevich V. and Semenov Yu. *Estimates for Fundamental Solutions of Second-order Parabolic Equations*, Journ. LMS, to appear.
7. Semenov Yu. *On perturbation theory for linear elliptic and parabolic operators; the method of Nash*, Contemporary Mathematics **221** (1999), 217-284.
8. Saloff-Coste L. *Parabolic Harnack inequality for divergence form second order differential operators*, Potential Analysis **4** (1995), 429-467.
9. Zhang Qi S. *On a parabolic equation with a singular lower order term, Part II: The Gaussian bounds*, Indiana Univ. Math. J., **43** (1997), 989-1020.

A Note on Nonlocal Equations in Mathematical Physics

Tadeusz Nadzieja[1]

Institute of Mathematics, Technical University of Zielona Góra
ul. Podgórna 50, 65-246 Zielona Góra, Poland

Abstract. The aim of this note is to present a simple proof of the existence of solutions of nonlocal equations appearing in the theory of self-interacting Brownian particles. A result on the nonexistence of solutions is also proved.

We will study the following problem

$$\varphi(x) = \frac{\lambda}{\int_\Omega e^{-\varphi}} \int_\Omega K(x,y) e^{-\varphi(y)}\, dy, \tag{1}$$

where $\varphi : \Omega \to \mathbb{R}$ is an unknown function from a bounded domain Ω of $\mathbb{R}^n$, $n > 1$, λ is positive parameter and $K(x,y)$ is a given symmetric kernel. The difficulty in the study of this problem lies not only in its nonlinearity but, above all, in the nonlocal term $\int_\Omega e^{-\varphi}$.

The problem (1) is motivated by the study of the stationary solutions of parabolic-elliptic systems describing the temporal evolution of a cloud of self-interacting Brownian particles [1–5].

The equation (1) can be derived also from the principle of maximum entropy. If U, ψ denote the density and the potential of a system of self-interacting Brownian particles confined to a container Ω, then

$$\psi(x) = \int_\Omega K(x,y) U(y)\, dy, \tag{2}$$

where the kernel K depends on the kind of interaction between the particles under consideration. For instance, assuming the existence of the Coulomb interaction $K = -E_n (= -G_n)$, where E_n (G_n) is the fundamental solution (the Green function) of the n-dimensional Laplacian. In the case of gravitational interaction it is reasonable to put $K = E_n$.

---

1) The preparation of this paper was supported by the KBN grant 2/P03A/011/19 and the Royal Society during the author's visit to King's College London.

CP553, *Disordered and Complex Systems*, edited by P. Sollich, et al.

If the system is in thermodynamical equilibrium then the density U maximizes the entropy $H(U) = -\int_\Omega U \log U$ under the constraints $\int_\Omega U = M$, $\frac{1}{2}\int_\Omega U\psi = E$. Here M and E are given mass (charge) and energy of the system. Thus U has the Boltzmann form

$$U = \frac{M}{\int_\Omega e^{-\psi/\Theta}} e^{-\psi/\Theta}, \tag{3}$$

where Θ is the temperature. Putting the Boltzmann form of the density (3) into (2) and denoting $\varphi = \psi/\Theta$, $\lambda = M/\Theta$, we get (1).

Note that if $K = \pm E_n$ or $K = \pm G_n$, then the solution of (1) satisfies the nonlocal differential equation

$$\Delta\varphi = \pm\lambda \frac{e^{-\varphi}}{\int_\Omega e^{-\varphi}}, \tag{4}$$

which is called the Poisson-Boltzmann(-Emden) equation in the case of electric (gravitational) interactions.

The Poisson-Boltzmann with zero boundary data was studied in [6–8]. It was proved that for all $\lambda > 0$ there exists a unique solution. Observe that (4) is the Euler-Lagrange equation for the functional $\lambda \log\left(\int_\Omega e^{-\varphi}\right) \mp \frac{1}{2}\int_\Omega |\nabla\varphi|^2$. We may use the variational methods for the proof of existence of solutions. These methods work well for the Poisson-Boltzmann equation and for the Emden problem in two-dimensional case [6]. In higher dimensional case or when we consider a more general nonlocal problem of the form

$$\varphi(x) = \frac{\lambda}{\left(\int_\Omega f(\varphi)\right)^p} \int_\Omega K(x,y) f(\varphi(y))\, dy, \tag{5}$$

where f is a given function and p is a positive parameter, we use an approach based on the Leray-Schauder theorem [9–11].

Here we give a simple proof of the existence of solutions of (1) under some natural assumptions on the kernel K.

Theorem 1. *Suppose that $K \geq 0$ and for some $r > 1$ $\sup_{x\in\Omega} \int_\Omega (K(x,y))^r\, dy < \infty$. Then, for each $\lambda > 0$ the problem (1) has a solution $\varphi \in L^\infty(\Omega)$.*

Proof. We define the nonlinear operator A on $L^\infty(\Omega)$

$$A\psi(x) := \lambda\mu(\psi) \int_\Omega K(x,y) e^{-\psi(y)}\, dy,$$

where $\mu(\psi) = \left(\int_\Omega e^{-\psi}\right)^{-1}$. We look for the fixed point of the operator A. A is a continuous and compact operator on $L^\infty(\Omega)$ [13]. To apply the Leray-Schauder theorem for the proof of existence of a fixed point of A we must have an *a priori* estimate for the possible solutions of the equation $\varphi = A\varphi$. Note that φ is positive, so

$$|\varphi|_\infty \leq \lambda\Gamma(\Omega)\mu(\varphi),$$

where $\Gamma(\Omega) = \sup_{x\in\Omega} \int_\Omega K(x,y)\,dy$. Hence we should estimate $\int_\Omega e^{-\varphi}$ from below. From the Jensen inequality we get

$$\exp\left(-\frac{1}{|\Omega|}\int_\Omega \varphi\right) \le \frac{1}{|\Omega|}\int_\Omega e^{-\varphi}.$$

Thus, to obtain the desired estimate, we must estimate $\int_\Omega \varphi$ from above.

We have

$$\int_\Omega \varphi = \lambda \frac{1}{\int_\Omega e^{-\varphi}} \int_\Omega \int_\Omega K(x,y) e^{-\varphi}\,dy\,dx$$
$$= \lambda \frac{1}{\int_\Omega e^{-\varphi}} \int_\Omega e^{-\varphi} \left(\int_\Omega K(x,y)\,dx\right) dy \le \lambda\Gamma(\Omega).$$

Hence we have an *a priori* estimate of $|\varphi|_\infty$, and as a consequence of the Leray-Schauder theorem we obtain the existence of solutions of (1).

Remark 1. The assumptions of Th. 1 are, of course, satisfied if $K = -E_n$ $(n \ge 3)$, or $K = -G_n$ [14].

Remark 2. Further regularity of solutions of (1) can be obtained under suitable assumptions on the kernel K. For example, assuming that for some $r > 1$ and $\beta \in (0,1]$ the kernel K satisfies

$$\sup_{x\in\Omega} \int_\Omega (K(x,y)|x-y|^{-\beta})^r\,dy < \infty,$$

$$\sup_{x\in\Omega} \int_\Omega (\nabla_x K(x,y)|x-y|^{1-\beta})^r\,dy < \infty,$$

the operator A transforms L^∞ into Hölder space $C^\beta(\Omega)$ [13]. Hence the solutions of (1) belong to $C^\beta(\Omega)$.

Remark 3. The regularity properties of φ near the boundary $\partial\Omega$ can be derived from the boundary behavior of the kernel K.

Remark 4. Using the same ideas as in the proof of Th. 1 we may obtain the existence of solutions of the problem (5) under some assumptions on the kernel K, function f and parameter p. Problems of this kind appear in the theory of thermistors [11].

Theorem 2. [9] *If K is positive definite then the equation (1) has at most one solution.*

Proof. Let φ_1 and φ_2 be two solutions of (1) and let $\nu_i = \log\mu(\varphi_i)$, $i = 1, 2$. We multiply the difference of the equation (1) written for φ_1, φ_2

$$\varphi_1(x) - \varphi_2(x) = \lambda \int_\Omega K(x,y)\left(e^{-(\varphi_1(y)-\nu_1)} - e^{-(\varphi_2(y)-\nu_2)}\right) dy$$

by $w(x) = e^{-(\varphi_1(x)-\nu_1)} - e^{-(\varphi_2(x)-\nu_2)}$ and integrate over Ω. We get

$$\int_\Omega (\varphi_1(x) - \varphi_2(x))w(x)\,dx = \int_\Omega ((\varphi_1(x) - \nu_1) - (\varphi_2(x) - \nu_2))w(x)\,dx$$
$$= \lambda \int_\Omega \int_\Omega K(x,y)w(x)w(y)\,dx\,dy \geq 0. \tag{6}$$

It follows from the monotonicity of the function e^{-x} that the left hand side of (6) is nonnegative if $\varphi_1 - \nu_1 = \varphi_2 - \nu_2$, so $\varphi_1 - \varphi_2 = const$. The left hand side of (1) is invariant with respect to transformation $\varphi \mapsto \varphi + c$, therefore $\varphi_1 = \varphi_2$.

It follows from Th. 1. and Th. 2. that the Poisson-Boltzmann equation has a unique solution for all $\lambda \geq 0$. It means that for the system of Brownian particles with electric self-interaction there exists a unique stationary state for each total mass (charge) of particles M and each temperature Θ.

If we consider the system with gravitational interaction (the Emden problem) the situation is more complicated. In star-shaped domains the solution exists only in some finite range of λ, i.e. if the total mass M of particles is sufficiently small or if the temperature Θ is sufficiently large. For instance, in the ball in $\mathbb{R}^2$, λ must be less than 8π and for this range of λ the solution is unique [12]. On the other hand, in an annulus or in a spherical shell the unique radially symmetric solution exists for all $\lambda > 0$ [15]. The problem whether the existence of solutions for all λ of the Emden problem depends on geometry or topology of Ω seems to be open and difficult [16]. The proofs of nonexistence results use the Pokhozaev identity (if $K = G_n$) [12] or the moment method [9]. In the proof of the next theorem we apply the latter.

We consider the problem (1) in a star-shaped domain with respect to the origin, i.e. such that for all $x \in \partial\Omega$, $x \cdot \nu(x) > 0$. Here $\nu(x)$ is the normal vector to the boundary at the point x. We prove

Theorem 3. *If Ω is a bounded, star-shaped with respect to the origin, and for all $x \in \Omega$, $\nabla_x K(x,y) \cdot x \geq \nabla_x E_n(x,y) \cdot x$, then for sufficiently large λ the problem (1) has no solution.*

Proof. If the density U has the Boltzmann form $U = M \frac{e^{-\varphi}}{\int_\Omega e^{-\varphi}}$ and φ satisfies (1) then

$$0 = \int_\Omega \nabla U \cdot x + \int_\Omega U \nabla\varphi \cdot x$$
$$= \int_\Omega \nabla U \cdot x + \int_\Omega U(x) \left(\int_\Omega \nabla_x K(x,y) \cdot x U(y)\,dy \right) dx.$$

Integrating by part the first integral and using the assumption on $\nabla_x K(x,y)$ we get

$$0 \geq \int_{\partial\Omega} U x \cdot \nu - n \int_\Omega U + \int_\Omega \nabla_x E_n(x,y) \cdot x U(y)\,dy\,dx.$$

After dropping the positive term and after symmetrization of the third term we have

$$0 \geq -nM + \frac{1}{2\sigma_n} \int_\Omega \int_\Omega U(x)U(y) \frac{(x-y) \cdot x + (y-x) \cdot y}{|x-y|^{-n}} \, dx \, dy$$

$$\geq -nM + \frac{1}{2\sigma_n} \int_\Omega \int_\Omega U(x)U(y)|x-y|^{-n+2} \, dx \, dy \geq -nM + \frac{1}{2\sigma_n} d^{2-n} M^2,$$

where d is the diameter of Ω. Therefore there are no solutions for $M > 2\sigma_n d^{2-n}$.

ACKNOWLEDGMENTS

The author wishes to thank Piotr Biler for interesting conversations.

REFERENCES

1. Aly J. J., *Physical Review E* **49**, 3771 (1994).
2. Bavaud F., *Rev. Mod. Phys.* **63**, 129 (1991).
3. Biler P., Krzywicki A., and Nadzieja T., *Rep. Math. Phys.* **42**, 359 (1998).
4. Streater R. F., *J. Stat. Phys.* **88**, 447 (1997).
5. Streater R. F., *J. Math. Phys.* **38**, 4570 (1997).
6. Caglioti E., Lions P.-L., Marchioro C., and Pulvirenti M., *Comm. Math. Phys.* **143**, 501 (1992).
7. Biler P., Hebisch W., and Nadzieja T., *Nonlinear Analysis T. M. A.* **23**, 1189 (1994).
8. Krzywicki A., and Nadzieja T., *Ann. Polon. Math.* **54**, 125 (1991).
9. Biler P., and Nadzieja T., *Colloq. Math.* **66**, 131 (1993).
10. Biler P., and Nadzieja T., *Nonlinear Analysis T. M. A.* **30**, 5343 (1997).
11. Krzywicki A., and Nadzieja T., *Banach Center Publ.* **52**, 147 (2000).
12. Krzywicki A., and Nadzieja T., *Zastosowania Matematyki* **21**, 265 (1991).
13. Kantorovich L. V., and Akilov G. P., *Functional Analysis*, Oxford, Pergamon Press, 1982.
14. Grüter M., and Widman K.-O., *Manuscripta Math.* **37**, 303 (1982).
15. Krzywicki A., and Nadzieja T., *Zastosowania Matematyki* **21**, 591 (1993).
16. Ding W., Jost J., Li J., and Wang G., *Ann. Inst. H. Poincaré Anal. Non Linéaire* **16**, 653 (1999).

PART V:
NEW DIRECTIONS
IN MATHEMATICAL FINANCE

Risk Avalanches in Financial Markets

Philippe Balland

Merrill Lynch International, 25 Ropemaker Street, London EC2Y 9LY, U.K.
Email ballaphi@ml.com

Abstract. Assuming simple trading rules for market-makers, short-term and long-term investors, and tight risk limits for the market-makers, we find that a financial market can self-organize in a critical state where risk avalanches and thus price changes of all sizes can take place.

FROM BACHELIER TO BLACK AND SCHOLES

In 1900, twenty years before the pioneering work of Norbert Wiener on Brownian motion, Louis Bachelier [1] first introduced the random walk to model the evolution of financial prices. Bachelier constructed the first Brownian motion from three assumptions: the path continuity, the independence and the stationarity of the increments. From these three assumptions, Bachelier proved using 'heuristic' arguments that price increments are normally distributed. The model introduced by Bachelier can be summarized as follows:

$$P_{t+dt} - P_t = \sigma \, dW_t \ ,$$

where σ is a constant, P_t is the accrued-adjusted price at time t of a perpetual bond issued by the French government and W_t is a Brownian motion. The work of Bachelier remained unnoticed until 1943 when Paul Levy [12] discovered that Bachelier had obtained some important results on Brownian motion at the begining of the century.

The use of random walks to model the evolution of prices originates from the efficient market hypothesis which asserts that all public information useful for making investment decisions is already incorporated in market prices. The efficient market hypothesis was first formulated by Bachelier who assumed that the expectation of trading gain is zero. The efficient market hypothesis is still the object of intense research although a substantial amount of empirical evidence in favour of this hypothesis has been accumulated over the past century. In 1970, Fama [5] claimed that returns are independent as a consequence of the efficient market hypothesis.

CP553, *Disordered and Complex Systems*, edited by P. Sollich, et al.

Fama argued that if the increments were correlated over a long time then short-term traders would take advantage of this dependence and as a consequence this dependence would disappear.

In 1963, Mandelbrot [13] rejected the assumption of continuous prices and Gaussian returns and proposed instead to model the price as a geometric symmetric Levy flight. Mandelbrot provided empirical evidence of the α-stability of return distributions.

In 1965, Samuelson [17] proposed to model stock prices using the following geometric Brownian motion:

$$dP_t/P_t = \mu\, dt + \sigma\, dW_t \ ,$$

where the 'real-world' drift μ and the volatility σ are assumed constant. Samuelson derived an expression for the price of a European stock option by simply calculating the expected value of the discounted payoff. The Samuelson price is not, however, the cost of replication as it depends on the 'real-world' drift μ.

In 1973, Black and Scholes [3] obtained an analytical expression for the cost of replicating a European stock option payoff when there are no dividends. The Black-Scholes option price is defined as the replication cost of the option payoff and it is independent of the drift μ. The Black-Scholes formula and the associated delta-hedging strategy are immediately adapted and extended to cover options on stocks with continuous dividend, on future contracts, on foreign exchange rates, on bonds and on swaps.

The lack of realism of the Black-Scholes assumptions is however well documented. Financial prices are not continuous and financial daily returns have fat tails and large kurtosis. Return distributions lack stationarity and the local volatility is not deterministic. There is volatility clustering in the sense that large changes in price tend to be followed by large changes and small changes tend to be followed by small changes.

These imperfections of the Black-Scholes model can be classified in two categories. The first type of imperfection relates to the observed fat-tails and high kurtosis of the conditional distribution of price returns. This corresponds to the 'Noe' effect as defined by Mandelbrot. The second type of imperfection relates to volatility clustering. This corresponds to the 'Joseph' effect as defined by Mandelbrot. Both the 'Noe' and 'Joseph' effects are supported by a large amount of empirical evidence.

In 1989, Bak [2] found that the energy released during earthquakes seems to exhibit the 'Noe' and 'Joseph' effects. Indeed, the distribution of the energy released during an earthquake has fat tails and periods of intense earthquake activity are usually followed by periods of intense earthquake activity. Bak found that both effects are in fact consequences of the self-organized critical behavior of the earth's crust. Financial markets also exhibit the 'Noe' and 'Joseph' effects. A natural question is therefore the following. Can financial markets exhibit self-organized critical behavior?

RISK AVALANCHES AND SELF-ORGANIZED CRITICALITY

We consider a trading economy for a single financial asset A. The market-players consist of a long-term investor, some short-term investors and a large number n of market-makers M_i. We assume that the market-makers publish after every transaction the last-traded price P for one unit of A.

The market-makers act as intermediaries between market-players having different needs or views. Market-makers have a tight risk-limit L. They buy and sell exclusively on 'demand' or when they have accumulated too much risk.

The long-term investor follows a long-term strategy. We assume that the long-term investor buys or sells with equal probability and at regular times $t = k\Delta t$, one unit of A by selecting at random a particular market-maker.

The short-term investors are speculators who look for trends in the market and who 'buy low' and 'sell high'.

The price elasticity is denoted ϵ and is assumed very small. An increase in demand/offer pushes the logarithm of the price up/down by an amount ϵ. If the current price is P then a market-player can buy one unit of A from a market-maker at a price Pe^{ϵ}. Following this transaction, the market-maker publishes Pe^{ϵ} as the last traded price and thus as the new indicative price.

A market-maker M_i with an accumulated position q_i of L (resp. $-L$) units of A transfers their risk by selling (resp. buying) $2L-1$ units of A to q market-makers chosen at random and to $l = 2L-1-q$ short-term investors who buy one unit of A as they believe that the current price is low (resp. high) compared to the historical level. Suppose the price of A is P before M_i transfers risk. As a result of the transactions, the logarithm of the price decreases (resp. increases) by an amount equal to $\epsilon(q-l)$. Some of these q market-makers may themselves need to sell (resp. buy) in order to adjust their positions. The initial transfer of risk can thus trigger an avalanche of risk which results in a large decrease (resp. increase) of the price. It is interesting to note that the short-term investors act as market-regulators as they limit the propagation of the risk avalanches. We assume that the life of a risk avalanche is much smaller than Δt, namely that the long-term investor does not update their strategy during an avalanche.

The cumulative number of transactions initiated by the long-term investor since $t = 0$ is denoted N and is referred to as business time. We denote by $N_t = [t^-/\Delta t]$ the business time at time t^-. The change in the logarithm of the price between two successive business times N and $N+1$ satisfies the following equation

$$\Delta \ln P_N = \delta_N \ (1 + (q-l)\Lambda_N) \ \epsilon \ ,$$

where δ_N takes values ± 1 with equal probability and Λ_N is the number of risk transfers following the Nth long-term transaction. We note that at time t, the last traded price satisfies $P_t = P_{N_t}$.

Suppose that the market-makers have an infinite risk limit. We assume that $\widehat{\epsilon} = \epsilon \Delta t^{-1/2}$ is independent of Δt. The characteristic function of $\ln P_t$ satisfies

$$E[P_t^{iz}] = P_0 \cos(z\widehat{\epsilon}\Delta t^{1/2})^{N_t} \overset{\Delta t \to 0}{\longrightarrow} P_0 \exp(-\frac{1}{2} z^2 \widehat{\epsilon}^2 t), \qquad \forall z.$$

The complex number $\sqrt{-1}$ is denoted i. The process $\ln P_t$ converges therefore to the Gaussian process $\widehat{\epsilon} W_t$ as Δt tends to zero. Note that this convergence is fast. In the absence of risk-limits and under the assumption of continuous and uniform trading, the model thus coincides with the Black-Scholes model. A constant drift can be incorporated by assuming that the long-term investor buys and sells with different probabilities.

Suppose that the risk limit L is finite. Denote by $q_{i,N}$ the position of market-maker M_i just before the Nth transaction. At business time N and for $0 \leq k \leq 2L-2$, the percentage of market-makers having a position equal to $k - L + 1$ is denoted $\chi_{k,N}$, and the number of risk-transfer following the Nth transaction is denoted Λ_N. We have to first order in $1/n$ and for all k such that $0 \leq k \leq 2L-2$:

$$\chi_{k,N+1} - \chi_{k,N} = \frac{q\Lambda_N}{n}(\chi_{\widehat{k\pm 1},N} - \chi_{k,N}) \ ,$$

where $\widehat{k}$ is equal to k modulo $2L-1$. Thus, the market evolves to a state satisfying $\chi_{\widehat{k\pm 1},+\infty} = \chi_{k,+\infty}$; this stationary state is characterized by $\chi_{k,+\infty} = 1/(2L-1)$ for all k such that $0 \leq k < 2L-1$. We suppose from now on that this stationary state has been reached by time $t=0$.

At business time N, the long-term investor can thus trigger a risk-transfer with a probability $1/(2L-1)$. The number of risk-transfers following this initial risk-transfer is a branching process with generating function:

$$f(z) = 1 - \frac{1}{2L-1} + \frac{1}{2L-1} z^q.$$

From this expression we derive the average number of new branches per node $\theta = f'(1) = q/(2L-1)$. Since this average of new risk-transfers is less than or equal to one, we conclude that risk-avalanches have almost surely finite size, i.e., $\Pr\{\Lambda_N < +\infty\} = 1$. Using a general result on branching process due to Harris [9], we obtain

$$\Pr\{\Lambda_N = k\} \overset{k \to +\infty}{\sim} \frac{C}{2L-1} \ k^{-3/2} \exp(-\lambda k) \ ,$$

where $\lambda \overset{\theta \to 1}{\sim} (1-\theta)^2$ and C is a constant.

It is remarkable that the power $3/2$ in the above formula is independent of the parameters L, q and l. It follows from the above equation that the density $d_{\Delta t}$ of $|\Delta \ln P_N|$ satisfies

$$d_{\Delta t}(s) \overset{s \to +\infty}{\sim} \frac{C \exp(\lambda/|q-l|)}{\epsilon|q-l|(2L-1)} (\frac{s}{\epsilon|q-l|})^{-3/2} \exp(-\lambda \frac{s}{\epsilon|q-l|}).$$

If the short-term investors do not intervene then we have $q = 2L - 1$. The system consisting of the market-makers is thus conservative as there is no dissipation of risk during the avalanches. The system is in fact equivalent to the random sandpile system studied by Christensen and Olami [4]. The density of $|\Delta \ln P_N|$ scales like $(s/\epsilon)^{-3/2}$ as shown by the above equation. This system thus exhibits self-organized critical behavior as defined by Bak [2] and Jensen [11]. The system evolves to a critical state where a local distortion can propagate throughout the entire market. Assume that $\widehat{\epsilon} = \epsilon \Delta t^{-2}$ is constant. The generalized central limit theorem as formulated by Gnedenko and Kolmogorov [8] shows that the process $\ln P_t$ converges to a symmetric 1/2-stable Levy flight with characteristic function

$$E[P_t^{iz}] \stackrel{\Delta t \to 0}{\longrightarrow} P_0 \exp(-c|z|^{1/2} t), \qquad \forall z.$$

In the presence of a risk limit and in the absence of short-term investors, the market exhibits self-organized critical behavior. The model coincides with the Mandelbrot model. The process $\ln P_t$ has very fat tails. The non-intervention of the short-term investors could correspond to a panic situation.

If the short-term investors intervene then we have $q < 2L - 1$. The size of the avalanches decays exponentially and the density $d_{\Delta t}$ has finite variance. We note however that this density scales like $s^{-3/2}$ for $s < \epsilon|q - l|/\lambda$. Hence, the density is similar to the truncated Levy distribution as defined by Mantegna and Stanley [14]. Assume that $\widehat{\epsilon} = \epsilon \Delta t^{-1/2}$ is independent of Δt. A direct application of the central limit theorem shows that the process $\ln P_t$ converges to a Gaussian process, in the limit of zero Δt. When Δt is of order $(1 - q/(2L - 1))^4$, the process $\ln P_t$ is like a stable Levy flight. The convergence of $\ln P_t$ to a Gaussian process can thus be as slow as the convergence of the truncated Levy flight introduced by Mantegna and Stanley [14].

CONCLUSION

Assuming simple trading rules for market-makers, short-term and long-term investors, and tight risk limits for the market-makers, we find that a financial market can self-organize in a critical state where risk avalanches and thus price changes of all sizes can take place. This state emerges when the short-term investors do not intervene. We found that the price process corresponding to this critical state of the market is a Levy stable process with exponent $\alpha = 1/2$. It is interesting to note that this exponent has a universal character as it is independent of the particular parameters defining the model.

ACKNOWLEDGMENT

The author would like to thank Professor L.P. Hughston, David Shelton, Messaoud Chibane and the participants of the ICMP 2000 Satellite Conference on Disordered and Complex Systems for stimulating discussions.

REFERENCES

1. Bachelier, L. (1900). *Theorie de la Speculation.* Ann. Sci. Ecol. Norm. Sup., 3:21.
2. Bak, P. (1996). *How Nature Works; the Science of Self-Organized Criticality.* Springer-Verlag.
3. Black, F. and Scholes, M.S. (1973). *The Pricing of Options and Corporate Liabilities.* J. Political Economy, 81:637.
4. Christensen, K. and Olami, Z. (1993). Phys. Rev. E, 48:3361.
5. Fama, E.F. (1970). *Efficient Capital Markets: A Review of Theoretical and Empirical Work.* J. Finance, 25:383.
6. Geman, H. and Ane, T. (1996). *Stochastic Subordination.* Risk, Sep:145.
7. Gikhman, I.I. and Skorokhod, A.V. (1969). *Introduction to the Theory of Random Processes.* Dover.
8. Gnedenko, B.W. and Kolmogorov, A.N. (1954). *Limit distributions for sums of independent variables.* Addison-Wesley.
9. Harris, T.E. (1963). *The Theory of Branching Processes.* Springer-Verlag.
10. Jacod, J. and Shiryaev, A.N. (1980). *Limit Theorems for Stochastic Processes.* Springer-Verlag.
11. Jensen, H.J. (1998). *Self-Organized Criticality.* Cambridge Lecture Notes in Physics.
12. Levy, P. (1937). *Theorie de l'Addition des Variables Aleatoires.* Gauthier-Villars.
13. Mandelbrot, B. (1962). *The Variations of Certain Speculative Prices.* J. Business, 36:349.
14. Mantegna, R.N. and Stanley, E.H. (1994). *The Truncated Levy Flight.* Phys. Rev. Lett., 73:2946.
15. Potters, M., Cont, R. and Bouchaud, J.P. (1996). *Financial markets as adaptive ecosystems.* Working paper.
16. Samorodnitsky, G. and Taqqu, M.S. (1994). *Stable Non-Gaussian Random Processes.* Chapman.
17. Samuelson, P.A. (1965). *Rational Theory of Warrant Pricing.* Industrial Management Review, 6:49

Hedging Contingent Claims on Defaultable Assets

Johan G.B. Beumee

ANFP Quant Group, Abbey National Treasury Services
Abbey House, 215-229 Baker Street, London NW1 6XL, U.K.
Email: johan.beumee@ants.co.uk

Abstract. Following the JLT model (Jarrow, Lando and Turnbull), this paper represents a defaultable asset as a continuous stochastic process plus a Poisson jump modelling the bankruptcy event. It is shown that if the recovery condition is known beforehand, a contingent claim can be hedged by a position in the defaultable asset and a risk-free instrument. In addition, the claim must satisfy a jump-diffusion equation and a risk-neutral representation is obtained using this equation. Examples include the price of a risky zero-coupon bond with a fixed recovery value and the prices of risky Call/Put options on corporate instruments (instrument terminates upon default).

INTRODUCTION

A considerable effort has been made in the last twenty-five years or so to model the observed reduction to an asset price due to the possibility of counterparty default. A very elegant approach was originally introduced by Black and Scholes [1], Merton [2], (see Longstaff and Schwartz [3] for a modern reference) where bankruptcy was modeled as the inability of a company to repay its assumed debt on a (series of) specific date(s). The value of the debt must always be less than the full asset value of the company to stop the bondholders from claiming the company. In this framework, bankruptcy occurs when the share value becomes negative and the recovery value of the debt becomes then the debt minus the remaining asset value of the company.

An alternative to the structural approach was introduced by R. A. Jarrow, D. Lando and S. M. Turnbull (henceforth referred to as JLT) who considered solely the price behavior of the defaultable instrument [4–7]. As long as there is no distress the asset price propagates as a continuous stochastic process and at the point of bankruptcy the price of the risky bond jumps to a recovery value. Typically, risk-neutrality is then introduced as a constraint and an appropriate probability density is chosen that is consistent with observed risky market prices (JLT, Zhou [8]). The main point of this paper is to reverse this procedure by first applying a classical

CP553, *Disordered and Complex Systems*, edited by P. Sollich, et al.

Black & Scholes hedging argument and then to derive a risk-neutral world as a consequence rather than as a starting point. This defines the continuous time hedges properly and shows that the price of a contingent claim must satisfy a Black & Scholes type equation with a jump term that reflects the profit/loss associated with default. This jump-diffusion type equation will be referred to as the BSP (Black-Scholes-Poisson) equation and is a special case of an equation previously introduced by Merton, see [2,9,10].

To construct the default hedge the recovery value/percentage for the instrument has to be known exactly. Then a precisely balanced short position on the underlying will hedge the default risk of a long contingent claim position, i.e., the short corporate bond position covers the potential default losses on the risky bond Call option exactly. Subsequently, any other default-free instrument can then hedge the remaining market risk in the portfolio so two hedges are always sufficient to provide a perfect protection against default and market risk simultaneously. The hedged portfolio is now risk-less and has to return the risk-free rate from which the BSP equation for the contingent claim price follows. Examples presented include the price of a risky zero-coupon bond with a fixed recovery value and the prices of risky Call/Put options on corporate instruments (instrument terminates upon default).

HEDGING MARKET AND DEFAULT EVENTS

The BSP equation is obtained by means of hedging a contingent claim from default and market risk and concluding that the combined portfolio must provide the risk-free rate of return. The risk-neutral property follows as a consequence.

Following [2,5,8,11–13] etc., the pricing process $S(t)$ can be represented as the solution to the following equation

$$dS(t) = \mu_S\, dt + \sigma_S\, dw(t) + \eta\, dN(t) \tag{1}$$

where the usual Brownian process $w(t)$ models the market risk and where the Poisson directed discontinuous process $N(t)$ models the default risk with hazard rate λ. Here μ_S, σ_S are the instantaneous return and volatility respectively and a typical choice of the recovery term is $\eta = X_d - S(t)$ where X_d is the recovery value of the asset. As long as the Poisson process $N(t) = 0$ the asset has not defaulted but $N(t)$ can jump into default $N(t) = 1$ with probability $\lambda\, dt$ in any time interval dt. Hence, up until the stopping time of the Poisson jump the price evolves as a regular Gaussian/log-normal process and upon default it jumps from $S(t)$ to X_d.

First consider hedging a contingent claim $C = C(S(t), t)$ with respect to the default event by means of a short position designed to offset the jump exposure, so let $\Pi = C(S(t), t) - \delta_J S(t)$. Then using (1) and Ito's equation (see Elliott [14]) it is clear that, if we define

$$\Gamma_\eta C = C(S(t) + \eta, t) - C(S(t), t) = C(X_d, t) - C(S(t), t),$$

then

$$d\Pi = d[C - \delta_J S(t)] =$$
$$= \left[\frac{\partial C}{\partial t} + \mu_S\left(\frac{\partial C}{\partial S} - \frac{\Gamma_\eta C}{\eta}\right) + \frac{\sigma_S^2}{2}\frac{\partial^2 C}{\partial S^2}\right] dt + \left(\frac{\partial C}{\partial S} - \frac{\Gamma_\eta C}{\eta}\right)\sigma_S\, dw(t)$$

if δ_J is chosen as

$$\delta_J = \frac{\Gamma_\eta C}{\eta} = \frac{C(X_d, t) - C(S(t), t)}{X_d - S(t)}\,. \tag{2}$$

Notice the considerable abuse of notation as neither the differentials or the body of the equation differentiate between the actual value of $S(t)$ (which includes the jump) and its left limit $S(t-)$. Hence δ_J is a proper default hedge as the short position in the underlying asset generates the same gain as the contingent claim loses upon default leaving the portfolio invariant to the default event.

The default-free portfolio Π is now driven entirely by the market process $w(t)$ hence it can be immunized from market risk by an additional hedge $\delta_c V(t)$ chosen such that

$$\delta_c = \frac{S(t)\sigma_S}{V(t)\sigma_V}\left(\frac{\partial C}{\partial S} - \frac{\Gamma_\eta C}{\eta}\right).$$

The portfolio $\Pi(t) = C(S(t), t) - \delta_J S(t) - \delta_c V(t)$ with this choice of market risk and δ_J chosen as in (2) becomes risk-free and should generate the risk-free rate of return. A fairly elaborate calculation then yields the main result.

Theorem 1 *Consider a contingent claim $C = C(S(t), t)$ depending on a risky asset $S(t)$ with recovery value $X_d = \alpha S(t)$ and a default-free asset $V(t)$ depending only on $w(t)$, then if*

$$\begin{aligned}\frac{dS(t)}{S(t)} &= \mu_S\, dt + \sigma_S\, dw(t) + (\alpha - 1)dN(t)\\ &= [\mu_S + \lambda(\alpha+1)]dt + \sigma_S\, dw(t) + (\alpha - 1)dM(t),\\ \frac{dV(t)}{V(t)} &= \mu_V\, dt + \sigma_V\, dw(t),\end{aligned}$$

the contingent claim C must satisfy the Black-Scholes-Poisson (BSP) equation

$$\frac{\partial C}{\partial t} + \theta S(t)\frac{\partial C}{\partial S} + \frac{1}{2}\sigma_S^2 S^2(t)\frac{\partial^2 C}{\partial S^2} + \frac{\theta - r}{1-\alpha}\Gamma_\eta C = rC, \tag{3}$$

where

$$\begin{aligned}\Gamma_\eta C &= C(S(t) + \eta, t) - C(S(t), t) = C(\alpha S(t), t) - C(S(t), t),\\ \theta &= \mu_S - (\mu_V - r)\frac{\sigma_S}{\sigma_V}\,.\end{aligned}$$

An immediate consequence of this theorem is the existence of a risk-neutral world. Assume that the price of a contingent claim $C(S(t), t)$ satisfies (3) and that the probability of default $\lambda = (\theta - r)/(1 - \alpha)$. Then

$$C(S(0), 0) = E\left[e^{-\int_0^T r(s)\,ds} C(S(T), T)\right], \tag{4}$$

where $r(s)$ is the instantaneous risk-free rate. The expectation ranges over an algebra of all events including possible default shocks as well as pricing fluctuations so the expression generalizes the classical Black (Scholes) risk-neutral world to include default events.

The risk-neutral expression is a very convenient tool for generating prices of many risky assets. As a first example, consider the zero-coupon bond, which provides a cash-flow of $N + cpn$ at maturity T and upon default recovers to $X_d = R(N + cpn)$. Using (4) this bond can be valued as

$$\begin{aligned} P_{BSP}(0) &= e^{-\lambda T} E_w\left[e^{-\int_0^T r(s)\,ds} P_{BSP}(T)\right] + \int_0^T dt\, \lambda e^{-\lambda t} E_w\left[e^{-\int_0^t r(s)\,ds} X_d\right] \\ &\approx \left[e^{-\lambda T}(N + cpn) + R(N + cpn)\lambda T\right] E_w\left[e^{-\int_0^T r(s)\,ds}\right]. \end{aligned}$$

Here the second term is simply the sum (integral) of all possible payoffs resulting from a default at time t with probability $\lambda e^{-\lambda t}$. Notice that in the second term the value entered is the post-jump-default value, in other words, it is the value of the bond at time t multiplied with the recovery value. In the special case that the recovery value were zero the risky zero coupon bond is proportional to the default-free zero coupon, which is a result often found in the literature, see for instance [6].

Obviously, the BSP equation can be used to find the exact price of a -maturity T and strike K-Call option C_{BSP} on a defaultable asset $S(t)$ with $S = S(0)$, which terminates upon default (after devaluing to $X_d = \alpha S(t)$). Using the risk-neutral representation (4) again, the Call option becomes

$$\begin{aligned} &\mathrm{Call}_{BSP}(S, 0, T) = \\ &= e^{-\lambda T} E_w\left[e^{-\int_0^T r(s)\,ds}\, \mathrm{Call}_{BSP}\left(S e^{(r+\lambda(1-\alpha)-\sigma_S^2/2)T + \sigma_S w(T)}, T, T\right)\right] \\ &\quad + \int_0^T dt\, \lambda e^{-\lambda t} E_w\left[e^{-\int_0^t r(s)\,ds}\, \mathrm{Call}_{BSP}\left(S e^{(r+\lambda(1-\alpha)-\sigma_S^2/2)t + \sigma_S w(t)}, t, T\right)\right] \\ &\approx e^{-\lambda T}\, \mathrm{Call}_{BS}(S, K, r, \sigma_S, q = -\lambda(1 - \alpha), 0, T)\ , \end{aligned}$$

where

$\mathrm{Call}_{BSP}(S, t, T) =$ T-maturity Call premium (BSP) at time t for defaultable underlying,

$\mathrm{Call}_{BS}(S, t, T) =$ T-maturity Black-Scholes Call premium at time t.

This approximation is sensible, as upon default the Call option is essentially valueless. The risky Call option is proportional to the BS option expression (the probability of not defaulting being the proportionality), but with a negative dividend term that compensates the holder of the option for taking the risk on default.

For the Put a different expression is obtained as upon default the Put option is essentially far in the money. Therefore

$$\begin{aligned}\mathrm{Put}_{BSP}(S,0,T) &= e^{-\lambda T}\,\mathrm{Put}_{BS}(S,K,r,\sigma_S,q=-\lambda(1-\alpha),T,T)\\ &\quad+\int_0^T dt\,\lambda e^{-\lambda t}E_w\left[e^{-\int_0^t r(s)\,ds}\,\mathrm{Put}_{BSP}\left(\alpha S(t),t,T\right)\right]\\ &\approx e^{-\lambda T}\,\mathrm{Put}_{BS}(S,K,r,\sigma_S,q=-\lambda(1-\alpha),0,T)+\lambda T(K-\alpha S)\end{aligned}$$

This again makes intuitive sense, as λT is approximately the probability of default over the interval time to maturity T at which point a payment of approximately $(K-\alpha S)$ is received. Notice that the extra term can have a noticeable effect on the delta of the option.

CONCLUSIONS

This paper follows the JLT representation of presenting default risk by means of a Poisson jump and modeling market fluctuations by means of a continuous stochastic process. The default exposure of a long contingent claim can be hedged by means of a short position in the underlying if the default conditions are known precisely. The remaining market risk exposure in this portfolio can then be hedged if there are tradable assets that depend on the same market factors. Under these circumstances the unique price for a contingent claim satisfies a jump-diffusion equation, which is a special case of Merton's original jump-diffusion equation. Using this equation it follows that all contingent claims can be priced as the discounted expectation over all future price fluctuations including defaults. In other words, a risk-neutral world has been recovered that incorporates all default events given sufficient knowledge of the recovery condition.

Examples included the premium of a zero coupon bond that retains a fixed recovery value upon default and risky Call and Put prices. It can be shown that default-risky option prices calculated from the BSP equation produce a volatility "frown" if fitted to the Black-Scholes volatilities. The risk-neutral representation can easily be adjusted for stochastic short-term interest rates and path dependent intensities hence there are many other interesting applications of the BSP methodology to interest rate related instruments. This approach is clearly very suitable for calculating prices of credit related structures like Default Swaps and Total Return Swaps and examples of those have been presented elsewhere.

REFERENCES

1. F. Black, M. Scholes, The pricing of options and corporate liabilities, Journal of Political Economy 81 (1973) 637-654.
2. R.C. Merton, Continuous-Time Finance, Chapter 12 (1990) Blackwell.
3. F.A. Longstaff, E.S. Schwartz, A simple approach to valuing risky and floating rate debt, Journal of Finance 50 (1995) 789-819.
4. R.A. Jarrow, S.M. Turnbull, Pricing Derivatives on Financial Securities Subject to Credit Risk, The Journal of Finance 1, March (1995) 53-85.
5. D. Lando, On Cox processes and credit risky securities, PhD thesis, Cornell University (1994).
6. D. Duffie, K. Singleton, Recursive Valuation of Defaultable Securities and the Timing of Resolution of Uncertainty, working paper, Stanford University (1995).
7. M.A.H. Dempster, S.R. Pliska, eds., Mathematics of Derivatives Securities (1997) Cambridge University Press.
8. C. Zhou, A Jump-Diffusion Approach to Modelling Credit Risk and Valuing Defaultable Securities, Federal Reserve Board working paper (March 1997).
9. R.C. Merton, Option Pricing when Underlying Stock Returns are Discontinuous, The Journal of Financial Economics 3 (1976a) 125-144.
10. S.P. Mason, S. Bhattacharya, Risky Debt, Jump Processes, and Safety Covenants, Journal of Financial Economics 9 (1981) 281-307.
11. M.H.A. Davis, T.C. Mavroidis, Valuation and Potential Exposure of Default Swaps, working paper, Tokyo-Mitsubishi International plc, London (1997).
12. T.C. Mavroidis, Pricing by Default, working paper, Tokyo-Mitsubishi International plc, London (1997).
13. M.A.H. Dempster, G.Ch. Gotsis, On the Martingale Problem for Jumping Diffusions, working paper, University of Cambridge (1998).
14. R.J. Elliott, Stochastic Calculus and Applications (1982) Springer Verlag, New York.

Hyperbolic and Semi-Parametric Models in Finance

N.H. Bingham* and Rüdiger Kiesel†

*Department of Mathematical Sciences, Brunel University, Uxbridge, Middlesex UB8 3PH, UK
E-mail: nick.bingham@brunel.ac.uk
†Department of Statistics, London School of Economics,
Houghton Street, London WC2A 2AE, UK
E-mail: r.t.kiesel@lse.ac.uk

Abstract. The benchmark Black-Scholes-Merton model of mathematical finance is parametric, based on the normal/Gaussian distribution. Its principal parametric competitor, the hyperbolic model of Barndorff-Nielsen, Eberlein and others, is briefly discussed. Our main theme is the use of semi-parametric models, incorporating the mean vector and covariance matrix as in the Markowitz approach, plus a non-parametric part, a scalar function incorporating features such as tail-decay. Implementation is also briefly discussed.

INTRODUCTION

The most basic ingredient of a mathematical model for equity markets is a distribution to model stock prices, or stock returns. Different behaviour is observed as the time-interval over which return are calculated is varied. For intervals longer than about 16 days—monthly returns, say—the Gaussian or normal model for log-prices (lognormal model for stock prices) provides a reasonable fit to observed data. This gives the benchmark Black-Scholes-Merton theory, based on geometric Brownian motion; see, e.g., Musiela & Rutkowski [15], Bingham & Kiesel [6] for textbook expositions. For intervals of the order of minutes, power-law or Paretian behaviour—distributions with tails decaying like a power—is observed; see Mandelbrot [14], Rachev & Mittnik [16] for background. For intermediate time-intervals—of the order of a day, say—one expects tail-behaviour intermediate between the log-quadratic decay of the Gaussian/normal case and the power-law decay of the Paretian case.

Following a suggestion of Barndorff-Nielsen, Eberlein & Keller [8] found empirically that the hyperbolic distributions provide a model suitable for financial returns over intervals of the order of a day. The hyperbolic distributions were introduced earlier by Barndorff-Nielsen [1] in the context of size distributions of sand particles and turbulence. The log-density is smooth, unimodal, and approaches linear

CP553, *Disordered and Complex Systems*, edited by P. Sollich, et al.

asymptotes at $+\infty$ and $-\infty$ ('log-linear tail-decay'). The simplest curve exhibiting such characteristics is the lower branch of a hyperbola, and this explains the origin of hyperbolic laws in this context.

HYPERBOLIC DISTRIBUTIONS

The normal distribution $N(\mu, \sigma^2)$ is completely specified by its mean μ and variance σ^2 (or in r dimensions, the multivariate normal $N_r(\boldsymbol{\mu}, \boldsymbol{\Sigma})$ is completely specified by its mean vector $\boldsymbol{\mu}$ and covariance matrix $\boldsymbol{\Sigma}$). Normal distributions are symmetric, and so are unable to model asymmetry, and have ultra-thin tails (log-quadratic tail-decay).

Since asymmetry and much fatter tails are commonly observed in financial return distributions, one seeks first a parametric model that can handle these aspects. The third and fourth moments—or third and fourth cumulants, or their affine invariant versions, the *skewness* and *kurtosis*—are typically used to model asymmetry and tail-thickness respectively. The hyperbolic distribution (log-density the lower branch of a hyperbola) has four parameters, and so provides a plausible four-parameter model for financial returns, which has been successfully used by Eberlein & Keller [8] and others.

The Bessel function K_λ of the third kind has the integral representation

$$K_\lambda(x) = \frac{1}{2}\int_0^\infty u^\lambda \exp\left\{-\frac{1}{2}(u + 1/u)\right\} du/u \qquad (x > 0)$$

(Watson [19], §6.22, (8)), which can be regarded as the definition of K_λ here. Observe that $K_{-\lambda} = K_\lambda$, and recall that $K_{\pm\frac{1}{2}}(x) = \sqrt{\pi/(2x)}e^{-x}$ (Watson [19], §3.71). So for $\lambda \in \mathbb{R}$, $\psi, \chi > 0$,

$$f(x) := \frac{(\psi/\chi)^{\frac{1}{2}\lambda}}{2K_\lambda(\sqrt{\chi\psi})}\, x^{\lambda-1} \exp\left\{-\frac{1}{2}(\psi x + \chi/x)\right\} \qquad (x > 0)$$

is a probability density on $\mathbb{R}_+$. From its functional form (or the theory of exponential families), its Laplace transform is

$$\zeta(s) = \left(\frac{\psi}{\psi + 2s}\right)^{\frac{1}{2}\lambda} \frac{K_\lambda(\sqrt{\chi(\psi + 2s)})}{K_\lambda(\sqrt{\chi\psi})} \qquad (s > 0),$$

which we call the *K_λ-formula.* The corresponding probability law is called the *generalized inverse Gaussian, GIG* or $GIG_{\lambda,\psi,\chi}$; the case $\lambda = 1$ is the *inverse Gaussian, IG* or $IG_{\psi,\chi}$.

In one dimension (we discuss the r-dimensional case later), we form the *normal mean-variance mixture* ($NMVM$), with *mixing distribution* $F = GIG$: we sample σ^2 randomly from GIG and then form the (conditionally) normal distribution $N(\mu + \beta\sigma^2, \sigma^2)$. The resulting (unconditional) distribution is called the *generalized hyperbolic*, GH. These laws have been studied in detail by Barndorff-Nielsen [1,3] and co-workers. They can model asymmetry and log-linear tails. The simplest case, $\lambda = 1$, has as log-density a hyperbola, whence the name.

ELLIPTICALLY CONTOURED DISTRIBUTIONS

While GH provides a better fit to daily stock returns than normality, it is unreasonable to expect any simple parametric model to capture the full complexity of financial reality. Accordingly, we now relax such parametric restrictions.

We return to the r-dimensional setting, motivated by the Markowitz mean-variance theory and diversification principle for portfolios. We diversify by holding a basket of assets, including some negative correlation; the returns are modelled by the mean vector $\boldsymbol{\mu}$ and the risk by the covariance matrix Σ. We wish to retain this parametric component $\theta := (\boldsymbol{\mu}, \boldsymbol{\Sigma})$, because of its interpretation in the Markowitz approach. So if we avoid a parametric model, we are working *semi-parametrically* (see Bickel *et al* [5] for background), with parametric component as above and a non-parametric component.

A density of the form

$$f(\boldsymbol{x}) := |\boldsymbol{\Sigma}|^{-\frac{1}{2}} g(Q), \qquad Q := (\boldsymbol{x} - \boldsymbol{\mu})^T \boldsymbol{\Sigma}^{-1} (\boldsymbol{x} - \boldsymbol{\mu}) \qquad (\boldsymbol{x} \in I\!R^r) \qquad (EC)$$

is called *elliptically contoured* (or *elliptically symmetric*), with *density generator* g. Here $g : I\!R_+ \to I\!R_+$ is the non-parametric component of our model, and handles features such as tail-decay. Note that (EC) can only handle elliptical symmetry, so we cannot model general asymmetry in this framework.

Example: the multivariate normal, with $g(u) = (2\pi)^{-\frac{1}{2}r} \exp(-\frac{1}{2}u)$.
The characteristic function of f is

$$\psi(\boldsymbol{t}) := I\!E \exp\left\{i\boldsymbol{t}^T \boldsymbol{X}\right\} = \exp\left\{i\boldsymbol{t}^T \boldsymbol{\mu}\right\} \cdot \phi(\boldsymbol{t}^T \boldsymbol{\Sigma} \boldsymbol{t}), \qquad (EC')$$

for some scalar function ϕ called the *characteristic generator* of ψ, or f; we write $f \sim EC_r(\boldsymbol{\mu}, \boldsymbol{\Sigma}; g)$ or $f \sim EC_r(\boldsymbol{\mu}, \boldsymbol{\Sigma}, \phi)$. For a textbook account, see Fang *et al.* [9], Ch. 2.

It is convenient here to restrict attention to f (or ψ) infinitely divisible. An elliptically contoured law is infinitely divisible iff $\phi = \phi_k^k$ for each $k = 1, 2, \ldots$ with each ϕ_k again a characteristic generator. Examples include the multivariate normal, and the generalized hyperbolic laws (Barndorff-Nielsen & Halgreen [2]).

Call a random r-vector $\boldsymbol{X}$ *self-decomposable* if for each $\rho \in (0, 1)$ it coincides in law with $\rho\boldsymbol{X} + \boldsymbol{Y}_\rho$, for some random r-vector $\boldsymbol{Y}_\rho$ independent of $\boldsymbol{X}$. The self-decomposable laws form a subset of the infinitely divisible ones; symbolically,

$$SD \subset ID.$$

The multivariate normal laws are self-decomposable, and it was shown by Halgreen [10] that so too are the generalized hyperbolic laws (see also Bingham & Kiesel [7]).

NORMAL VARIANCE MIXTURES

Given a law F on $(0, \infty)$, we sample U from F and then sample $\boldsymbol{X}$ from $N_r(\boldsymbol{\mu}, U\boldsymbol{\Sigma})$:

$$\boldsymbol{X} \,|(U = u) \sim N_r(\boldsymbol{\mu}, u\boldsymbol{\Sigma}). \qquad (NVM)$$

The resulting law is called a *normal variance mixture*, with *mixing law* F, $NVM(\boldsymbol{\mu}, \boldsymbol{\Sigma}; F)$. Writing $\phi_m(s) := \int_0^\infty e^{-su} dF(u)$ $(s > 0)$ for the Laplace transform of F ('m for mixture'), $\boldsymbol{X}$ has characteristic function

$$\psi(\boldsymbol{t}) = \exp\left\{i\boldsymbol{t}^T \boldsymbol{\mu}\right\} \cdot \phi_m(\tfrac{1}{2}\boldsymbol{t}^T \boldsymbol{\Sigma} \boldsymbol{t}). \qquad (NVM')$$

Comparing with (EC'), we see that *all normal variance mixtures are elliptically contoured*, with characteristic generator $\phi(s) = \phi_m(s/2)$. Normal variance mixtures combine the advantages of being elliptically contoured with those of being normal mean-variance mixtures; symbolically,

$$NVM = NMVM \cap EC.$$

Both infinite divisibility and self-decomposability transfer from the mixing law F through to the mixture law; symbolically,

$$IDNVM = NMVM \cap IDEC, \qquad SDNVM = NMVM \cap SDEC.$$

We now restrict ourselves to self-decomposable normal variance mixtures—the class $SDNVM$ above—for four main reasons:

1. Self-decomposable distributions are absolutely continuous and unimodal (Sato [17]), Th. 28.4 and Th. 53.1). So for $SDEC$, the density generator *g exists and is decreasing.* Such a g, and its estimation, is the key to the model proposed here.
2. Self-decomposable laws are exactly those that can arise as marginal laws in autoregressive time-series models $\boldsymbol{X}_t = \rho \boldsymbol{X}_{t-1} + \boldsymbol{\epsilon}_t$ (where as usual, the innovations $\boldsymbol{\epsilon}_t$ are independent of $\boldsymbol{X}_s$ for $s < t$).
3. Self-decomposability transfers from the mixing law to the normal variance mixture law.
4. Self-decomposability, rather than the specifics of the parametric (normal inverse Gaussian) model, is the key to the important recent work on financial time-series by Barndorff-Nielsen *et al.* [4].

IMPLEMENTATION

We begin with a financial time-series of n r-vectors $\mathbf{x}_i$, recording the returns of the r assets in our portfolio. We then
(i) estimate the parameters $\boldsymbol{\mu}$, $\boldsymbol{\Sigma}$, obtaining estimators $\boldsymbol{\mu}_n$, $\boldsymbol{\Sigma}_n$;
(ii) use non-parametric density estimation to estimate f, by f_n;
(iii) use f_n, $\boldsymbol{x}_n$, $\boldsymbol{\Sigma}_n$ as 'plug-in' values in (EC), plus monotone regression (as in e.g. Härdle [11], p.217) to estimate g.

Our output is then a non-decreasing non-parametric estimate of g. Note that the 'curse of dimensionality'—explosion of the computational complexity with r—affects only the 'f step' but not the 'g step', which is one-dimensional.

The simplest case, $r = 2$, is also the most easy to visualize. We implemented our method for $r = 2$ on 1,000 daily returns of IBM and HP stock from December 1987 to December 1991. Plots of the two time-series, contour and relief plots of our estimate of the density (obtained by using the function kde2d from Venables & Ripley [18] in S-Plus), and our estimate of g are as shown in Figures 1 and 2.

Related work. In one dimension, Korsholm [13] treats normal mean-variance mixtures—and so can handle asymmetry—using mixture models for both theory and algorithms. Hodgson, Linton & Vorkink [12] treat general elliptical symmetry in r dimensions, with emphasis on the parametric component; their motivation is to test the capital asset pricing model.

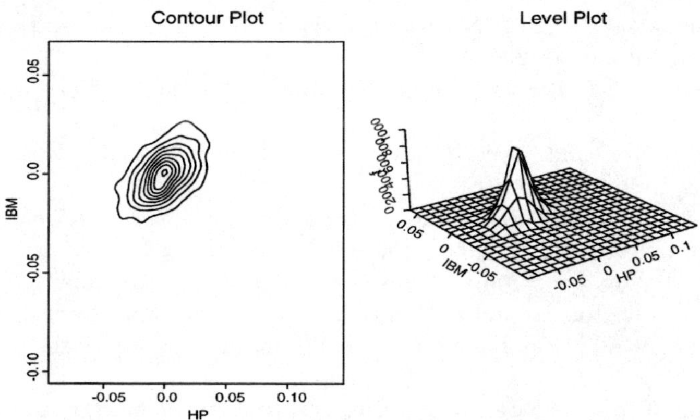

FIGURE 1. Contour and level plots for IBM/HP data.

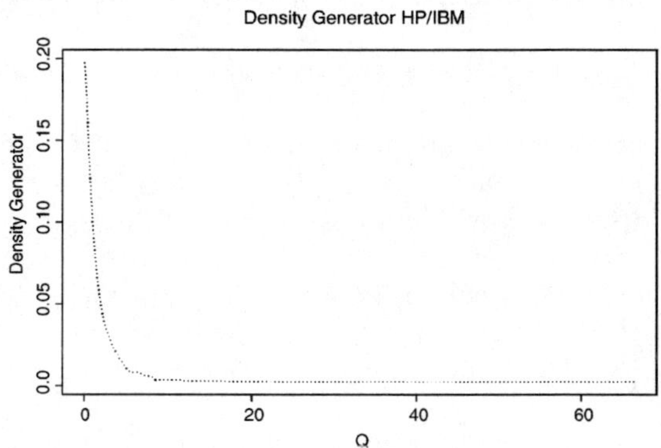

FIGURE 2. Density generator for IBM/HP data.

REFERENCES

1. O.E. Barndorff-Nielsen. Exponentially decreasing distributions for the logarithm of particle size. *Proc. Roy. Soc. London A*, 353:401–419, 1977.
2. O.E. Barndorff-Nielsen and O. Halgreen. Infinite diversibility of the hyperbolic and generalized inverse Gaussian distributions. *Z. Wahrschein.*, 38:309–312, 1977.
3. O.E. Barndorff-Nielsen, J.L. Jensen and M. Sørensen. Wind shear and hyperbolic distribution functions. *Boundary Layer Meteorology*, 49:417 – 431, 1989.
4. O.E. Barndorff-Nielsen, J.L. Jensen and M. Sørensen. Some stationary processes in discrete and continuous time. *Adv. Appl. Probab.*, 30:989 – 1007, 1998.
5. P.J. Bickel, C.A.J. Klaasen, Y. Ritov and J.A. Wellner. *Efficient and adaptive estimation for semi-parametric models.* Springer, New York, 1998. 1st edition 1993, Johns Hopkins University Press.
6. N.H. Bingham and R. Kiesel. *Risk-neutral valuation.* Springer Finance. Springer London, 1998; corrected reprinting, 2000.
7. N.H. Bingham and R. Kiesel. Semi-parametric modelling in finance. J. Royal Statist. Soc. B, to be submitted, 2000+.
8. E. Eberlein and U. Keller. Hyperbolic distributions in finance. *Bernoulli*, 1:281–299, 1995.
9. K.-T. Fang, S. Kotz and K.-W. Ng. *Symmetric multivariate and related distributions.* Chapman & Hall, London, 1990.
10. C. Halgreen. Self-decomposability of the generalized inverse Gaussian and hyperbolic distribution functions. *Z. Wahrschein.*, 47:13–18, 1978.
11. W. Härdle. *Applied nonparametric regression.* Cambridge University Press, 1990.
12. D.J. Hodgson, O. Linton and K. Vorkink. Testing the capital asset pricing model efficiently under elliptical symmetry: a semi-parametric approach. Preprint, LSE, 2000.
13. L. Korsholm. The semi-parametric normal mean-variance mixture model. *Scand. J. Statist.*, 27:227–261, 2000.
14. B.B. Mandelbrot. *Fractals and scaling in finance. Discontinuity, concentration, risk.* Springer, 1997. Selected works of Benoit B. Mandelbrot.
15. M. Musiela and M. Rutkowski. *Martingale methods in financial modelling.* Springer, New York, 1997.
16. S. Rachev and S. Mittnik. *Stable Paretian models in finance.* John Wiley & Sons, Chichester, 2000.
17. K.-I. Sato. *Lévy processes and infinite divisibility.* Cambridge University Press, 1999.
18. W.N. Venables and B.D. Ripley. *Modern Applied Statistics with S-PLUS.* Statistics and Computing. Springer, New York, 3rd edition, 1999.
19. G.N. Watson. *A treatise on the theory of Bessel functions.* Cambridge University Press, 2nd edition, 1944. 1st edition 1922.

Applications of Information Geometry to Interest Rate Theory

Dorje C. Brody* and Lane P. Hughston†

*Blackett Laboratory, Imperial College, London SW7 2BZ, UK
and Centre for Mathematical Science, Cambridge University,
Wilberforce Road, Cambridge CB3 0WA, UK
†Department of Mathematics, King's College London,
The Strand, London WC2R 2LS, UK

Abstract. There is an information geometry associated with the theory of interest rates. This arises from the fact that the system of smooth yield curves is isomorphic to the convex space of density functions on the positive real line. The arbitrage-free interest rate dynamics of Heath, Jarrow & Morton (1992) can be represented as a process on this space. Properties of this process are investigated and some special cases examined.

Recently there has been a growing interest in modelling the interest rate term structure as a dynamical system (see, e.g., Musiela 1993, Björk & Svensson 1999, Björk & Christensen 1999, Björk & Gombani 1999, Björk 2000, Brody & Hughston 2000, Filipović 2000). The idea is a simple one: we treat the yield curve as a mathematical object in its own right, a "point" ρ lying in a space $\mathfrak{M}$, which we identify as "the space of all possible smooth yield curves". Then with the specification of an initial yield curve ρ_0 we wish to understand the resulting dynamics, which we model as a random trajectory ρ_t in $\mathfrak{M}$. The following questions then arise: How do we best model $\mathfrak{M}$? What natural structures exist on $\mathfrak{M}$ that might assist us in determining the dynamical laws governing ρ_t? These questions are interesting because the standard HJM theory (Heath, Jarrow & Morton 1992) of arbitrage-free interest rate dynamics makes no use of specific structures on $\mathfrak{M}$. The thought is that by bringing the structure of $\mathfrak{M}$ into play it will be possible both to clarify the status of existing interest models, and to devise new interest rate models that might form a more adequate representation of the properties of observed interest rates. In what follows we shall sketch some tentative examples of new models that can be developed by following this line of argument.

The key point we emphasise here is that there is a natural information geometry associated with the space of yield curves. This comes about as follows. Let $t = 0$ denote the present, and P_{0T} a family of discount bond prices satisfying $P_{00} = 1$,

CP553, *Disordered and Complex Systems*, edited by P. Sollich, et al.

where T is the maturity date ($0 \leq T < \infty$). We impose the economic condition that interest rates should always be positive by means of the following criterion:

Definition. *A term structure will be said to be admissible if the discount function* P_{0T} *is of class* C^2 *and satisfies* $0 < P_{0T} \leq 1$, $\partial_T P_{0T} < 0$, *and* $\lim_{T\to\infty} P_{0T} = 0$.

The remarkable fact about an admissible discount function P_{0T} is that it can be viewed as a complementary probability distribution: we think of the maturity date as an abstract random variable X, and for its distribution we write $\Pr[X < T] = 1 - P_{0T}$. Then the associated density function $\rho(T) = -\partial_T P_{0T}$ clearly satisfies $\rho(T) > 0$ for all T, and $\int_T^\infty \rho(u)du = P_{0T}$. Now suppose we define the positive real line $\mathbb{R}^1_+ = [0, \infty)$. We shall say a density function is *smooth* if it is of class C^2 on $\mathbb{R}^1_+$. Then we have the following characterisation of the space $\mathfrak{M}$:

Proposition 1. *The system of admissible term structures is isomorphic to the convex space* $\mathcal{D}(\mathbb{R}^1_+)$ *of smooth density functions on the positive real line.*

The requirement that P_{0T} should be C^2 is to some extent arbitrary, and we may consider strengthening or weakening this condition slightly. It is reasonable, however, to insist that the forward short rate curve $f_{0T} = -\partial_T \ln P_{0T}$ should be continuous and nonvanishing for all $T < \infty$.

The space of probability distributions on a given sample space has natural geometric properties. In particular, given a pair of term structure densities $\rho_1(x)$ and $\rho_2(x)$ we can define a distance function ϕ_{12} on $\mathfrak{M}$ by the formula:

$$\phi_{12} = \cos^{-1} \int_0^\infty \xi_1(x)\xi_2(x)dx, \tag{1}$$

where $\xi(x) = \sqrt{\rho(x)}$. We call this 'angle' the Bhattacharyya distance between the given yield curves. The geometrical interpretation of ϕ_{12} is as follows. The map $\rho(x) \to \xi(x)$ associates to each point of $\mathfrak{M}$ a point in the positive orthant $\mathcal{S}^+$ of the unit sphere in the Hilbert space $L^2(\mathbb{R}^1_+)$, and ϕ is the resulting spherical angle on $\mathcal{S}^+$. We note that $0 \leq \phi < \frac{1}{2}\pi$ and that orthogonality can never quite be achieved on account of the requirement that forward rates are always nonvanishing.

Example 1. The discount bond family $P_{0T} = \exp(-rT)$ for constant r determines a 'flat' term structure with a continuously compounded rate of interest r for each value of the maturity date T. The spherical distance between two such yield curves is $\phi_{12} = \cos^{-1}\left(2\sqrt{r_1 r_2}/(r_1 + r_2)\right)$. ◇

We note that in this case the distance function is determined by the ratio of the geometric and arithmetic means of the corresponding rates.

Example 2. The discount bond family $P_{0T} = (1 + \kappa^{-1}rT)^{-\kappa}$ for constant r and κ corresponds to a 'flat' term structure with a constant annualised rate of interest r assuming a compounding frequency κ over the life of each bond (κ need not be an integer). In particular, the spherical distance between two such yield curves for $\kappa = 1$ is $\phi_{12} = \cos^{-1}\left(\ln(r_1/r_2)\sqrt{r_1 r_2}/(r_1 - r_2)\right)$. ◇

The spherical distance computation is applicable when we wish to make a comparison in nonparametric situations. In the case of a parametric family of yield curves we consider the corresponding submanifold of $\mathfrak{M}$ and the Fisher-Rao geometry induced on this submanifold (cf. Brody & Hughston 1998).

Now suppose we write P_{tT} for the random value at time t of a discount bond that matures at time T, where $T \in \mathbb{R}^1_+$ and $0 \leq t \leq T$. We assume for each value of T that P_{tT} is an Ito process defined on the interval $t \in [0,T]$, for which the dynamics are given by $dP_{tT} = m_{tT}dt + \Sigma_{tT}dW_t$. The process W_t, to which P_{tT} is assumed to be adapted, is a standard Brownian motion taking values in a separable Hilbert space $\mathcal{H}$ (cf. Da Prato & Zabczyk 1992, Filipović 2000). The absolute drift m_{tT} thus defined, along with the absolute volatility process Σ_{tT}, which takes values in the dual Hilbert space $\mathcal{H}^*$, are assumed to satisfy regularity conditions sufficient to ensure that $\partial_T P_{tT}$ is also an Ito process. For interest rate positivity we require $0 < P_{tT} \leq 1$ and $\partial_T P_{tT} < 0$. Additionally we impose the boundary conditions $P_{TT} = 1$, $\lim_{T\to\infty} P_{tT} = 0$, and $\lim_{T\to\infty} \partial_T P_{tT} = 0$.

We define the short rate $r_t = \rho_{tt}$, and the forward short rate $f_{tT} = \rho_{tT}/P_{tT}$. Because P_{tT} is positive, f_{tT} is an Ito process iff ρ_{tT} is an Ito process. We note that f_{tT} is the forward rate fixed at time t for the short rate at time T. The formula $\int_t^\infty \rho_{tu}du = 1$ says that the value at time t of a continuous cash flow that pays the small amount $f_{tu}du$ at time u is unity. For no arbitrage we require the existence of an $\mathcal{H}^*$-valued process λ_t, independent of T, such that $m_{tT} = r_t P_{tT} + \lambda_t \Sigma_{tT}$. We do not assume the bond market is complete.

To proceed further, we introduce the Musiela parameterisation $x = T - t$ for time to maturity and write $B_{tx} = P_{t,t+x}$ for the bond price thus parameterised. Thus B_{tx} represents the price at time t of a bond for which the time left until maturity is x, and we have the dynamical relation

$$dB_{tx} = (r_t - f_{t,t+x})B_{tx}dt + \Sigma_{t,t+x}(dW_t + \lambda_t dt), \tag{2}$$

which has a simple intuitive interpretation. Now suppose we write $\rho_t(x) = -\partial_x B_{tx}$. Because $\rho_t(x) > 0$ and $\int_0^\infty \rho_t(x)dx = 1$ for all $t > 0$, it follows that $\rho_t(x)$ is a density valued process. Then writing $\omega_{tx} = -\partial_x \Sigma_{t,t+x}$ we obtain the following dynamics:

$$d\rho_t(x) = (r_t\rho_t(x) + \partial_x\rho_t(x))dt + \omega_{tx}(dW_t + \lambda_t dt). \tag{3}$$

The process ω_{tx} is subject to the constraints $\int_0^\infty \omega_{tx}dx = 0$ and $\lim_{x\to\infty}\omega_{tx} = 0$, which implies that it can be expressed in the form $\omega_{tx} = \rho_t(x)(\nu_t(x) - \bar{\nu}_t)$, where $\bar{\nu}_t = \int_0^\infty \rho_t(u)\nu_t(u)du$ and $\nu_t(x)$ is unconstrained. The process $\nu_t(x)$ plays a role similar to the HJM forward short rate volatility. However, if the volatility structure is specified arbitrarily in the HJM theory there is no guarantee that the interest rate system will be admissible, whereas that feature is built into the present dynamics. The resulting bond volatility structure Σ_{tx} is invariant under transformations of the form $\nu_t(x) \to \nu_t(x) + \alpha_t$. This freedom can be used to set $\lambda_t = -\bar{\nu}_t$ without loss of generality. Then, both λ_t and Σ_{tx} are determined by $\nu_t(x)$.

Proposition 2. *The general admissible term structure evolution based on the information set generated by a Brownian motion W_t on a Hilbert space $\mathcal{H}$ is given by a measure valued process $\rho_t(x)$ in $\mathcal{D}(\mathbb{R}^1_+)$ satisfying*

$$\frac{d\rho_t(x)}{\rho_t(x)} = \left(\rho_t(0) + \partial_x \ln \rho_t(x)\right) dt + \left(\nu_t(x) - \bar{\nu}_t\right)\left(dW_t - \bar{\nu}_t dt\right), \tag{4}$$

where $\bar{\nu}_t = \int_0^\infty \rho_t(u)\nu_t(u)du$. The process $\nu_t(x)$ can be specified exogenously along with the initial term structure density $\rho_0(x)$.

It follows from the relation $\rho_t(0) = r_t$, that the process for the short rate satisfies

$$dr_t = \left(r_t^2 + \partial_x \rho_t(x)|_{x=0}\right) dt + r_t(\nu_t(0) - \bar{\nu}_t)(dW_t - \bar{\nu}_t dt). \tag{5}$$

The appearance of r_t^2 in the drift might seem counterintuitive. This is compensated by the second term appearing in the drift. For example, if we consider the CIR model (Cox et al. 1985), a calculation shows that the r_t^2 term cancels a similar term arising from $\partial_x \rho_t(x)|_{x=0}$, leaving the correct mean-reverting behaviour.

Proposition 3. *The solution of the dynamical equation for $\rho_t(x)$ in terms of the volatility structure $\nu_t(x)$ and the initial term structure density $\rho_0(x)$ is*

$$\rho_t(T-t) = \rho_0(T) \frac{\exp\left(\int_{s=0}^t V_{sT} dW_s - \frac{1}{2}\int_{s=0}^t V_{sT}^2 ds\right)}{\int_{u=t}^\infty \rho_0(u) \exp\left(\int_{s=0}^t V_{su} dW_s - \frac{1}{2}\int_{s=0}^t V_{su}^2 ds\right) du}, \tag{6}$$

where $V_{tu} = \nu_t(u-t)$. The corresponding formula for the bond price process is

$$P_{tT} = \frac{\int_{u=T}^\infty \rho_0(u) \exp\left(\int_{s=0}^t V_{su} dW_s - \frac{1}{2}\int_{s=0}^t V_{su}^2 ds\right) du}{\int_{u=t}^\infty \rho_0(u) \exp\left(\int_{s=0}^t V_{su} dW_s - \frac{1}{2}\int_{s=0}^t V_{su}^2 ds\right) du}, \tag{7}$$

and for the unit-initialised money market account $B_t = \exp\left(\int_0^t r_s ds\right)$ we have

$$B_t = \frac{\exp\left(\int_{s=0}^t \bar{\nu}_s dW_s - \frac{1}{2}\int_{s=0}^t \bar{\nu}_s^2 ds\right)}{\int_{u=t}^\infty \rho_0(u) \exp\left(\int_{s=0}^t V_{su} dW_s - \frac{1}{2}\int_{s=0}^t V_{su}^2 ds\right) du}. \tag{8}$$

The formulae for $\rho_t(x)$, P_{tT} and B_t indicated in Proposition 3 are surprisingly simple, given the nonlinearities of the dynamics of $\rho_t(x)$. These results can be used as a starting point for the derivation of new interest rate models, some of which we remark on briefly below.

Example 3. If $\nu_t(x)$ is chosen to be a deterministic function of t and x then we obtain a class of 'quasi-lognormal' models. In such a model the bond prices are given by ratios of superpositions of lognormally distributed random variables. ◇

Example 4. Martingale representations for discount bonds. It follows from Proposition 3 that an alternative expression for $\rho_t(x)$ is given by

$$\rho_t(T-t) = \frac{\rho_0(T) M_{tT}}{\int_{u=t}^\infty \rho_0(u) M_{tu} du}, \tag{9}$$

where for each value of T the process M_{tT} is a martingale over the interval $0 \le t \le T$, such that $M_{tT} > 0$ and $M_{0T} = 1$. We see that M_{tT} is the exponential martingale associated with V_{tT}. This expression for $\rho_t(T-t)$ arises also in the Flesaker-Hughston model (Flesaker & Hughston 1996, 1997, 1998; Rutkowski 1997; Musiela & Rutkowski 1997; Brody 2000; Hunt & Kennedy 2000), for which we have

$$P_{tT} = \frac{\int_{u=T}^{\infty} \rho_0(u) M_{tu} du}{\int_{u=t}^{\infty} \rho_0(u) M_{tu} du} \tag{10}$$

as a representation for the bond price. $\diamondsuit$

Example 5. Quasi-linear models. Suppose we write $\rho_0(u) = \int_0^\infty e^{-uR}\phi(R)dR$, where $\phi(R)$ is the inverse Laplace transform of the initial term structure density. Then for certain choices of the martingale family M_{tu} the integration $\int_T^\infty e^{-uR} M_{tu} du$ can be carried out explicitly. For instance, let M_t be an arbitrary martingale $(0 \le t < \infty)$ and Q_t the associated quadratic variation, so $(dM_t)^2 = dQ_t$, and set $M_{tT} = \exp\left((\alpha+\beta T)M_t - \frac{1}{2}(\alpha+\beta T)^2 Q_t\right)$. This model is obtained by putting $\nu_t(T-t) = (\alpha+\beta T)\sigma_t$ in Proposition 3, where $dM_t = \sigma_t M_t dW_t$. Then the u-integration can be carried out explicitly in the expressions for $\rho_t(x)$ and P_{tT}, and the results expressed in closed form:

$$\begin{aligned} \int_{u=T}^{\infty} e^{-uR} M_{tu} du = {} & \frac{1}{|\beta|\sqrt{Q_t}} \exp\left(\frac{1}{2}\frac{(M_t - \beta^{-1}R)^2}{Q_t} + \alpha\beta^{-1}R\right) \\ & \times \mathcal{N}\left(\pm\frac{(M_t - \beta^{-1}R)}{\sqrt{Q_t}} \mp (\alpha+\beta T)\sqrt{Q_t}\right), \end{aligned} \tag{11}$$

where $\mathcal{N}(x) = (2\pi)^{-1/2}\int_{-\infty}^{x} \exp(-\frac{1}{2}\xi^2)d\xi$ is the normal distribution function, and the $\pm$ sign is chosen according to the sign of β (cf. Brody 2000). These models are 'quasi-linear' in the sense that we have a closed form solution for a simple flat-rate initial term structure, and more general solutions can be constructed by taking linear superpositions of the resulting 'simple' martingale families. $\diamondsuit$

Example 6. Canonical exponential models. If we let $\rho_t(x)$ be of the form

$$\rho_t(x) = \frac{\exp\left(g_t(x) - \theta_t h_t(x)\right)}{\int_{u=0}^{\infty} \exp\left(g_t(u) - \theta_t h_t(u)\right) du}, \tag{12}$$

where θ_t is a one-dimensional diffusion and the processes $g_t(x)$ and $h_t(x)$ are deterministic, then at each time t the term structure belongs to an exponential family parameterised by the value of θ_t. The no-arbitrage condition implies that θ_t is a time inhomogeneous square-root process, and that $g_t(x)$ and $h_t(x)$ satisfy Riccati-type equations. $\diamondsuit$

Example 7. State-variable models. Here we consider the structure of admissible term structure dynamics driven by a stationary Markov process. In this case we write $\rho_t(T-t) = N_{tT}/\int_{u=t}^{\infty} N_{tu}du$, and we assume that the martingale family N_{tT} is of the form $N_{tT} = F(\psi_t, T-t)$, where the vector state variable ψ_t is a homogeneous diffusion process satisfying $d\psi_t = \alpha(\psi_t)dt + \beta(\psi_t)dW_t$. The condition that N_{tT} has no drift is $\partial_T F = \Delta F$, where $\Delta = \alpha\partial_\psi + \frac{1}{2}\beta^2\partial_\psi^2$ is the generator of the diffusion. The solution is formally $F(\psi_t, T-t) = e^{(T-t)\Delta}\Phi(\psi_t)$, where $\Phi(\psi_t)$ is a positive bounded function of class C^∞. We assume that Δ is negative and possesses an inverse Δ^{-1}. The solution for $\rho_t(x)$ is

$$\rho_t(x) = \frac{e^{x\Delta}\Phi(\psi_t)}{\int_{u=0}^{\infty} e^{u\Delta}\Phi(\psi_t)du}, \tag{13}$$

or, equivalently, $\rho_t(x) = -e^{x\Delta}\Phi(\psi_t)/\Delta^{-1}\Phi(\psi_t)$. It follows, further, on account of the Markov property, that $N_{tT} = E_t[\Phi(\psi_T)]$. The stationarity property ensures that $\alpha(\psi_t)$, $\beta(\psi_t)$ and $\Phi(\psi_t)$ determine the term structure at any time t as a function of the state variable ψ_t. This applies in particular to the initial term structure. The short rate is given by $r_t = -\Delta Z(\psi_t)/Z(\psi_t)$, where $Z(\psi_t) = \Delta^{-1}\Phi(\psi_t)$, and for the bond prices we have $B_{tx} = e^{x\Delta}Z(\psi_t)/Z(\psi_t)$. The forward short rates are then given by $r_{tx} = -e^{x\Delta}\Delta Z_t/e^{x\Delta}Z_t$, where $r_{tx} = f_{t,t+x}$, and we have $\rho_t(x) = -e^{x\Delta}\Delta Z_t/Z_t$. The significance of the process $Z(\psi_t)$ is that it represents the state-price density (cf. Rogers 1997, Hunt & Kennedy 2000), and a calculation shows that $dZ_t/Z_t = -r_t dt - \lambda_t dW_t$, where $\lambda_t = (\partial_\psi \ln Z)\beta_t$ is the risk premium process. ◇

Let us turn now to the problem of the relative movement of term structures, which can be given a simple characterisation by the use of information geometry. Given a density process $\rho(x)$ we form the associated square root process $\xi(x)$. Letting $\nu(x)$ and λ be exogenous and writing $\bar{\nu} = \int_0^\infty \rho(x)\nu(x)dx$, we have

$$d\xi = \left(\tfrac{1}{2}r\xi(x) + \partial_x\xi(x) - \tfrac{1}{8}(\nu(x) - \bar{\nu})^2\xi(x)\right)dt + \tfrac{1}{2}\xi(x)(\nu(x) - \bar{\nu})(dW + \lambda dt), \tag{14}$$

where for convenience here we suppress the time index. Given a pair of such processes $\xi_1(x)$ and $\xi_2(x)$ we form the process for the cosine of the spherical distance ϕ_{12} between the corresponding yield curves given in formula (1). An increase in $\cos\phi_{12}$ means a decrease in the distance. Writing $\sigma(x) = \nu(x) - \int_0^\infty \xi^2(x)\nu(x)dx$, for the dynamics of $\cos\phi_{12}$ we obtain

$$\begin{aligned} d\cos\phi_{12} = &\left(\tfrac{1}{2}(r_1 + r_2)\cos\phi_{12} - \sqrt{r_1 r_2} - \tfrac{1}{8}\int_0^\infty (\sigma_1(x) - \sigma_2(x))^2\xi_1(x)\xi_2(x)dx\right)dt \\ &+\tfrac{1}{2}\left(\int_0^\infty (\sigma_1(x) + \sigma_2(x))\xi_1(x)\xi_2(x)dx\right)(dW + \lambda dt). \end{aligned} \tag{15}$$

A special case arises when we consider the relative dynamics of two yield curves with differing initial conditions but governed by the same volatility and risk premium $\nu(x)$ and λ. Defining $\bar{\nu}_{12} = \int_0^\infty \xi_1(x)\xi_2(x)\nu(x)dx$, we have:

Proposition 4. *The Bhattacharyya distance process for two yield curves subject to the same underlying interest rate dynamics satisfies*

$$d\cos\phi_{12} = \left(\tfrac{1}{2}(r_1+r_2)\cos\phi_{12} - \sqrt{r_1r_2} - \tfrac{1}{8}(\bar{\nu}_1-\bar{\nu}_2)^2\cos\phi_{12}\right)dt + \left(\bar{\nu}_{12} - \tfrac{1}{2}(\bar{\nu}_1+\bar{\nu}_2)\cos\phi_{12}\right)(dW+\lambda dt). \quad (16)$$

The distance process is clearly invariant under transformations of the form $\nu(x) \to \nu(x) + \alpha$. It is interesting to note that the ratio of the geometric and arithmetic means of the short rates plays a critical role in determining the behaviour of $\cos\phi$. In particular, in a risk neutral world the two curves will necessarily tend to diverge once the distance is sufficiently great to ensure that $\cos\phi \leq 2\sqrt{r_1r_2}/(r_1+r_2)$.

Acknowledgements. The authors are grateful to P. Balland, T. Björk, D. Filipović, B. Flesaker, P. Glasserman, R. A. Jarrow, Y. Jin, T. Knudsen, D. Madan, and R. F. Streater for helpful remarks. DCB acknowledges the support of The Royal Society. LPH acknowledges the hospitality of the Finance Department of the Graduate School of Business of the University of Texas at Austin.

REFERENCES

1. Björk, T. & Christensen, B. J., *Mathematical Finance* **9**, 323 (1999).
2. Björk, T. & Gombani, A., *Finance and Stochastics* **3**, 413 (1999).
3. Björk, T., A Geometric View of Interest Rate Theory. In *Handbook of Mathematical Finance*, Cambridge: Cambridge University Press (2000).
4. Brody, D. C., *Mathematical Theory of Finance*, Tokyo: Nihon-Hyoronsya (2000).
5. Brody, D. C. & Hughston, L. P., *Proc. Roy. Soc. London* **454**, 2445 (1998).
6. Brody, D. C. & Hughston, L. P., Interest Rates and Information Geometry (Preprint, Imperial College London and King's College London, 2000).
7. Cox, J. C., Ingersoll, J. & Ross, S., *Econometrica* **53**, 385 (1985).
8. Da Prato, G. & Zabczyk, J., *Stochastic Equations in Infinite Dimensions*, Cambridge: Cambridge University Press (1992).
9. Filipović, D., Consistency Problems for HJM Interest Rate Models, PhD thesis, Swiss Federal Institute of Technology (2000).
10. Flesaker, B. & Hughston, L. P., *Risk Magazine* **9**, 46 (1996).
11. Flesaker, B. & Hughston, L. P., *Net Exposure* **3**, 55 (1997).
12. Flesaker, B. & Hughston, L. P., Positive Interest: An Afterword, in *Hedging with Trees*, ed. Broadie, M. & Glasserman, P., London: Risk Publications (1998).
13. Hunt, P. J. & Kennedy, J. E., *Financial Derivatives in Theory and Practice*, Amsterdam: North-Holland (2000).
14. Heath, D., Jarrow, R. & Morton, A., *Econometrica* **60**, 77 (1992).
15. Musiela, M., Stochastic PDEs and Term Structure Models, Journées Internationales de Finance, La Baule (1993).
16. Musiela, M. & Rutkowski, M., *Martingale Methods in Financial Modelling*, Berlin: Springer-Verlag (1997).
17. Rogers, L. C. G., *Math. Finance* **7**, 157 (1997).
18. Rutkowski, M., *Applied Math. Finance* **4**, 151 (1997).

Measures of Dependence for Multivariate Lévy Distributions[1]

J. Boland*, T. R. Hurd*, M. Pivato† and L. Seco†

*Dept. of Mathematics and Statistics, McMaster University
Hamilton, Canada, L8S 4K1
†Dept. of Mathematics, University of Toronto
Toronto, Canada, M5S 3G3

Abstract. Recent statistical analysis of a number of financial databases is summarized. Increasing agreement is found that logarithmic equity returns show a certain type of asymptotic behaviour of the largest events, namely that the probability density functions have power law tails with an exponent $\alpha \approx 3.0$. This behaviour does not vary much over different stock exchanges or over time, despite large variations in trading environments. The present paper proposes a class of multivariate distributions which generalizes the observed qualities of univariate time series. A new consequence of the proposed class is the "spectral measure" which completely characterizes the multivariate dependences of the extreme tails of the distribution. This measure on the unit sphere in M–dimensions, in principle completely general, can be determined empirically by looking at extreme events. If it can be observed and determined, it will prove to be of importance for scenario generation in portfolio risk management.

INTRODUCTION

Much research over the past forty years has examined statistical properties of stock returns with an aim to find models which improve over the Brownian motion models which began with Bachelier's work in the early 1900s. Bachelier's theory of speculation, adapted to logarithmic returns, assumes that successive increments of the logarithmic returns $X(t, \delta t) = \log[S(t + \delta t)/S(t)]$ of a time series of prices $S(t)$ are (a) random, (b) statistically independent, (c) identically distributed, and (d) Gaussian with zero mean. Bachelier himself pointed to assumption (d) as the weakest conceptual link, and subsequent researchers have noted that by replacing Gaussians with more general distributions the remaining three assumptions stand up relatively well. Amongst the models which have gained some popularity in the literature are stable–Lévy processes [11], [13], truncated stable–Lévy processes

[1] Research supported by the Natural Sciences and Engineering Research Council of Canada and by MITACS, Canada.

CP553, *Disordered and Complex Systems*, edited by P. Sollich, et al.

[3], fractional Brownian motion [12], hyperbolic distributions [7], [8] and Bessel distributions [10], [9].

Of the shortcomings of geometric Brownian motion, the most egregious is the fat tail problem: it is generally agreed that empirical time series for financial returns are leptokurtic, i.e., they have tails which are thicker than those of any Gaussian. Amongst the fat–tailers who regard this fact as very important we can identify two camps: the "exponential tailers" and the "power law tailers" who see a probability density function which is asymptotically proportional to a negative power of the event size.

The theoretical justification for power law tails rests on the experience of physicists, geologists and others who study the mathematics of complex systems which consist of an enormous number of units connected by finite range interactions. For example, highly turbulent fluids are described qualitatively by a cascade of energy from large distance scales through to a short distance dissipative scale. The spectrum of excitations exhibits "Kolmogorov scaling" which is a clear power law. The concept of "self–organized criticality" has been proposed as an explanation for the ubiquity of power laws in such complex non-equilibrium systems [1]. A key consequence of this concept is the "universality" of the exponents of the power laws in question, i.e., their invariance under continuous changes in the microscopic interaction laws.

An equity market can be viewed as a large system of traders in interaction with each other and the external economy. Thus it is natural to imagine that prices, being a measure of the state of the system, might exhibit self–organized criticality and consequently the type of universal power law seen in some physical systems. Of course, finance is not physics, and any tidy picture from physics will be severely muddied by problems of changing market conditions (stemming, for example, from evolving technology), irrationality and other psychological factors, the wide variation in trading styles, etc. Nevertheless, if self–organized criticality is applicable in finance it leads naturally to an optimistic prediction, consequences of which we will explore in this paper:

Scaling Prediction: *The log returns within equity markets should exhibit power law tails with a universal exponent* α. *This exponent will the same in different sectors within a given stock market, and will be the same in different stock markets around the world.*

In section 2 we review some recent work on the statistics of financial time series. The evidence does appear to offer some support to the validity of the scaling prediction: the log returns of a great number of different equities and equity indices, over time intervals from one minute to several days, show power law tails all with exponents $\alpha \approx 3.0$. Furthermore, there is some evidence for universal behaviour beyond that asserted in the scaling prediction: the probability density functions for returns over time intervals in the range from five minutes to several days change only by an overall scale, having the same universal shape.

The present review takes this empirical evidence seriously, and explores the consequences of the scaling prediction. The main difficulty in considering non-Gaussian distributions is that the dependence structure is not easily characterized in a unique way. In the Gaussian case, the correlation matrix determines the multivariate distribution once the marginals are known. Moreover, despite the fact that determining the full correlation structure exactly from market data is often difficult, the Gaussian distribution is simple enough that determining a few of its principal components is often sufficient for risk management purposes, and these can easily be obtained from market data.

We will characterize a general class of multivariate distributions consistent with the scaling prediction for its marginals, and derive their associated asymptotic dependence structure ("tail dependence"). This tail dependence will turn out to be entirely determined by a measure on a high-dimensional sphere, and we will obtain estimates for the impact of this measure on the conditional correlations in the tails of the marginals. In a precise sense, this "spectral measure" is the natural analog for Lévy distributions of the Gaussian correlation matrix. The results presented here are proven in the working paper [2]. The issue of the determination of the dependence structure from historical data will be addressed in future work.

REVIEW OF EMPIRICAL STUDIES

Here we briefly summarize some empirical studies done over the past forty years which have been important for researchers studying the distributions of price log-returns $X(t, \delta t)$. For an extended and readable discussion, see [5].

Benoit Mandelbrot [11] was an early pioneer in studying the returns of cotton price fluctuations, and proposed nearly forty years ago that the distribution of log returns $X(t, \delta t)$ can be modelled by a stable Lévy distribution. More precisely, he proposed that the probability that $X(t, \delta t)$ is bigger than x is asymptotically $Cx^{-\alpha}$ as $x \to \infty$, where $\alpha \approx 1.7$. His proposal was based on two elements: first, that the distribution of price returns is decidedly non-Gaussian, and second, that the functional form of these distributions doesn't depend on the increment size δt. Models based on this stable ($\alpha < 2$) Lévy hypothesis have the difficulty that their distributions do not have a finite second moment, but stability guarantees that there is a well-developed theory generalizing many features of Gaussian distributions [13].

But we now know this picture is too simplistic, since as δt grows large, the distributions converge to a Gaussian [3]. Two papers [6] examine the stock price log-returns of individual companies and the log-returns of the S&P 500 index. They present evidence that this crossover to Gaussian behavior occurs for $\delta t \approx 4$ days in the case of the S&P 500 index, and for $\delta t \approx 16$ days in the case of individual stocks. They emphasize however, that for δt less than these crossover points, the distributions are essentially independent of δt.

This brings us to the second departure from Mandlebrot's hypothesis. While the return distributions are power laws as he suggested, the work just cited shows that for a great range of x values (from 3 to 50 standard deviations in the case of the

S&P 500, and from 2 to 80 standard deviations in the case of individual companies), the exponent α is approximately 3, not 1.7. This is true for any of the 1000 stocks chosen, as well as the Hang-Seng and NIKKEI indices, and does not depend on the size of δt (below the crossover mentioned above). Dacorogna et al [4] have found $\alpha \gtrsim 3$ power laws in foreign exchange markets as well. These findings suggest that return distributions for financial fluctuations should not be modelled with Lévy stable laws, but with more general Lévy distributions characterized by $\alpha > 2$. Our main goal in this paper is to extend this idea to the multivariate setting, where we assume that there is a fixed $\alpha \approx 3$ so that in every direction, the cumulative density function has power law decay with exponent α.

MODELLING ASSUMPTIONS

In the remainder of this paper we focus on a collection of M different equity prices. We let $S_t = \{S_t^k\}_{k=1,\dots,M}$ be the vector stock price process where S_t^k is the price of the kth equity at time $t > 0$. We focus on the sequence of returns sampled at N discrete times $t_j = j\delta t,\ j = 0, 1, \dots, N$, where $\delta t > 0$ is a fixed time interval. We define the vector of (log) returns from t_{j-1} to t_j to be $X_j = \{X_j^k\}_{k=1,\dots,M}$ with $X_j^k = \log(S_{t_j}^k / S_{t_{j-1}}^k)$. In general, and in the remainder of the paper, we are concerned with determining a set of mathematically natural conditions on the process X_t which are compatible with the observed regularities in the data.

Remark: We defer to a subsequent paper the question of how the distribution of the returns vary with the value of δt. In the present paper, δt should be taken to be a value intermediate between 1 minute and 4 days.

We list a set of modelling assumptions which will characterize a certain class of multivariate distributions. We state these assumptions as applicable directly to the time series of returns: more realistically, one should think of these assumptions as applicable to the set of underlying risk factors in the model. Observed time series would then be modelled as a process driven by these factors. Nonetheless, when applied directly to observed time series our modelling assumptions do a surprisingly reasonable job.

H1 *The random variables X_j are independent and identically distributed. We call the underlying M-dimensional distribution X, thus $X_j \stackrel{d}{=} X$ for all j. X has finite covariance and is infinitely divisible.*

We assume finite covariance since this does seem to be true in observed time series. Infinite divisibility is a consequence of thinking of X as the increment $X_{\delta t}$ of an underlying continuous time process X_t. From this condition the Lévy–Khintchine theorem implies the following representation for the log–characteristic function $\Psi_X(u) = -\log(E[e^{i(u,X)}])$ of X:

$$\Psi_X(u) = -i(u,m) + (u,Qu)/2 + \int_{\mathbf{R}^M} [e^{i(u,a)} - 1 - i(u,a)\chi_{|a|<1}(a)]d\mu(a) \quad (1)$$

Here $u, m \in \mathbf{R}^M$, Q is a positive semi-definite quadratic form on $\mathbf{R}^M$, the Lévy–Khintchine (LK) measure μ is a positive measure on $\mathbf{R}^M$ which satisfies a certain integrability condition, and for any set $A \subset \mathbf{R}^M$, χ_A denotes the indicator function for that set.

The LK–measure has an interpretation which we find very useful in guiding our work. A Poisson process with jump size $a \in \mathbf{R}^M$ and rate $\lambda > 0$ has log–characteristic function $\lambda[e^{i(u,a)} - 1]$. Roughly, the last term of (1) is interpreted as a (continuous) compounding of Poisson jump processes in which jumps of size in the set $[a, a + da]$ occur at the rate $d\mu(a)$.

The second assumption restricts the form of the log–characteristic function to be that of a "purely discontinuous" process:

H2 *We assume* $Q = 0$.

The final assumption embodies and abstracts the observation that observed equities as well as numerous *stock indices* seem to have power law tails with the same exponent.

H3 *There is a constant* $\alpha > 0$ *such that every generalized (univariate) marginal* $Y = \xi^T X$, $\xi \in \mathbf{R}^M$ *of the* $\mathbf{R}^M$*–valued random variable* X *has a probability density function* ρ_Y *with power law asymptotics:*

$$\rho_Y(y) \sim g_\pm(\xi)|y|^{-1-\alpha} \quad \text{as } y \to \pm\infty$$

for some constants $g_\pm(\xi) \geq 0$. *We say that* X *has universal exponent* α.

A CLASS OF MULTIVARIATE DISTRIBUTIONS

In this section we discuss a general family which satisfies **H1-H3**.

Definitions: Given $\alpha > 2$, we say an M–dimensional random variable is of class α if its log–characteristic function has the form (1) with $Q = 0$ and such that the Lévy-Khintchine measure $d\mu_X(a)$ is asymptotically in separated form:

$$d\mu_X(a) \sim r^{-1-\alpha}\, dr\, d\Gamma(\hat{a}), \quad r \to \infty \tag{2}$$

where $r = |a|$ and $\hat{a} = a/r$. The "spectral measure" Γ is a general (positive, finite) measure on the unit sphere $S \in \mathbf{R}^M$. Let $\mathcal{H}_{M,\alpha}$ denote the family of all M–dimensional class α random variables, and let $\mathcal{H}_\alpha = \cup_{M>0}\mathcal{H}_{M,\alpha}$.

The main structural result concerning the family $\mathcal{H}_\alpha$ is that it is closed under linear transformations.

Theorem 1 *Let* $X \in \mathcal{H}_{M,\alpha}$, *and let* $B : \mathbf{R}^M \to \mathbf{R}^k$ *be a non–zero linear transformation. Then* $B(X) \in \mathcal{H}_{k,\alpha}$.

Theorem 1 implies that **H3** is true for $X \in \mathcal{H}_{M,\alpha}$. To see this, we need the following univariate result:

Proposition 2 *Suppose the univariate X has LK measure μ which satisfies*

$$\mu(a) \sim g_{\pm}|a|^{-1-\alpha} \quad \text{as } a \to \pm\infty \tag{3}$$

Then the probability density function of X has the same asymptotics as μ.

A direct asymptotic calculation now shows that class α distributions satisfy **H3**:

Corollary 3 *Let $X \in \mathcal{H}_{M,\alpha}$. Then for any $\xi \in \mathbf{R}^M$, the generalized marginal of X in the direction of ξ, $Y = (\xi, X)$, has power law asymptotics with exponent α and constants*

$$g_{\pm}(\xi) = \int ((\xi, \hat{a}))_{\pm}^{\alpha} d\Gamma(\hat{a}) \tag{4}$$

where $(x)_{\pm} = \max(\pm x, 0)$.

We can use a multivariate version of Proposition 2 to strengthen the connection between the LK measure and the probability density function of X.

Theorem 4 *Let $X \in \mathcal{H}_{M,\alpha}$. Then the probability density function of X has the same asymptotics as the LK measure of X.*

Thus the behaviour of large values of a random vector $X \in \mathcal{H}_{M,\alpha}$ is determined by Γ. We introduce a set of "large events" parameterized by a radius $R > 0$: $B_R = \{X : |X| > R\}$, where $|\cdot|$ denotes Euclidean norm. Now let$X_R = X|R$ be X conditioned on the event B_R. The probability density function of X_R is $\rho(x)\chi_R(x)/\int \rho(x)\chi_R(x)\, dx$. To emphasize that the following result is very different from what will happen with Gaussians, observe that if X is a multivariate Gaussian, then for R large X_R becomes concentrated on the maximal eigenspace of the covariance matrix of X. Thus the dependences possible for large Gaussian events are very restricted.

Theorem 5 *Suppose $X \in \mathcal{H}_{\alpha,M}$ with spectral measure Γ and vanishing mean. Then the correlation matrix of X_R is asymptotic to the correlation of Γ:*

$$C_{R,ij} = \frac{E[(X_{Ri} - \bar{X}_{Ri})(X_{Rj} - \bar{X}_{Rj})]}{\sqrt{var(X_{Ri})var(X_{Rj})}} \sim \frac{E_\Gamma[(a_i - \bar{a}_i)(a_j - \bar{a}_j)]}{\sqrt{var_\Gamma(a_i)var_\Gamma(a_j)}}$$

Example: A well-studied fat–tailed distribution in economics is the univariate t–distribution with 3 degrees of freedom which has probability density function $C(1+x^2)^{-2}$ and log–characteristic function $\Psi(u) = |u| - \log(1+|u|)$. We can prove that this distribution satisfies **H1**, **H2**. Now we use this distribution to generate a family of multivariate t–distributions satisfying **H1**–**H3** with tail exponent $\alpha = 3$. Let $X = (X_1, X_2, \ldots, X_M)$ be a random vector whose components are i.i.d. random variables with this distribution. Then define the family to consist of all $Y = B(X)$ for any value M and any linear transformation B. Such distributions have the asymptotic behaviour (2). The spectral measure of Y consists of a finite number of "atoms" (delta functions) at the points $B(e_i)$ with weights $|B(e_i)|^\alpha$, where e_i is the i–th standard basis vector in $\mathbf{R}^M$.

CONCLUSIONS

We have seen that the scaling prediction leads for each M to a broad class of M–dimensional distributions which are characterized by an asymptotic scaling in both the LK measure and in the probability density function. For such random vectors, the joint distribution of large events is determined by an exponent $\alpha > 2$ and the spectral measure Γ, an arbitrary finite measure on the unit sphere in $\mathbf{R}^M$. Normalized moments of the conditional random vector X_R are given by the moments of Γ if R is large enough.

Such behaviour is completely different in character from what is possible in the Gaussian universe.

It remains to be seen whether the spectral measure can really be observed in real data. In principle, this will require processing huge amounts of data to sift out the large events. One is looking for the normalized moments of X_R to stabilize as R gets large. Equivalently, one can project X_R radially onto the unit sphere: the resulting measure should be approximated by Γ for R large. If the spectral measure does exist in real equity data, this will be a strong verification that some form of self–organized criticality is true in financial markets.

REFERENCES

1. P. Bak. *How nature works: the science of self–organized criticality.* Copernicus, Springer-Verlag, 1996.
2. J. Boland, T. R. Hurd, M. Pivato and L. Seco. Extreme value dependences in multivariate financial time series. Working paper, 2000. `http://www.phimac.org`
3. J. P. Bouchaud and M. Potters. *Theory of financial risk.* To appear.
4. M. Dacorogna, U. Müller, O. Pictet and C. de Vries. Olsen & Associates preprint, 1998. `http://www.olsen.ch/library/research/oa_working.html`
5. J. Farmer. *Computing in Science and Engineering (IEEE)*, Nov.-Dec.:26–39, 1999.
6. P. Gopikrishnan, V. Plerou, L. Amaral, M. Meyer, H. Stanley. *Phys. Rev.* E 60:5305, 1999; and
   V. Plerou, P. Gopikrishnan, L. Amaral, M. Meyer, H. Stanley. *Phys. Rev.* E 60:6519, 1999.
7. U. Keller and E. Eberlein. *Bernoulli*, 1:281–299, 1995.
8. R. Kiesel and N .H. Bingham. Modelling asset returns with hyperbolic distributions. Preprint, 1999.
9. F. Longstaff. Stochastic volatility and option valuation. UCLA School of Management preprint, 1995.
10. D. B. Madan, P. Carr, and E. Chang. *European Finance Review*, 2:79–105, 1998.
11. B. Mandelbrot. *J. of Business*, 36:394–419, 1963.
12. B. Mandelbrot and J. W. Van Ness. *SIAM Review*, 10:422–437, 1968.
13. G. Samorodnitsky and M. S. Taqqu. *Stable non-Gaussian random processes.* Chapman & Hall, New York, 1994.

Scaling and Multiscaling in Financial Markets

Giulia Iori

Department of Mathematics,
King's College London,
The Strand, London WC2R 2LS, U.K.

Abstract. This paper reviews some of the phenomenological models which have been introduced to incorporate the scaling properties of financial data. It also illustrates a microscopic model, based on heterogeneous interacting agents, which provides a possible explanation for the complex dynamics of market returns. Scaling and multiscaling analysis performed on simulated data is in good quantitative agreement with empirical results.

INTRODUCTION

The probability distribution of returns of many market indexes and exchange rates, at a given (but not too long) time scale, seems consistent with an asymptotic power law decay $P(r) \sim r^{-(1+\mu)}$. A crossover between a Levy stable regime with $\mu < 2$ [1], and a power law with an exponent $2 < \mu < 4$ [2] has been reported [3]. Power law distributions are self-similar, therefore the distribution of returns $r_\tau(t) = \log(P(t+\tau)/P(t))$ at different time scales τ, rescaled by a lag-dependent factor $\eta(\tau)$, satisfies

$$P(r_\tau) = \frac{1}{\eta(\tau)} \mathcal{F}(\frac{r_\tau}{\eta(\tau)}) \tag{1}$$

where $\mathcal{F}(u)$ is a time independent scaling function. For a self-affine function $\eta(\tau) = \tau^H$, with $H = 1/2$ for a Gaussian process and $H = 1/\mu$ for a Levy process. If scale consistency holds the time-scale at which the process is observed becomes irrelevant. The situation in financial markets is more complicated than this and one observes that when τ increases beyond a certain value (a few days for market indices and a few weeks for individual stocks) the scaling breaks down and the shape of a Gaussian, as predicted by the random-walk hypothesis, is recovered. This implies that it is not possible to find a unique real number H such that the statistical properties of the rescaled variables $r_\tau(t)/\tau^H$ do not depend on τ. The notion of multiaffinity is introduced to characterize a stochastic process $X(t)$ which satisfies

CP553, *Disordered and Complex Systems*, edited by P. Sollich, et al.

$$\langle |X(t+\tau) - X(t)|^q \rangle = c(q)\tau^{\phi_q} \tag{2}$$

where ϕ_q is the scaling function. Self-affine processes are characterized by a ϕ_q which is linear and fully determined by its index H: $\phi_q = Hq$. Multi-affine processes instead are characterized by a non-linear scaling function ϕ_q. Anomalous scaling, or multiscaling, appears in a wide class of phenomena where global dilatation invariance fails. For example, intermittent behaviour in dynamical system, i.e., strong time dependence in the degree of chaoticity, is accompanied by anomalous scaling with respect to time dilatations in the trajectory space. If the non-linear shape of the scaling exponents ϕ_q is a consequence of intermittent behaviour, multiaffinity involves multifractality of an appropriately defined probability measure. A measure of the degree of intermittency could then be provided in terms of an infinite set of exponents associated with the geometrical structure of this probability measure [4].

Multiscaling in financial markets is an indication that returns, even if uncorrelated, are not independent stochastic variables, and it reveals the presence of wild fluctuations. The wilder the fluctuations, the larger the difference of ϕ_q from a linear behaviour in q. Indeed, it is well known that while stock market returns are uncorrelated, the autocorrelation function of a measure of volatility, such as the absolute value of returns, is positive and slowly decaying, indicating long memory effects. This phenomenon, known in the literature as volatility clustering, implies temporal dependencies, with periods of large price changes alternating with periods of smaller changes. Empirical analysis shows that the decay of the autocorrelations of absolute returns is hyperbolic over a large range of time lags (from one day to one year), with an exponent $0.1 < \gamma < 0.4$ [5].

Considerable attention has been devoted to detecting comovements of volatility with other economic variables in an attempt to interpret and capture the source of clustering effects in returns. In particular, much effort has been devoted to the analysis of correlations between volatility of returns and trading volume. Empirical evidence has been provided [2] for a positive cross correlation between these two quantities. Furthermore, a lack of regularity has been observed in the trading or business time. Stochastic trading time models [6–8] have also been proposed as a possible explanation for the emergence of persistence in volatility.

PHENOMENOLOGICAL MODELS

A simple model proposed by Mandelbrot and van Ness [9] which can generate volatility persistence is Fractional Brownian Motion. FBM is a random process with stationary, self-similar increments which grow locally at a rate τ^H, where H is the self-affinity exponent. Even though this model accounts for correlated volatility fluctuations, these are generated through dependent, and hence predictable, price increments with negative autocorrelation when $0 < H < 1/2$ and positive correlation when $1/2 < H < 1$.

An earlier model introduced by Mandelbrot [10], which has become very popular in the physics literature, is the Levy Flight (LF) model. An LF is a random walk in which the step length is chosen from a Levy distribution. Since Levy distributions are stable under convolution the LF process exhibits exact self-similarity. The exact scaling of both FBM and LF is not consistent with the breakdown of scaling observed in financial market data.

To overcome the many limitations of LF (not least the difficulties of dealing with a process with a diverging second moment) Truncated Levy flights (TLF) have been introduced. A TLF is a process based on a truncated Levy stable distribution with a cut-off in its power law tail. The truncation can be introduced as a sharp cut-off as in Mantegna and Stanley [1] or, to preserve the infinitely divisible property of the pdf, through a smoother exponential decay in the tail as in Koponen [11]. Because of the cut-off, the distribution is no longer self-similar when convoluted and has finite variance. Even though the TLF pdf belongs to the basin of attraction of a Gaussian, it converges to the Gaussian very slowly and the process exhibits Levy scaling in a wide range of sampling intervals. Nakao [12] has shown that TLFs à la Koponen exhibit the simplest form of multiscaling, i.e., bi-fractality. Nonetheless his results, i.e., $\phi_q = q/\alpha$ for $0 < q < \alpha$ and $\phi_q = 1$ for $q > \alpha$ are not consistent with the empirical findings. Similar results have been found by Chechkin [13] who analysed a finite sample of a simulated ordinary LF.

Indeed, even though Levy flights have stationary, statistically self-affine and stably distributed increments, the finiteness of the sample size violates self-affinity, giving rise to spurious multi-scaling. Consequently, while the moments of stable Levy distributions with $\alpha < 2$ diverge, together with the q-th order structure function, for $q > \alpha$ both quantities are finite for finite sample size. A rough estimate of finite sample effects gives [13] a linear τ-dependence of the q-th order structure functions with an exponent which does not depend on q for $q > \alpha$. So both ordinary and truncated Levy flights can generate multiaffinity, but the shape of the scaling function ϕ_q is not consistent with the empirical one.

Further aspects remain unexplained by the TLF: the first is that the TLF describes only the central part of distribution of returns well, but not the far tails which decay with an exponent μ well outside the Levy regime. Second the crossover to the Gaussian regime occurs at much larger times than would be expected from the TLF and, finally, correlated variance fluctuations are not accounted for.

In the economic literature the modelling of financial time series has developed in quite different directions, and leptokurtotic distributions have been introduced through a variance conditionally dependent on its past (heteroschedasticity), as in the ARCH/GARCH models [14]. GARCH models nonetheless, even if fat-tailed, only achieve weak memory effetcs, manifested in an exponential (and not hyperbolic) decay of the volatility autocorrelation function. The recently developed FIGARCH process [15] achives long memory still mantaining the martingale property of asset returns. But unlike GARCH, FIGARCH does not converge to a Gaussian process over long sampling intervals and fails to describe the scaling properties of pdfs at different time horizons.

As a generalization of ARCH models, Heteroskedastic Levy Flight (HLF) processes have recently been introduced. Podobnik *et al.* [16] show that due to correlations in the variance the process generates power-law tails in the distribution of returns whose exponents can be controlled through the way in which the correlations in the variance are introduced. The crossover between the two power-law regimes ($\mu < 2$ and $2 < \mu < 4$) can also be generated in these models. Moreover, Santini [17] has shown that in the HLF the Levy scaling of the pdf persists for times of order of magnitude larger than for uncorrelated variance fluctuations. It would be interesting to analyse whether this model also reproduces the multiscaling behaviour of financial data.

An alternative exactly soluble model that mimics the long range volatility correlations has been introduced by Bouchaud *et al.* [18]. Although their model is uni-affine by construction, it shows apparent multiscaling, in good agreement with empirical data, as a result of very long transient effects, induced by the long range nature of the volatility correlations.

This scenario seems to indicate that multiscaling in financial markets is not a trivial effect of tail truncation, but is a consequence of correlated variance fluctuations.

Following a completely different approach, an altenative model, the Multifractal Model of Asset Return (MMAR), has been introduced by Mandelbrot [19]. Fluctuations in volatility are introduced in MMAR by a random trading time, generated as the cumulative distribution function of a random multifractal measure. Trading time is assumed highly variable and contains long memory. Both these characteristics are passed on to the price process through compounding. Subordinate stochastic processes have been extensively used in the economic literature, where either the trading volume [6] or the trading time [7,8] has been chosen as the directing process.

Note that in general the distributions of subordinated stochastic processes do not possess scaling properties. In the MMAR model the return process is a compound process $X(t) = B_H[\theta(t)]$ where $B_H(t)$ is a fractional Brownian Motion with self-affinity index H, and $\theta(t)$ is a stochastic trading time. In particular $\theta(t)$ is a multifractal process with continuous, non-decreasing paths and stationary increments. If $B_H(t)$ is a Brownian motion ($H = 1/2$) without drift the MMAR generates, together with multiscaling, uncorrelated increments and persistence in volatility.

Aside from the phenomelogical characterization of the scaling properties of financial data, the microscopic origins of the complexity of financial markets need to be investigated. From a physicist's point a view the market is a perfect example of a complex system, with a large number of heterogeneous agents interacting in an intricate way. It is then tempting to describe the behaviour of market players using models developed in the context of the statistical mechanics of disordered systems, as proposed in the following section.

A MICROSCOPIC MODEL

It is still an open question whether the emergence of power law fluctuations and volatility clustering observed in financial data is due to external factors, like the arrival of new information, or to the inherent interaction among market players, i.e., the trading process itself.

Iori [20] proposed a model where large fluctuations in returns arise purely from communication and imitation among traders. The key element in the model is the introduction of a trade friction (representing transaction costs) which, by responding to price movements, creates a feedback mechanism on future trading and generates volatility clustering. In [20] (to which the reader is referred for more details and for references to alternative agent based models), the market consists of a market maker plus a number of noise traders. Traders buy from or sell to the market maker and respond to a signal which incorporates idiosyncratic preferences and the influence of the traders nearest to them. Only one kind of stock is traded, whose price is set by the market maker on the basis of the observed order flow and the overall trading volume.

The model is studied through numerical simulations. It reproduces correctly the scaling of the distribution of returns, with a power law decay with $\mu \sim 3$ at short time, and slow convergence to a Gaussian distribution at larger time scales. Moreover, the model generates volatility clustering and positive cross-correlation between volatility and trading volume. The autocorrelation function of the volatility decays hyperbolically with an exponent $\gamma \sim 0.3$ consistent with empirical observations.

The analysis of the moments of the distribution of the simulated data also reveals multiscaling (fig. 1a). By using a step-wise linear regression, the slopes in the structure function are estimated as 0.47 for $q < 3$ and 0.14 for $q > 3$. These results are in good agreement with the empirical analysis performed by Baviera *et al.* [21]

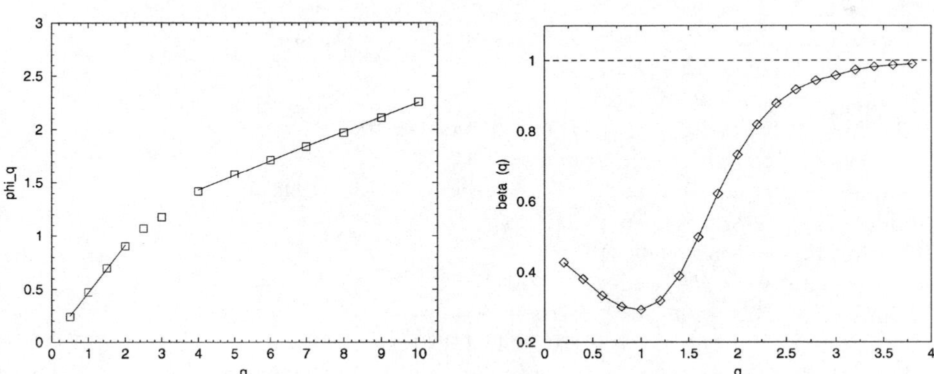

FIGURE 1. Exponent ϕ_q versus q (left). The slopes are 0.47 for $q < 3$ and 0.14 for $q > 3$. Scaling exponent $\beta(q)$ as a function of q (right). Anomalous scaling with $\beta(q) < 1$ is shown.

on the DEM-USA exchange rate quotes.

Baviera *et al.* [21] also detected multiscaling in the volatility autocorrelations by analyzing the generalized correlation functions:

$$C_q(L) = \langle |r(t)|^q |r(t+L)|^q \rangle - \langle |r(t)|^q \rangle \langle |r(t+L)|^q \rangle. \quad (3)$$

If the absolute return series $|r(t)|^q$ has long term memory we would find $C_q(L) \sim L^{-\beta(q)}$ with $\beta(q) < 1$ while if $|r(t)|^q$ is an uncorrelated process $\beta(q) = 1$. Multiscaling would be signaled by a non-linear shape of $\beta(q)$. The scaling exponent $\beta(q)$ measured from simulated data is shown in fig. 1b for $0 < q < 4$. In analogy with the NYSE index and the DEM-USA exchange rate [21,22], $\beta(q)$ is not a constant function of q revealing the presence of different anomalous scales. The convergence of $\beta(q)$ to one reveals that large fluctuations are practically independent. This observation might justify the convergence of the distribution of returns to a Gaussian even though returns are not independent random variables.

This simple model can reproduce many of the stylized facts of stock market returns and has outlined a mechanism which can explain the emergence of the observed power-law fluctuations.

REFERENCES

1. Mantegna, R. and Stanley, H.E., *Phys. Rev. Lett.* **73**, 2946 (1994).
2. Pagan, A., *Journal of Empirical Finance*, **3**, 15102 (1996).
3. Gopikrishnan, P. *et al.*, Phys. Rev. E **60**, 5541 (1999).
4. Calvet, L., http://cowles.econ.yale.edu/P/au/d-c.htm.
5. Ding, Z. *et al.*, *Journal of Empirical Finance* **1**, 83 (1993).
6. Clark, P.K., *Econometrica* **41**, 135-156 (1973).
7. Dacorogna, M.M. *et al.*, *Journal of International Money and Finance* **12**, 135-156 (1993).
8. Scalas, E. *et al.*, cond-mat/0001120 and cond-mat/0006454.
9. Mandelbrot, B.B., and van Ness, J.W., *SIAM review* **10**, 422-437 (1968).
10. Mandelbrot, B.B., *Journal of Business* **40**, 394-419 (1963).
11. Koponen, I., *Phys. Rev. E* **52**, 1197 (1995).
12. Nakao, H., cond-mat/0002027.
13. Chechkin, A.V. and Gonchar, V.Y., cond-mat/9907234.
14. Bollerslev, T., *Journal of Econometrics* **31**, 307-327 (1986).
15. Baillie, R.T. *et al.*, *Journal of Econometrics* **74**, 3-30 (1996).
16. Podobnik, B. *et al.*, cond-mat/9910433.
17. Santini, P., cond-mat/9906413.
18. Bouchaud, J.P. *et al.*, cond-mat/9906347.
19. Mandelbrot, B.B. *et al.*, http://cowles.econ.yale.edu/P/au/d-c.htm
20. Iori, G., *Int. J. of Mod. Phys. C* **10**, 1149-1162 (1999) and Iori, G., adap-org/9905005, *Journal of Economic Behaviour and Organization*, forthcoming.
21. Baviera, R. *et al.*, cond-mat/9811066
22. Pasquini, M. and Serva, M., *Physica A* **269**, 140-147 (1999).

Diagrammatic Approach to Real Options

Steve Leppard

Research Group, Enron Europe Ltd.,
40 Grosvenor Place, London SW1X 7EN, U.K.

Abstract. Real option valuation requires the combination of financial option pricing methods with business-based optimization techniques. In this article a diagrammatic representation of real option deals is introduced, which allows the full complexity of real option formulations to be discussed with non-technical practitioners. Examples from finance and the energy industry are presented and discussed.

INTRODUCTION

The term "real options" is used to refer to a whole range of techniques intended to assign a value to management flexibility in project valuation or the valuation of physical assets (see, for example, [1]). The proper evaluation of real option deals requires an appropriate synthesis of business operational research and financial derivative pricing techniques. This paper describes a diagrammatic method which allows the user to express the optionality present in a deal, define how this optionality impacts value, and to construct automatically the appropriate equations needed to price the deal.

With the synthesis offered by the diagrammatic method we see OR optimization and tree-based derivative pricing as special cases of real option valuation. Simple examples from finance and the energy industry are presented to illustrate the technique.

A useful image to keep in mind is that of the "state space" on which real option valuations are conducted (see Figure 1). The state space is usually formed from the cross product of price and physical/contractual state spaces. Movements through price space are (usually) considered to be random, while movements through physical/contractual state space are (usually) considered to arise from management decision making. Real option valuation is obtained by evaluating the cost implications of optimal decision making in the face of these uncertain price movements.

CP553, *Disordered and Complex Systems*, edited by P. Sollich, et al.

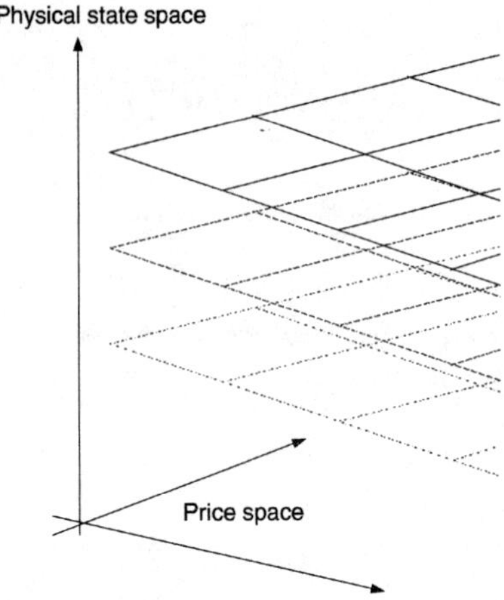

FIGURE 1. State space

INTRODUCING THE DIAGRAMMATIC NOTATION

The tree-like diagrammatic notation was originally conceived as an aid to communication between managers and quantitative analysts. The intention was to combine consistently the best of the diagrammatic methods used to express option pricing trees, decision trees and discounted cash flow diagrams. A diagrammatic grammar has been defined which ensures that the (dynamic programming) pricing algorithm follows automatically from grammatically-correct diagrams.

To motivate the notation, a simple binomial tree valuation of an American put option is examined. The financial term "derivative" of course arises from the fact that the deal derives its value from the underlying. We extend this notion to recognize that management decision-making during the lifetime of the contract will also affect value.

In the diagrammatic notation we distinguish:

- The "states" of the system, which are the values of the underlying(s), and any contractual/physical states reached through management decisions. States are denoted by a box (containing annotation) attached to the underside of the value it impacts.
- The "values", which are derived through a decision-making process based on these states. Values are denoted by circles, containing some symbol representing the value at that node.
- "Cash flows" which are parameters whose values are not resolved in terms of subsequent "values". Cash flows are denoted by upward or downward directed arrows.

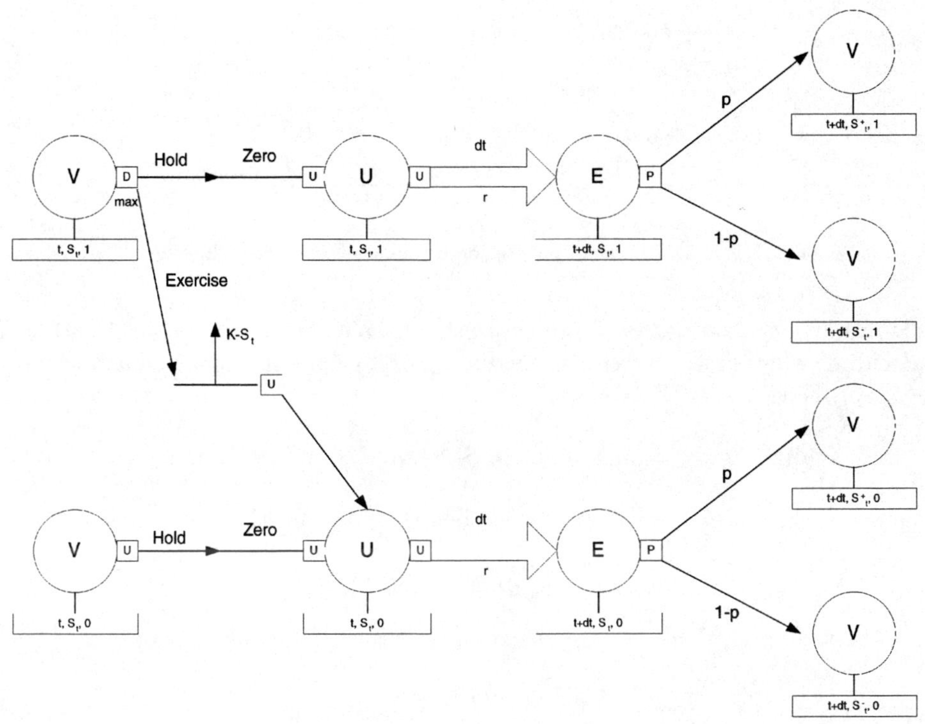

FIGURE 2. American put option diagram

- Occurrences which are the outcome of decision making, probabilistic events, or compulsory events. Such occurrences emanate from (respectively) "decision", "probability" and "unconditional transitions". These three types of transition are shown as small boxes containing the initial letters "D","P" or "U".
- The passing of time through the use of solid "time arrows", along which discounting takes place.

For the case of an American put option the system's states are given by the cross product of a binomial price space and a two-state contractual space corresponding to the states "option exercised" and "option not exercised". The income from making the one-way transition in contractual space is the intrinsic value of the option at that time/price combination.

The diagram to express this valuation, together with the end condition is given in Figures 2 and 3.

As Figures 2 and 3 indicate, the states of the system may be described by the pair (S, u) consisting of the underlying price S, and how many "units of exercise" remain $u \in \{0, 1\}$.

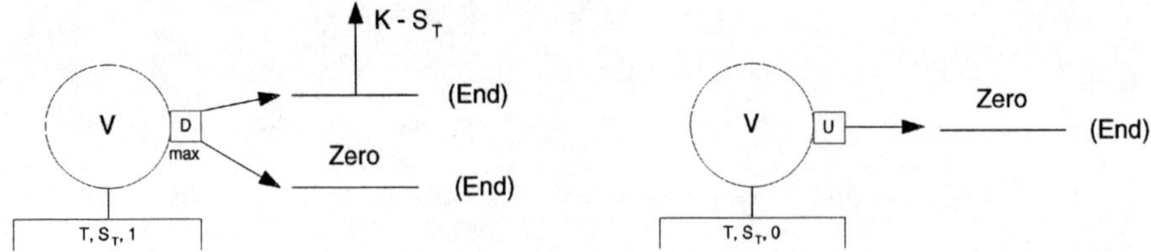

FIGURE 3. American put option termination condition diagram

The value of the derivative where the option hasn't been exercised $(S, 1)$ is found by deciding whether to exercise immediately or to hold the option for one more time step. Thus

$$V(t, S_t, 1) = \max\left(0 + U(t, S_t, 1), K - S_t + U(t, S_t, 0)\right).$$

If the option has already been exercised then there is only the unconditional transition giving

$$V(t, S_t, 0) = U(t, S_t, 0).$$

We have that $U(t, S_t, u)$ is the discounted future value of the option $E(t + \Delta t, S_{t+\Delta t}, u)$:

$$U(t, S_t, u) = e^{-r\Delta t} E(t + \Delta t, S_{t+\Delta t}, u).$$

Finally $E(t, S_t, u)$ is the expected value of the value of the option under the binomial process up and down ticks:

$$E(t, S_t, u) = pV(t, S_t^+, u) + (1 - p)V(t, S_t^-, u),$$

where p and $1 - p$ are the risk neutral up and down tick probabilities.

Of course we have the termination conditions:

$$V(T, S_T, 1) = \max(K - S_T, 0), \qquad V(T, S_T, 0) = 0,$$

the second of which implies that $V(t, S_t, 0) = 0$ for all t. Substitution of these equations for U and E into V give the recursive formulas

$$V(t, S_t, 1) = \max\left(e^{-r\Delta t}(qV(t, S_t^+, 1) + (1 - q)V(t, S_t^-, 1)), K - S_t\right),$$

and

$$V(T, S_T, 1) = \max(K - S_T, 0).$$

Some readers may recognize this as the backward-recursive Bellman equation of stochastic dynamic programming.

POWER PLANT VALUATION WITH OPTIMIZED MAINTENANCE

We now consider the real option valuation of a merchant power plant, which is free to sell power into the market whenever it is economical to do so. Such assets are usually valued in terms of a series of "spark spread" options, the payoff of which is given by the difference between power and fuel prices, with variable plant operation costs as the strike price:

$$\max(S_{\text{power}} - S_{\text{fuel}} - K_{\text{variable}}, 0) \times \text{plant capacity}.$$

We only run the plant if fuel and variable costs are more than counterbalanced by income from power prices.

We consider a monthly model in which the plant has an output that degrades over time. We need to consider our maintenance and running options at each time step, for each system state:

Don't run, where the plant stays at the same stage of its maintenance cycle, and only incurs fixed costs.

Run, where the plant moves one month further through its cycle, reaping the power-fuel spread.

Maintain, where the plant moves back to the beginning of its cycle, incurring turnaround costs, but earning no income.

The system state is parameterized by the price state, power-fuel spread $S_{\text{power-fuel},t}$, and the physical state, number of months since the plant was last maintained y. For the sake of simplicity, we assume that the power-fuel spread follows a binomial price evolution.

Figure 4 shows the real option diagram for this valuation. If we denote the monthly fixed costs by f, the plant capacity as a function of maintenance state by $c(y)$, and turnaround maintenance costs by M, then

$$\begin{aligned} V(S_{\text{power-fuel},t}, y) = \max(&-f + 0 + U(S_{\text{power-fuel},t}, y), \\ &-f + (S_{\text{power-fuel},t} - K) \times c(y) + U(S_{\text{power-fuel},t}, y-1), \\ &-f - M + U(S_{\text{power-fuel},t}, 0), \end{aligned}$$

with the terminating condition for one month after the project time horizon T:

$$V(S_{\text{power-fuel},T+1}, y) = 0,$$

for all $S_{\text{power-fuel},T+1}, y$. It is easy to see how the remainder of the recursive Bellman equation can be constructed in a similar manner to that above.

CONCLUSION

The diagrammatic technique introduced above allows quantitative analysts and non-mathematical business practitioners to discuss the valuation of fixed assets.

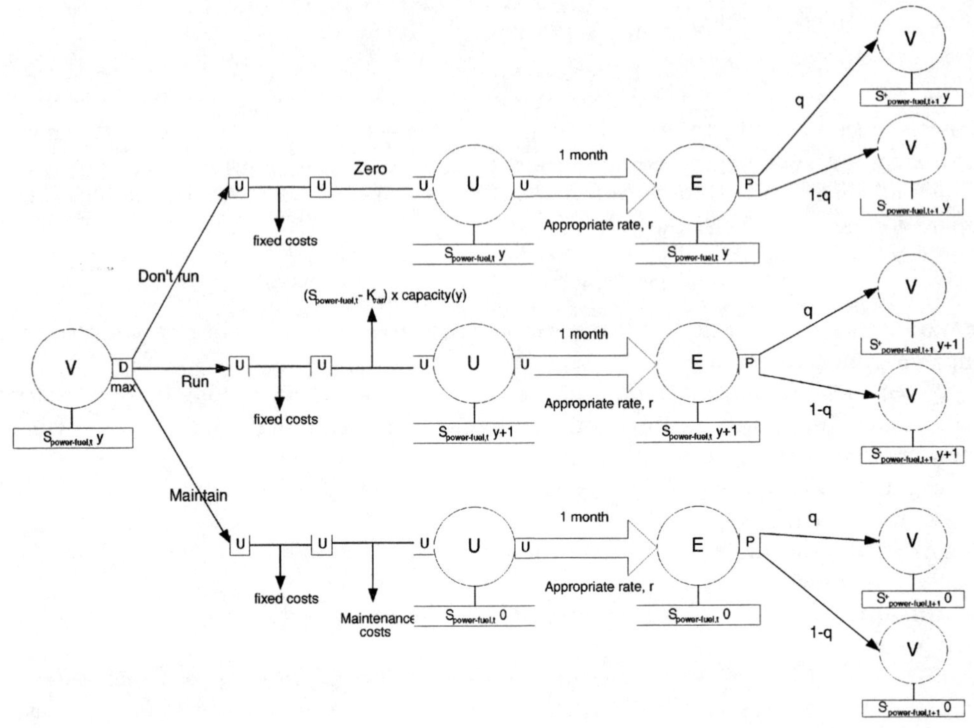

FIGURE 4. Power plant with optimized maintenance diagram

If the state space is chosen appropriately then the recursive Bellman equation of stochastic dynamic programming naturally arises. It is extremely simple to implement these recursive equations as a computer model. Tree-based derivative pricing techniques and the dynamic programming optimization technique may both be considered as special cases of real option valuation, and can be expressed naturally in the notation.

ACKNOWLEDGEMENTS

The author would like to acknowledge the advice and assistance of Stinson Gibner, Olivier Herbelot, Vince Kaminski, Zimin Lu and Grant Masson in the production of this work.

REFERENCES

1. Trigeorgis, L., *Real Options*, The MIT Press (1996).

Instability of Implied Volatility, Fictitious Skews and Smiles and the Hazards of Exotics: The Importance of the Inverse Function Theorem in Mathematical Finance

William T. Shaw

Quantitative Analysis Group, Nomura International plc
Nomura House, 1 St. Martin's-le-Grand, London EC1A 4NP
william.shaw@nomura.co.uk

Abstract. The extraction of the parameters of a model from observational data requires the inversion of the mapping from model parameters to observables. In physical systems this mapping is often simple and linear and the inversion presents no difficulties. In mathematical finance the relevant mappings can be non-linear and a proper understanding of the inversion requires an appreciation of the *inverse function theorem*. The computation of *implied volatility* represents a situation where the inverse function theorem either nearly fails or fails absolutely, resulting in, respectively, extreme instability or multi-valuedness of the implied volatility. Under such circumstances even tiny amounts of noise in the price can result in massive changes in the implied volatility. This note explains how this phenomenon can arise in an analysis of vanilla and simple exotic options, and argues that in many circumstances, the implied volatility may be meaningless, and that complex modelling of volatility skews and smiles may be misguided.

INTRODUCTION

The journals and magazines that deal with derivatives are currently fascinated by volatility modelling. Models of extraordinary detail are now being built to explain observed volatility skews and smiles. These include 2+-factor models, GARCH models and their relatives, as well as older elasticity-type models. An excellent survey is given by [1]. Such exotic structures need to be built on firm foundations. This note explores the foundations, and the whole picture which emerges is, to my mind, and in the case of *implied* volatility, a house of cards built on quicksand.

Why do I say this? One of our primary goals is to understand and exploit the way volatility data is implied by market pricing information. The construction of a volatility surface, and/or its cross-sections in the form of skews or smiles,

CP553, *Disordered and Complex Systems*, edited by P. Sollich, et al.

is fundamental to this process. This requires that the computation of volatility surfaces from pricing data is a "meaningful" process. The problem is that the inversion of option-pricing formulae is, depending on parameters, potentially a highly unstable process. When the inversion is unstable, a tiny change in a market price can cause a huge change in the "implied volatility" or IV. Any noise in the price or in the valuation algorithm can be reflected in massive shifts in volatility, so that the volatility that is calculated is essentially meaningless. "Noise" in the price can arise from, for example,

- bid-offer spread;
- post-theory price adjustments based on psychology;
- tick-size (price quantization) effects,

and any error in the valuation algorithm will also produce noise that may amplified in the volatility computation. For this last reason this note will focus on European style instruments, for which we can give clean and ultra-high precision direct computations with *Mathematica*—the American case raises further complications in the numerical detail. This discussion gets a grip on some important situations when market data is actually un-usable for calculating a vol surface. Throughout this discussion the term "vol surface" will refer to the surface (or discretely, matrix) of implied term volatilities obtained from market prices. The matter of calculating a corresponding surface of local volatilities will not be addressed here. A *Mathematica* notebook containing a more detailed exposition is available from the author by e-mail.

The Inverse Function Theorem

The problems I will discuss are associated with the failure, or near failure, of the *inverse function theorem.* Informally speaking, this theorem states that the mapping $x \to f(x)$ is smoothly invertible in a neighbourhood of a point $y_0 = f(x_0)$ provided $f'(x_0) \neq 0$. In higher dimensions the derivative mapping must have full rank. In simple financial option calculations, where the inversion mapping is the calculation of implied volatility from a price, the relevant mathematical derivative is lambda (also known as vega). Option pricing theorists need to pay rather more attention to the consequences of the failure of this theorem, or indeed its near failure. What happens when lambda is zero, or just very small, or when the space of unknown or poorly-specified parameters is of dimension greater than one, is of critical importance. I am not claiming any originality for these observations—my role here is to communicate an issue which I have found to be important, and which seems to be rather critical in volatility modelling: *the mapping from observables to theoretical parameters is an unstable or multi-valued one.*

INSTABILITY OF IMPLIED VOLATILITY AND VANILLA OPTIONS

Absolute and Relative Instability

In situations where implied volatility is to be computed routinely, we need to impose a stability criterion, and to impose this we need a measure of stability—in fact, we shall define the *instability*. This measure is closely linked to the sensitivity of the option price with respect to volatility, usually known as lambda or vega. In our analysis this quantity (which I denote by Λ) is given by

$$\Lambda = \frac{\partial P}{\partial \sigma} \tag{1}$$

with P, the option value, measured in currency units, and σ, the volatility, measured in absolute terms, so $\sigma = 0.25$ corresponds to an annualized volatility of 25%. It is convenient to define an absolute and a relative measure of instability. We define the absolute instability, $\mathcal{I}_A$, of the implied volatility by

$$\mathcal{I}_A = \frac{1}{\frac{\partial P}{\partial \sigma}} = \frac{1}{\Lambda} \tag{2}$$

and the relative instability, $\mathcal{I}_R$, of the implied volatility by

$$\mathcal{I}_R = \frac{P}{\frac{\partial P}{\partial \sigma}} = \frac{P}{\Lambda} \tag{3}$$

The relative instability gives the shift, in percentage terms, of the implied volatility, caused by a 1% relative change in the option price—this is the measure of instability that we shall focus on most in our analysis of vanilla instruments. If the relative instability is unity, a change in price from 99.5 to 100.5 will cause the vol to increase by 1%. If the relative instability is 10, for example, an increase in price from 49.5 to 50.5 might cause an increase in vol from 25% to 45%. Clearly a large value of the relative instability implies that the implied volatility has no meaning, or at least that it is pointless trying to interpret any detail in the volatility surface. First we look at the relative instability. To explore this concept we use some *Mathematica* graphics functions, using the routines in [2]. Let's take a first look at this measure of instability. We consider a simple call struck at 100, with underlying price varying from 70 to 130. We cap the displayed instability at 20. In Figure 1 we show the instability for a simple call with volatility 20%, with an interest rate of 5% and yield of 2%, for times ranging from maturity, $t = 0$ to one year, $t = 1$. This plot reveals that there is massive instability for shorter-dated in-the-money options.

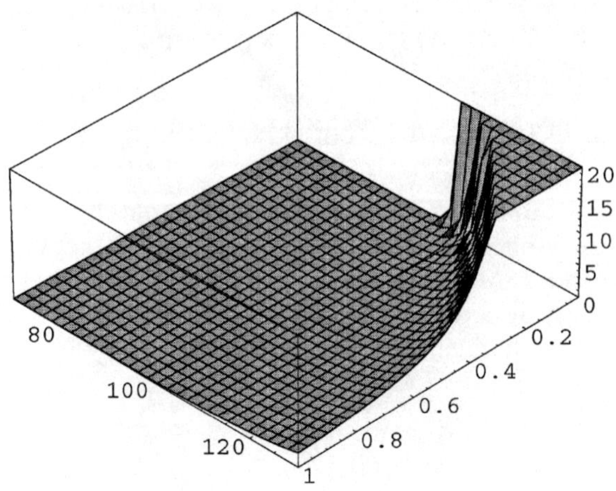

FIGURE 1. Instability Surface for a Simple Call

FAKE SKEW/SMILES I—IMPACT OF POOR SPREAD MANAGEMENT

The instability issue introduced above impacts the computation of implied volatility in several independent and serious ways. The first issue to consider is the impact of failing to clean market data of any spread effects. We can introduce some implied volatility computations to illustrate this.

Suppose market participant AAA prices a call option cleanly using an accurate Black-Scholes model and makes NO other psychological adjustments to the price other than a spread adjustment. That is, the price AAA quotes to the market is

$$1 + \frac{\text{spread}}{100} \tag{4}$$

times the clean Black-Scholes value with a given fixed volatility. Participant BBB is equipped with an accurate Black-Scholes model and gives AAA's raw price to her implied volatility calculator. The implied volatility reported is easily calculated by running the Black-Scholes model forward, adjusting the price by a spread, then running it backward. Let's plot the implied volatility, for an asset price of 100, for various strikes ranging from 70 to 130, but with parameters otherwise the same as used for our instability plot Figure 1. The result is as shown in Figure 2. Figure 2 demonstrates a massive fictitious skew! It also raises the question as to whether we can clean market prices sufficiently to obtain an implied volatility that is at all meaningful for shorter dated and/or deep in the money vanilla options. This is a direct consequence of the high value of the instability, or the smallness of lambda (vega).

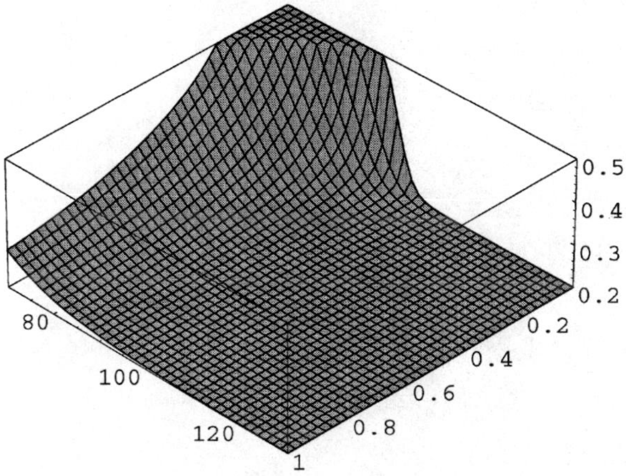

FIGURE 2. IV for Price NOT Cleaned of Spread with varying strike and maturity

FAKE SKEW/SMILES II—FLAT CHARGES AND ABSOLUTE INSTABILITY

We have seen that the impact of a spread charge as a proportion of the premium is to produce a one-sided skew. The impact of a *flat rate* charge per contract exposes the absolute or $1/\Lambda$ measure of implied volatility instability $\mathcal{I}_{\mathcal{A}}$ and easily leads to a fake smile. It's easy to define a function to compute implied volatility in the presence of such a charge and, as an example, let's charge 0.001 currency units per contract. The result is shown in Figure 3. This is rather an extreme smile (guffaw perhaps!) but indicates the impact of a leftover component of a flat charge in the price. By combining spread and flat charges we can build up various combinations of skews and smiles.

Comments

These informal calculations reveal the impact of implied volatility instability on the calculation of volatility surfaces in the presence of small adjustments to the market price of options associated with very simple models of transaction costs. These results suggest that at least some component of the smiles and skews present in surfaces of implied volatility may have as much to do with instability effects as with sophisticated volatility models, and that extremely careful cleaning of market prices is needed prior to computations of implied volatility. These results need generalization to the American case to handle many of the instruments commonly used in volatility surfaces.

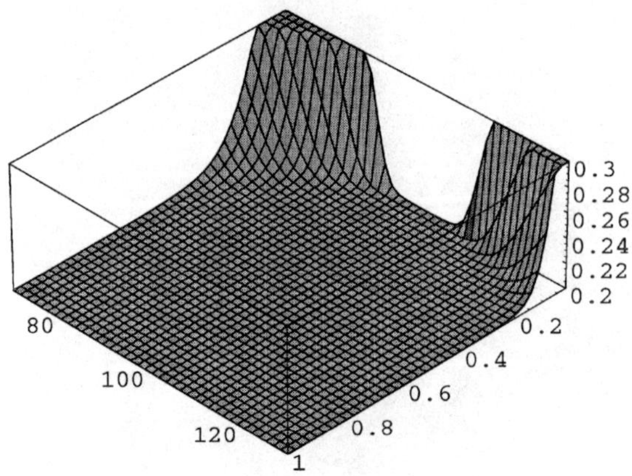

FIGURE 3. IV for Price NOT Cleaned of Flat Charge with varying strike and maturity

Simple Exotics

The examples presented above explore the near-failure of the inverse function theorem. It is easy to find examples of absolute failure amongst the simple exotics. For example, Chapter 1 of [2] gives some examples for Barrier Options where Λ has one, two, or even infinitely many zeroes. The simpler case of Binary Options can be analyzed analytically using the elementary theory of quadratic equations, which reveals that depending on parameters, there may be no (real) IV, one value, or two values, in accordance with the pattern of real solutions of simple quadratics (details can be found in a notebook available from the author). The case of convertible bonds is also very interesting, in that the spread associated with the risk of the issuer defaulting on the cash flows potentially becomes a second unknown parameter along with the volatility, so that armed only with a market price one can only generate spread-IV contours, as has been demonstrated in a series of calculations by my colleague R. Wilson, which show that one can then have a range of IV depending on your view on the spread. Finally we observe that in interest-rate modelling, the matter is still more intricate, in that it is common practice to calibrate such models by extracting local IVs from term IVs, by a process akin to differentiation. This can introduce a second layer of instability on top of the one elucidated here.

REFERENCES

1. Jarrow, R., *Volatility—New Estimation Techniques for Pricing Derivatives*, RISK Publications (1998).
2. Shaw, W.T., *Modelling Financial Derivatives with Mathematica*, Cambridge University Press (1999).

Valuation of Financial Models with Non-Linear State Spaces

Nick Webber

University of Warwick, Coventry CV4 7AL, U.K.
Email: Nick.Webber@Warwick.ac.uk

Abstract. A common assumption in valuation models for derivative securities is that the underlying state variables take values in a linear state space. We discuss numerical implementation issues in an interest rate model with a simple non-linear state space, formulating and comparing Monte Carlo, finite difference and lattice numerical solution methods. We conclude that, at least in low dimensional spaces, non-linear interest rate models may be viable.

INTRODUCTION

Economic and financial state variables are usually thought of as taking values in a linear state space. While this has resulted in some useful models that can approximate behaviour observed in the market, particularly near equilibrium, there is no reason to suppose that the 'true' world is linear.

A number of papers have explored the possibility of defining models on non-linear state spaces. Hughston [4] shows how a no-arbitrage model of asset prices can be formulated when the underlying state variables take values in a Riemannian manifold. He shows how geometric objects such as the Riemannian connection, torsion and curvature can be interpreted in financial terms.

Numerical results can also be achieved. Nunes and Webber [7] show how it is possible to construct interest rate models on simple 2-manifolds, fitting one such model to empirical data.

While numerical methods for extracting prices from financial models defined on linear state spaces are well developed, this is not the case for non-linear state spaces. In this paper we discuss how each of the main methods used in the linear case might be applied to the non-linear case, comparing their implementations for a particular non-linear model.

The main part of this paper discusses the numerical pricing strategies and how these might be applied to the non-linear case. The numerical methods are then applied to calculate term structures in a simple interest rate model.

CP553, *Disordered and Complex Systems*, edited by P. Sollich, et al.

THE VALUATION FRAMEWORK

Our fundamental assumption is that the states of our economic system take values in an N-dimensional manifold $\mathcal{M}$. We shall immediately assume that $\mathcal{M}$ is Riemannian, so that it possesses a symmetric tensor field g of type $(0,2)$, positive definite at each $x \in \mathcal{M}$. g defines a unique affine connection ∇. An affine connection determines the map $\exp_x : U_x \to \mathcal{M}$, at each $x \in \mathcal{M}$, that maps a region U_x of the origin in T_x diffeomorphically onto a region of x in $\mathcal{M}$.

The fundamental option pricing relationship, due to Harrison and Kreps [2] and Harrison and Pliska [3], expresses the current value c_t of an option as an expected discounted payoff, where the discounting factor is determined by a numeraire,

$$c_t = \mathbb{E}_t \left[c_T \frac{P_t}{P_T} \right], \tag{1}$$

and the expectation is taken under a measure in which c_t/P_t is a martingale. Often the numeraire is taken to be the accumulator account, $P_t = \exp\left(-\int_0^t r_s \mathrm{d}s\right)$, where r_t is the instantaneous short rate at time t, and (1) becomes

$$c_t = \mathbb{E}_t \left[\exp\left(-\int_t^T r_s \mathrm{d}s \right) c_T \right],$$

where the expectation is taken with respect to the so-called risk neutral measure. Equation (1) can be evaluated if we are able to define martingales and compute the expectation.[1]

The state X_t of the system at time t evolves stochastically in $\mathcal{M}$ is directly observable. Instead there is a set of observables $S = (S_1, \dots, S_M) : \mathcal{M} \to \mathbb{R}^M$. We shall write $S_t = S(X_t)$ for the state of the observables at time t. S_t may include observations of asset prices and interest rates. The expectation (1) may only occasionally be solved explicitly. Standard numerical methods in a linear space include numerical integration by Monte Carlo methods, solving a related partial differential equation (PDE) by finite difference methods, and discretising time and the state space in a lattice method.[2] We consider each of the three main numerical methods in turn to determine how it might extend to a non-linear state space. For concreteness we consider the application of each method to a particular interest rate model.

A Non-linear Model of Interest Rates

Consider an interest rate model with state space $\mathcal{M} = \mathcal{S}^\infty$. $\mathcal{S}^1$ is covered by $\mathbb{R}$, and is endowed with a metric tensor inherited from $\mathbb{R}$ via the covering map,

---

1) See Emery [1]. The existence of a Riemannian structure and the associated connection allows local martingales to be defined on $\mathcal{M}$, and hence the extension to $\mathcal{M}$ of the no-arbitrage framework for Itō processes in the linear case.

2) These are reviewed in the interest rate case by, for instance, James and Webber [5].

$\pi : x \mapsto x \bmod 2\pi$. Processes taking values in $\mathbb{R}$ project down onto processes with values in S^1. Suppose that $\bar{X}_t$ is a Markovian Itō process in $\mathbb{R}$ with $\mathrm{d}\bar{X}_t = \alpha\left(t, \bar{X}_t\right) \mathrm{d}t + \rho\left(t, \bar{X}_t\right) \mathrm{d}z_t$. It projects down to a Markov process $X_t = \pi \bar{X}_t$ on S^1 if for all $s \in S^1$ and $x, y \in \pi^{-1}(s)$, we have $\alpha(t, x) = \alpha(t, y)$ and $\rho(t, x) = \rho(t, y)$, so that $\alpha\left(t, \bar{X}_t\right) = \alpha\left(t, \bar{X}_t + 2\pi\right)$ and $\rho\left(t, \bar{X}_t\right) = \rho\left(t, \bar{X}_t + 2\pi\right)$ for all t and $\bar{X}_t$. Our state variable is $X_t = \pi \bar{X}_t$ with α and ρ constants.

We define the short interest rate $r : S^1 \to \mathbb{R}$ as $r(\theta) = \mu + \kappa \sin(\theta)$, so that the short rate process is $r_t = r(X_t) = \mu + \kappa \sin\left(\bar{X}_t\right)$. r_t is confined to the range $[r_{\min}, r_{\max}]$, $r_{\min} = \mu - \kappa$, $r_{\max} = \mu + \kappa$. Because we have an explicit construction, it is easy to simulate the process r_t. If α is non-zero, r_t cycles between high and low values. r_t is less volatile near $r_{\min}$ and $r_{\max}$.

We assume that processes are defined under the risk neutral measure. The value at time t of a pure discount bond $B_t(T)$ with payoff 1 at time T is $B_t(T) = \mathbb{E}_t\left[\exp\left(-\int_t^T r_s \mathrm{d}s\right)\right]$. The term structure is the set of spot rates $r_t(T) = -\frac{1}{T-t} \ln B_t(T)$. Solving for term structures in this model (see below) we find they can be damped waves or they can be monotonic.

Solving a PDE

In a linear state space the Feynman-Kac formula allows us to express c_t in (1) as the solution to a certain PDE. Hughston [4] shows how the same PDE arises covariantly on a Riemannian manifold. The price process c_t satisfies the covariant equation

$$\frac{\partial c}{\partial t} + \frac{1}{2} g^{ij} \nabla_i \nabla_j c = \lambda^i \nabla_i c + rc \tag{2}$$

where g^{ij} is the inverse of the metric tensor and λ^i is a price-of-risk vector field. If c is a function of real valued asset prices S_i, $i = 1, \ldots, N$, then one obtains a generalization of the usual Black-Scholes PDE,

$$\frac{\partial c}{\partial t} + \frac{1}{2} g^{ij} \frac{\partial^2 c}{\partial S_i \partial S_j} + r S^i \frac{\partial c}{\partial S_i} = rc. \tag{3}$$

The inverse metric g^{ij} has a natural interpretation as a variance matrix. The PDE (3) is defined on the manifold $\mathcal{M}$. To solve the PDE numerically the manifold must be discretized. Both in the linear case and on Riemannian manifolds one may use finite element techniques (see Zvan [9]).

If the manifold admits a covering by $\mathbb{R}^N$ then a grid on $\mathbb{R}^N$ may project down consistently onto a grid on $\mathcal{M}$. For the illustrative model, option prices satisfy the PDE

$$\frac{1}{2} \rho^2 c_{X,X} + \alpha c_X + c_t = r(X_t) c, \tag{4}$$

where subscripts denote partial differentiation, with the boundary condition $c(t, X_t) = c(t, X_t + 2\pi)$, and for a pure discount bond $c(X_T, T) = 1$. Since everything is periodic in X_t, solutions are indeed defined on the circle S^1.

Nunes and Webber [7] numerically solve the PDE. The covering space is discretised, ensuring that the grid has period 2π. For the illustrative example, essentially a one dimensional problem, no special problems are encountered.

Monte Carlo Methods

There are a number of ways in which processes can be simulated on a Riemannian manifold $\mathcal{M}$. One general method uses the *exp* mapping. If $\mathcal{M}$ has a covering space or imbeds into $\mathbb{R}^N$ then other methods can be employed.

Suppose $\mathcal{M}$ is an N-dimensional manifold and that $\{y_i \in I^N\}_{i=1,\dots,M}$ is a uniform sample of M points in the N-dimensional unit hypercube. Let $F_{\Delta t}$ be the distribution function of the N-dimensional normal distribution with mean 0 and variance Δt. Then $F_{\Delta t}^{-1}(y_i) \in \mathbb{R}^N$ is distributed as an N-dimensional normal variate with mean 0 and variance Δt. An isometry $\iota_{x_0} : \mathbb{R}^N \to T_{x_0}$ onto the tangent space T_{x_0}, $x_0 \in \mathcal{M}$, extends to an isometry ι_y for each $y \in \mathcal{M}$ via the connection. Consider the map

$$x_i \longmapsto x_{i+1} = \exp_{x_i}\left(\iota_{x_i} F_{\Delta t}^{-1}(y_i)\right), \; i = 1, \dots, M. \tag{5}$$

The sequence $\{x_i\}_{j=1,\dots,M}$ is a sample path of a Brownian motion X_t on $\mathcal{M}$.[3,4]

If $\mathcal{M}$ imbeds into $\mathbb{R}^N$ then processes on $\mathcal{M}$ become processes in $\mathbb{R}^N$, restricted to $\mathrm{Im}(\mathcal{M})$, under the imbedding. The processes can then be simulated as processes in $\mathbb{R}^N$ (Rogers and Williams [8]).

If $\mathcal{M}$ admits a covering space, compatible with the metric structure, then a sample path in the covering space projects down onto a sample path in $\mathcal{M}$. We use this method for Monte Carlo valuation in the illustrative S^1 model.

Lattice Methods

Lattices can be constructed via the *exp* map, or via a covering space or an imbedding if these are available.

Choose a set of directions $a_i \in \mathbb{R}^N$, $i = 1, \dots, L$, and for each i choose a probability p_i, $0 < p_i < 1$, with $\sum p_i = 1$. Specify a scale factor $s(\Delta t)$ such that the discrete branching

$$z_t \to z_{t+\Delta t}^i = z_t + s(\Delta t)\, a_i, \quad i = 1, \dots, L \tag{6}$$

---

[3] The map is only defined if $\iota_{x_i} F_{\Delta t}^{-1}(y_i)$ lies within a small enough neighbourhood of $0 \in T_{x_i}$, but in practise, with a small enough time step Δt, this is not a computational objection.

[4] Sample paths can be determined from a low discrepancy sequence. Other regularisation techniques, such as stratified sampling and antithetic variate methods, also carry over from the linear case.

TABLE 1. Spot rates for maturities T: Comparison of numerical results

20 steps per year				100 steps per year			
T	PDE	Lattice	M.C.	T	PDE	Lattice	M.C.
5	0.13487	0.13415	0.13404	1	0.11203	0.11219	0.11218
10	0.10764	0.10761	0.10779	2	0.12238	0.12246	0.12258
15	0.10488	0.10465	0.10448	3	0.12986	0.12985	0.12998
20	0.10825	0.10769	0.10779	4	0.13383	0.13374	0.13385
25	0.10129	0.10144	0.10139	5	0.13416	0.13401	0.13408

converges to a Wiener process as $\Delta t \to 0$. (See McCarthy and Webber [6] for details.) Given an isometry of $\iota : \mathbb{R}^N \to T_{x_0}$ and a connection on $\mathcal{M}$, the exponential map $\exp : (x, T_x) \to \mathcal{M}$ maps branching on tangent planes T_x to a branching on $\mathcal{M}$ itself. In general, this branching won't recombine.

For the illustrative model we construct a lattice via the covering space $\mathbb{R}$. S^1 is discretised into M levels, $l_i = i\Delta r$, $i = 1, \dots, M$, where $\Delta r = \frac{2\pi}{M}$. Fix a time step Δt. At time $t_j = j\Delta t$ the state variable takes values in the set $\left\{X_{t_j} = X_0 + j\alpha\Delta t + i\rho\sqrt{k\Delta t}\right\}_{i=1,\dots,M}$ for a parameter k. Branching at time t_j is given by

$$X_{t_{j+1}} = X_{t_j} + \alpha\Delta t + \begin{cases} \rho\sqrt{k\Delta t} & \text{with probability } 1/2k, \\ 0 & \text{with probability } (k-1)/k, \\ -\rho\sqrt{k\Delta t} & \text{with probability } 1/(2k), \end{cases} \tag{7}$$

where level $(M+1)\,\rho\sqrt{k\Delta t}$ is identified with level $\rho\sqrt{k\Delta t}$. For $k > 1$ the branching specified by (7) is well defined and has identical first and second moments to X_t. k is chosen so that the branching is compatible with the projection map. In fact we require $\rho\sqrt{k\Delta t} = \Delta r$, so set $k = \rho^{-2}\Delta r^2/\Delta t$. The time and space step must be chosen so that $k > 1$.

Pricing on the lattice now proceeds using forward induction to construct Arrow-Debreu security prices at each node on the lattice, at each time step.

TERM STRUCTURE VALUATION IN THE S^1 MODEL

The three methods described above were used to compute the term structure of interest rates in the S^1 model. Model parameter values were $\mu = 0.1$, $\kappa = 0.05$, $\alpha = 0.5$ and $\rho = 0.5$.

Term structures were computed out to 25 years at 20 time steps per year, and out to 5 years at 100 time steps per year. The PDE was solved with an explicit finite difference method with 100 space steps. The Monte Carlo method used antithetic variates and 2000 sample paths. The lattice method used 120 space steps, giving a value of $k = 1.37$. Table 1 summarises the results for selected maturities, T, in years.

With our parameter values and numerical specifications the three methods normally agree to with a few basis points, even for the coarser time step. The lattice solution and the Monte Carlo solution normally agree to within 1 to 2 basis points. The PDE solution is greater than the other two by 7 basis points at 5 years and at 18.5 years. With the finer time step the lattice method and the Monte Carlo method are within a basis point of one another. Now, though, the PDE method is in much closer agreement, within 1 basis point of the other two methods.

In both cases the PDE method was the fastest, closely followed by the lattice method. The Monte Carlo method was significantly slower. The lattice method appears to give the best trade off between speed and accuracy, at these parameter values and numerical specifications.

CONCLUSIONS

In the context of Hughston [4], we have investigated how numerical algorithms for the linear case can be adapted to non-linear manifolds. We have compared the pricing methods, finding that each is able to adequately price a simple instrument.

We conclude that there seems to be no numerical objection to setting up models of derivative securities on non-linear state spaces, at least in simple examples of the sort considered in this paper.

REFERENCES

1. M. Emery. Stochastic Calculus on Manifolds. *Springer-Verlag*, 1989.
2. J.M. Harrison and D.M. Kreps. Martingales and Arbitrage in Multiperiod Securities Markets. *Journal of Economic Theory*, 20:381–408, 1979.
3. J.M. Harrison and S.R. Pliska. Martingales and stochastic integrals in the theory of continuous trading. *Stochastic Processes and their Applications*, 11:215–260, 1981.
4. L.P. Hughston. Stochastic Differential Geometry, Financial Modelling, and Arbitrage-Free Pricing. *Working Paper, Merrill Lynch*, pages 1–22, 1994.
5. J. James and N.J. Webber. Interest Rate Modelling. *Wiley*, 2000.
6. L.A. McCarthy and N.J. Webber. An Icosahedral Lattice Method for Three-Factor Models. *Working paper, University of Warwick*, pages 1–45, 1999.
7. J. Nunes and N.J. Webber. Low Dimensional Dynamics and the Stability of HJM Term Structure Models. *Working Paper, University of Warwick*, pages 1–27, 1997.
8. L.C.G. Rogers and D. Williams. Diffusions, Markov Processes and Martingales: Vol. 2, Ito Calculus. *Wiley*, 1987.
9. R. Zvan, P.A. Forsyth, and K.R. Vetzal. A General Finite Element Approach for PDE Option Pricing Models. *Working paper, University of Waterloo*, pages 1–26, 1997.

Natural Hazard Metaphors for Financial Crises

Gordon Woo[1]

EQECAT, London

Abstract. Linguistic metaphors drawn from natural hazards are commonly used at times of financial crisis. A brewing storm, a seismic shock, etc., evoke the abruptness and severity of a market collapse. If the language of windstorms, earthquakes and volcanic eruptions is helpful in illustrating a financial crisis, what about the mathematics of natural catastrophes? Already, earthquake prediction methods have been applied to economic recessions, and volcanic eruption forecasting techniques have been applied to market crashes. The purpose of this contribution is to survey broadly the mathematics of natural catastrophes, so as to convey the range of underlying principles, some of which may serve as mathematical metaphors for financial applications.

INTRODUCTION

'There is no crash at the moment, only a slight premonitory movement of the ground under our feet'. The context of this warning was a mid-nineteenth century crisis in England; not a seismic one, although tremors are felt from time to time, but one involving the Bank of England, soon after it assumed the responsibilities and liabilities of a central bank. In the parallel context of an American banking crisis enveloping President Andrew Jackson, an economic commentator remarked: 'A general collapse of credit, however short the time, is worse than the most terrible earthquake'. One and a half centuries later, this apparently straightforward statement has bcome partly self-referential in that a massive Tokyo earthquake might destroy the property collateral for many loans, thus triggering banking failures, and hence a general collapse of credit. The graphic language of natural hazards provides a rich vocabulary to express the most dire of financial predicaments, signifying the interplay of complex dynamical market processes appearing to be under little better human control than many arising in Nature. Attempts to ward off an economic recession may be no more successful than attempts to divert a lava flow. The scientific understanding of natural hazards is founded on equations rather than verbal hyperbole, and a financial mathematician may well enquire how far the vocabulary borrowed from the realm of natural hazards can be extended from the

[1] Contact address: 11, Woodsford Square, London W14 8DP, U.K.

CP553, *Disordered and Complex Systems*, edited by P. Sollich, et al.

literary to the mathematical domain. To the extent that this is possible, quantitative ideas may profitably be imported. Departures from the convenient yet perhaps simplistic Brownian motion paradigm come readily to mind; a training in diffusive processes hardly prepares a physicist for the wild and dangerous world of natural catastrophes. A time series for a financial index may be trend-reinforcing, indicative of dynamical persistence, rather than displaying signs of Brownian motion. The long-range memory embedded in the time series may be gauged by a high value of the Hurst exponent, which is a classic measure used by civil engineers to assess the flood or drought tendency of a river. As Mandelbrot would remark with multi-disciplinary zeal, financial engineers have much to gain from familiarity with dam reservoir engineering, as well as with the diverse topics surveyed below.

SCALING

Hyperbolic distributions have attracted increasing attention in financial mathematics, an application long advocated by Mandelbrot and co-workers. The practical origins of scaling distributions and fractal concepts are obscured within a geographical study by the English physicist Lewis Fry Richardson into the lengths of the boundaries between adjacent countries. Richardson's 1951 paper [4] had a seminal influence on Mandelbrot's investigations into the fractal geometry of Nature. Richardson was an early pioneer of numerical weather forecasting, and he also worked in a seismological observatory; professional experience which foreshadows the pervasive influence of notions of fractal scaling in modern Atmospheric and Earth Science. As Mandelbrot has emphasized, the natural world provides a multitude of common examples of scaling. In photographing rocks, geologists are taught to include within each image an absolute size measure, such as a coin or geological hammer. Otherwise a picture of a large fault segment may not be distinguishable from that of a small rock fragment. From geological field observations, the number of faults above a given length has been observed to fall off as a power of fault length. The magnitude of an earthquake is a function of fault length, so this scaling relation helps explain the statistics of the occurrence of earthquakes of different sizes. Most disturbingly for the risk averse, great earthquakes, associated with ruptures several hundred kilometres long, tend to occur far more often than might be presumed naively under a Gaussian distribution. Those accustomed only to small tremors may have their peace of mind shattered by an earthquake larger than any in living memory. The probabilistic counterpart of a power law for fault length is a hyperbolic distribution, examples of which abound in the study of natural hazards.

EARTHQUAKES AND RECESSIONS

The death toll from earthquakes is a stark measure of the difficulty of earthquake prediction. The social consequences of recessions may also be tragic, and the challenge to economists of predicting recessions makes their predicament also precarious. As with seismologists, there are economists who are perpetually warning of doom, but where medium-term forecasts are made, what accuracy is achievable? Both earthquakes and recessions are discrete, comparatively rare events, possessing a set of observable indicators which may have occasional value as predictive precursors. Information on these indicators establishes a time-series pattern which may not be readily discernible to a statistician's eye or computer software for multivariate analysis, but potentially could be recognized using special artificial intelligence techniques. Following the methodology of Gelfand, the Russian theorist Keilis-Borok [2] has formulated an approach to earthquake prediction, which is based on a number of observable seismic traits, such as the current level of seismic activity. It was for California in the 1980's that the most celebrated Russian medium-term forecast was made; a forecast that was sufficiently important to be relayed in person to Ronald Reagan by Mikhail Gorbachev. The destructive Loma Prieta earthquake occurred in October 1989, during a period forecast as having a Temporary Increase in Probability (TIP) of an earthquake occurring. Although the overall success of this approach has been mixed, the idea of using Gelfand's pattern recognition algorithm in an economic setting has been alluring enough for Keilis-Borok to have developed it for application to forecasting recessions in the USA. In this economic context, the algorithm warns of a recession according to the values of a number of chosen economic indicators. For the USA, six indicators were used by Keilis-Borok et al [2]. Apart from indices of monthly economic activity and employment advertising, weekly claims for unemployment insurance were included, as well as the level of business inventories. The other two indicators were financial: the interest rate on 90-day US treasury bills; and the difference between the interest rate on 10-year US Treasury Bond and the federal funds interest rate. Ideally, an alarm would be raised with conviction if there were unanimity among the indicators. In practice, an alarm might be raised if all but a few indicators signal a recession. Application of the six-indicator model to historical USA recessions within the period from January 1959 to April 1996 has proved reasonably encouraging. All five recessions were preceded by continuous alarms, and there were no continuous alarms not ending in a recession. This positive result needs to be qualified by a caveat: there is no guarantee that the six economic indicators used will continue to be reliable predictors in the future, so the actual practical value of this pattern recognition approach remains conjectural.

TAKING THE STRAIN

To the extent that a stock market crash is a type of cooperative phenomenon, governed by collective interactions between traders, the crash hazard rate (i.e., the probability per unit time of a crash happening) might be viewed as an analogue of a strain rate in a brittle material. One of the popular visual metaphors of a financial crash is of a bow being bent back to snapping point. The paradigm of material failure offers a rich source of ideas for representing sudden cataclysms, whether they be physical or economic. Indeed, the standard nonlinear equation describing the rate dependence of material failure has been suggested as a basic preliminary model for stock market crashes [1]. A number of simple volcanological applications of the material failure equation have been attempted, with the aim of forecasting the time of occurrence of volcanic eruptions. There have been some modest successes, e.g. [5], in instances where strain deformation on the slopes of a volcano has proved to be a suitable proxy for a measure of eruption threat. However, the correspondence between strain deformation and events interior to the volcano may be very complex, as happens when the nonlinear dynamics of multi-phase flow play a significant role in eruptive behaviour. This dynamical complexity inevitably erodes the time resolution of eruption forecasts. It might be expected that superior forecasting results using the strain paradigm would be obtained in modelling hazard phenomena where there was a more direct observable linkage between strain and failure. Indeed, accurate forecasts are achievable for the time of occurrence of the breakage of ice and rock masses in avalanche-prone alpine areas. The success of such forecasts owes much to the practical feasibility of accurate instrumental monitoring of the deformation. Some consideration has also been given to the use of crustal strain monitoring in earthquake forecasting [3]. However, the necessary resolution of strain needed to justify and apply such forecasting methods is not yet available, although satellite interferometry holds some promise for the future.

REFERENCES

1. Johansen, A., and Sornette, D., *Risk* **January**, 91–95 (1999).
2. Keilis-Borok, V., Stock, J.H., Soloviev, A., and Mikhalev P., *Journal of Forecasting* **19**, 65–80 (2000).
3. Main, I.G., *Geophysical Journal International* **139**, F1–F6 (1999).
4. Richardson, L.F., *General Systems Yearbook* **6**, 580–627 (1951).
5. Voight, B., *Nature* **332**, 200–203 (1988).
6. Woo, G., *The Mathematics of Natural Catastrophes*, Imperial College Press, London, 1999.

Martingale Approach to Real Options

Lane P. Hughston and Mihail Zervos

Department of Mathematics, King's College London,
The Strand, London WC2R 2LS, UK

Abstract. We formulate a general mathematical model for investments in real assets from the perspective of the real options approach. We then derive an analytic expression for its price under a market completeness assumption. This expression is the solution of a stochastic optimisation problem. The generality of the model is such that it can also provide a framework for the study of financial options.

In contrast with traditional economic approaches to the problem of investment pricing, the real options approach recognises the value of managerial flexibility. By incorporating managerial decision strategies into appropriate models, the new approach aims at correcting a well-documented discrepancy between observed and theoretical prices. The new research area has already been studied extensively in the economics literature, and the interested reader can consult the books by Dixit & Pindyck [2] and Trigeorgis [5].

Up to now, the real options approach has been developed within the framework of specific investment models. Such models typically involve a small number of managerial decisions such as the decision to activate a project or the decision to abandon it at discretionary times. Overall, apart from offering a generic framework of a qualitative nature, the new theory appears to be rather fragmented. The motivation for this paper is to develop and analyse a general model that can provide the beginning of a unifying quantitative as well as qualitative framework.

Adopting an abstract point of view, we can state that the constituent elements of every investment consist of a set of managerial decision flows and an uncertain flow of profits and losses associated with each managerial decision strategy. This is the model that we study. More specifically, we identify every investment with a set of finite variation processes modelling cumulative profits and losses parametrised by an abstract set of managerial decision flows. The model is so general that it can also be used as a framework for the study of financial derivatives such as European and American options. To motivate our definition, we consider an investment model that has been studied by several authors and leads to results of an explicit nature (see Duckworth & Zervos [1], and the references therein). The model itself, as well as several issues pertaining to its pricing, first appears in Knudsen, Meister &

CP553, *Disordered and Complex Systems*, edited by P. Sollich, et al.

Zervos [4]. However, these authors develop their analysis within the framework of specific investment models, and under strong regularity assumptions.

In this paper we assume that all of the random payoffs that can possibly result from the management of any investment are perfectly correlated with the prices of a finite number of actively traded assets. This is the usual market completeness assumption often made in finance. Also, it is commonly made in the literature of real options. Such an assumption may hold reasonably well for a number of important cases which include investments associated with commodities for which there exist liquid futures markets. In situations where there is less market liquidity, e.g. real estate acquisition, or corporate investment in fine art, further considerations are required that go beyond the scope of the present analysis. With regard to the market of the actively traded assets, we consider the standard stochastic calculus model for a frictionless market of continuous trading (see e.g. Karatzas & Shreve [3]).

The main result of this paper is that, under the market completeness assumption discussed above, the prices at which a seller and a buyer would accept a deal are given in terms of the solution of a stochastic optimisation problem. Moreover, the existence of a price that both parties would accept is equivalent to the existence of an optimal decision strategy in this optimisation problem.

THE MARKET

Fix a constant $T > 0$, and consider a complete probability space $(\Omega, \mathcal{F}, P)$ carrying a standard, n-dimensional Brownian motion $(W_t,\ t \in [0,T])$. Let $(\mathcal{F}_t,\ t \in [0,T])$ be the augmentation of the natural filtration of W by the P negligible sets in $\mathcal{F}$, and assume that $\mathcal{F} = \mathcal{F}_T$. We denote by $\mathcal{C}$ the set of all $(\mathcal{F}_t)$-adapted, finite variation, càdlàg processes $(C_t,\ t \in [0,T])$.

We consider a market where $n+1$ assets are traded continuously. One of them (an "ideal money market account") is locally riskless, and has a price S_t^0, at time t, that evolves according to the differential equation

$$dS_t^0 = r_t S_t^0\, dt, \qquad S_0^0 = s^0 > 0.$$

For notational simplicity, we introduce the discount factor $R_t = \exp\left(-\int_0^t r_s\, ds\right)$, so $S_t^0 = s^0/R_t$. The remaining n assets are risky, and their prices are modelled by the following stochastic differential equations:

$$dS_t^i = b_t^i S_t^i\, dt + S_t^i \sum_{j=1}^{n} \sigma_t^{ij}\, dW_t^j, \qquad S_0^i = s^i > 0, \qquad i = 1, \ldots, n.$$

We assume that the *interest rate* process r, the *instantaneous rates of return* $b = (b^1, \ldots, b^n)'$, and the *volatility* matrix $\sigma = (\sigma^{ij})$ satisfy the following assumption.

Assumption 1. r, b, σ are $(\mathcal{F}_t)$-progressively measurable processes, and there exists a constant K such that $\sum_{i=1}^n (|r_t| + |b_t^i|) + \sum_{i,j=1}^n |\sigma_t^{ij}| \leq K$ and $\xi'\sigma_t\sigma_t'\xi \geq K^{-2}|\xi|^2$, $\forall \xi \in \mathbb{R}^n$, $\forall t \in [0,T]$, P-a.s.

Now, consider a "small investor", i.e. an agent whose actions cannot influence the prices, who starts with an *initial capital* x and invests in the $n+1$ market assets. This investor selects a *portfolio process* π and a *cumulative consumption process* $C \in \mathcal{C}$. More specifically, at every instant t, the investor decides how much money π_t^i to invest in the i-th risky asset and what the cumulative consumption C_t should be. By assuming that C is a finite variation process, and not simply an increasing process, we allow for the possibility that the "consumption" $C_{t+\Delta t} - C_t$ in the time interval $]t, t+\Delta t]$ can be negative, in which case an amount $-C_{t+\Delta t} + C_t$ is actually infused into the portfolio.

To ensure that stochastic integrals are well defined, we impose the following condition on the portfolio process:

Assumption 2. The portfolio process $\pi = (\pi^1, \ldots, \pi^n)'$ is $(\mathcal{F}_t)$-progressively measurable and satisfies $\int_0^T |\pi_s|^2\, ds < \infty$, P-a.s.

We can then show that, if $X^{(x,\pi,C)}$ is the investor's total wealth at time t, we have

$$X_t^{(x,\pi,C)} = R_t^{-1}\left(x - \int_{[0,t]} R_s\, dC_s + \int_0^t R_s\,(\sigma_s'\pi_s)\cdot d\tilde{W}_s\right), \quad t \in [0,T],$$

where the $(\mathcal{F}_t, P)$-semimartingale $\tilde{W}$ is defined by

$$\tilde{W}_t = W_t + \int_0^t \sigma_s^{-1}(b_s - r_s\underline{1}_n)\, ds, \quad t \in [0,T],$$

and $\underline{1}_n$ denotes the n-dimensional vector with each component unity.

Now consider the uniformly integrable martingale $(L_t,\ t \in [0,T])$ defined by

$$L_t = \exp\left(-\int_0^t \lambda_s \cdot dW_s - \frac{1}{2}\int_0^t |\lambda_s|^2\, ds\right),$$

where the *relative risk* process λ is defined by $\lambda_t = \sigma_t^{-1}(b_t - r_t\underline{1})$, and let $\tilde{P}$ be the unique probability measure on $(\Omega, \mathcal{F})$ defined by $d\tilde{P}/dP = L_T$. Girsanov's theorem implies that $\tilde{W}$ is a $(\mathcal{F}_t, \tilde{P})$-Brownian motion. Furthermore, when discounted by R, the prices of the traded assets are all martingales.

To exclude arbitrage opportunities, we have to restrict our portfolios to lie within the following class.

Definition 1. A portfolio π is *tame* if it satisfies Assumption 1 and, if M is the $(\mathcal{F}_t, \tilde{P})$-local martingale defined by $M_t = \int_0^t R_s\,(\sigma_s'\pi_s)\cdot d\tilde{W}_s$, then the process M^- is of class (D) with respect to $\tilde{P}$. We denote by $\mathcal{A}$ the set of all tame portfolios.

Note that this assumption implies that the local martingale M is a supermartingale. As a consequence, the market offers no arbitrage opportunities if tame portfolios are the only admissible ones. In the absence of such an assumption, we can construct trading strategies which yield arbitrage opportunities (see Karatzas & Shreve [3, Example 1.2.3]). Also, note that this assumption is weaker than corresponding assumptions in the literature.

REAL OPTIONS

To motivate our definition of a real option, consider an investment that produces a single commodity, and suppose that the price Y_t of one unit of the commodity at time t satisfies

$$dY_t = c_t Y_t \, dt + Y_t e_t \cdot dW_t, \qquad Y_0 = y > 0.$$

Here, we assume that c and $e = (e^1, \ldots, e^n)'$ are $(\mathcal{F}_t)$-progressively measurable processes such that $\int_0^T [|c_t| + |e_t|^2] \, dt < \infty$, P-a.s.

The investment can operate in two modes, say "open" and "closed". The transition from one operating mode to the other one is immediate and forms a sequence of decisions made by the management. We model these decisions by a process $Z \in \mathcal{Z}$, where $\mathcal{Z}$ is the family of all adapted, finite variation, càglàd processes Z with values in $\{0, 1\}$. Specifically, $Z_t = 0$ means that the investment is "closed" at time t, whereas $Z_t = 1$ means that the investment is "open" at time t. Also, the stopping times at which the jumps of Z occur are the intervention times at which the investment's mode is changed.

Now, given any choice of a switching strategy Z, the management faces the cumulative payoff $C^Z \in \mathcal{C}$ defined by

$$C_t^Z = \int_0^t e^{-rs}[h(Y_s)Z_s - C(1 - Z_s)] \, ds - \sum_{0 \le s \le t} e^{-rs}[K_1(\Delta Z_s)^+ + K_0(\Delta Z_s)^-],$$

for $t \le T$, where $\Delta Z_t = Z_{t+} - Z_t$ and $(\Delta Z_t)^{\pm} = \max\{\pm \Delta Z_t, 0\}$. Here, h models the running costs and payoffs resulting from an "open" investment. If in a short time interval $[t, t + \Delta t]$ the investment is "open" and the commodity price is y, then, depending on whether $h(y)$ is positive or negative, the investment yields a profit or incurs a loss equal to $h(y)\Delta t$. The constant C models running "standby" costs resulting from a "closed" investment. Also, the constants K_1, K_0 are the costs resulting from "switching" the investment from its "closed" mode to its "open" one, and vice versa, respectively.

The problem is to *price this investment taking into account the managerial flexibility provided by the choice of a switching strategy* $Z \in \mathcal{Z}$.

With a view to constructing a general model, we observe that the key components of the model outlined above are (i) a set of managerial decision strategies, namely the set of all admissible switching strategies $\mathcal{Z}$, and (ii) a random flow of profits and losses associated with each decision strategy Z, which is given by the process C^Z. This observation gives rise to the following abstract model.

Definition 2. A *real option* $\{C^\partial,\ \partial \in \mathcal{D}\}$ is an economic entity that (i) offers its owner a set of decision strategies $\mathcal{D}$, and (ii) presents its owner with a cumulative payoff $C^\partial \in \mathcal{C}$ if they adopt decision strategy $\partial \in \mathcal{D}$.

We do not impose any specific structure on the decision space $\mathcal{D}$ itself. Of course, as far as the modelling of actual problems is concerned, every decision strategy should be non-anticipative. However, this requirement is captured by the general assumption that C^∂ is $(\mathcal{F}_t)$-adapted, for every $\partial \in \mathcal{D}$.

Financial derivatives can also be considered as special cases:

Example 1. Given $i = 1, \ldots, n$ and $T_1 \in]0, T]$, a *European call option* on asset i with maturity T_1 and strike price K is the special case of the above definition which arises if we define $\mathcal{D} = \mathcal{F}_{T_1}$, and, given $\partial \in \mathcal{D}$, $C_t^\partial = (S_T^i - K)\,1_\partial 1_{\{T_1 \leq t\}}$, for all $t \geq 0$. According to these definitions, if the holder follows decision strategy $\partial \in \mathcal{D}$, then they exercise the option if the event $\partial \in \mathcal{F}_{T_1}$ occurs, and only then.

Example 2. Given $i = 1, \ldots, n$ and $T_1 \in]0, T[$, an *American call option* on asset i with maturity T_1 and strike price K is the special case which arises if we define $\mathcal{D} = \mathcal{S}$, where $\mathcal{S}$ is the set of all $(\mathcal{F}_t)$-stopping times, and, given any stopping time $\partial \in \mathcal{D}$, $C_t^\partial = (S_\partial^i - K)\,1_{\{\partial \leq T_1 \wedge t\}}$, for all $t \geq 0$. Observe that, given a choice of an exercise time $\partial \in \mathcal{D}$, the option is exercised on the event $\{\partial \leq T_1\}$ and expires without being exercised on the event $\{\partial > T_1\}$.

THE PRICE OF A REAL OPTION

Fix any real option $\{C^\partial,\ \partial \in \mathcal{D}\}$. We define the set of the *seller's prices* by

$$\mathcal{V}^s = \{p \in \mathbb{R} \mid \forall \partial \in \mathcal{D},\ \exists x \leq p,\ \exists \pi \in \mathcal{A} : X_T^{(x,\pi,C^\partial)} \geq 0,\ P\text{-a.s.}\}$$

and the set of the *buyer's prices* by

$$\mathcal{V}^b = \{p \in \mathbb{R} \mid \exists \partial \in \mathcal{D},\ \exists x \leq -p,\ \exists \pi \in \mathcal{A} : X_T^{(x,\pi,-C^\partial)} \geq 0,\ P\text{-a.s.}\}$$

Clearly, as long as they are not empty, $\mathcal{V}^s$ is an interval which is a neighbourhood of ∞, and $\mathcal{V}^b$ is an interval which is a neighbourhood of $-\infty$.

Definition 3. The *lowest seller's price* of the real option is $V^s = \inf \mathcal{V}^s$, whereas the *highest buyer's price* of the real option is $V^b = \sup \mathcal{V}^b$. Moreover, if $\mathcal{V}^s \cap \mathcal{V}^b \neq \emptyset$, then any $V \in \mathcal{V}^s \cap \mathcal{V}^b$ is a *rational price* of the real option.

The ideas behind this definition are the following. Suppose first that the price of the real option is set at any $p \in \mathcal{V}^s$. In this case, the decision of the owner to sell the real option is rational because the owner can find a trading strategy that is at least as profitable as any decision strategy. Indeed, consider any $\partial \in \mathcal{D}$, and let $x \leq p$, $\pi \in \mathcal{A}$ be such that $X_T^{(x,\pi,C^\partial)} \geq 0$. If the owner sells the real option for p and invests x to maintain the portfolio-consumption pair (π, C^∂), then the owner makes a riskless profit equal to $p - x$ at time 0, and the portfolio's value is eventually nonnegative. This is a situation which is preferable to maintaining the real option's ownership and following decision strategy ∂.

On the other hand, suppose that the price of the real option is set at any $p \in \mathcal{V}^b$, and let $\partial \in \mathcal{D}$, $x \leq -p$, $\pi \in \mathcal{A}$ be such that $X_T^{(x,\pi,-C^\partial)} \geq 0$. In this case, consider the strategy consisting of: (i) investing an amount p to buy the real option and adopt the decision strategy ∂, (ii) investing an amount x to maintain the portfolio-consumption pair $(\pi, -C^\partial)$, and (iii) infusing the profits resulting from the investment into the portfolio, and vice versa. Clearly, this strategy, which yields a riskless profit equal to $-p-x$ at time 0 and a total wealth that is eventually nonnegative, is at least as good as doing nothing.

Now, if $\mathcal{V}^s \cap \mathcal{V}^b \neq \emptyset$, then both parties would be keen to accept a deal at a price $V \in \mathcal{V}^s \cap \mathcal{V}^b$. Therefore, V being a price at which the deal is satisfactory for both parties, it represents a rational price for the real option.

The following result shows that, under the assumptions that we have made, the lowest seller's price coincides with the highest buyers's price, and the existence of a rational price is equivalent to the existence of an optimal control strategy.

Theorem 1 *Consider a real option $\{C^\partial,\ \partial \in \mathcal{D}\}$ as described by Definition 2, and suppose that* $\sup_{\partial \in \mathcal{D}} \tilde{E} \int_{[0,T]} R_s \, d\check{C}_s^\partial < \infty$. *Then*

$$V^s = V^b = \sup_{\partial \in \mathcal{D}} \tilde{E} \left[\int_{[0,T]} R_s \, dC_s^\partial \right].$$

Moreover, if there exists a $\partial^* \in \mathcal{D}$ such that $V^s = \tilde{E}\left[\int_{[0,T]} R_s \, dC_s^{\partial^*}\right]$, then V^s is the unique rational price of the real option.

REFERENCES

1. Duckworth K. & Zervos M., A Model for Investment Decisions with Switching Costs, *The Annals of Applied Probability*, to appear (2000).
2. Dixit A. K. & R. S. Pindyck R. S., *Investment under Uncertainty*, Princeton University Press (1994).
3. Karatzas I. & Shreve S. E., *Methods of Mathematical Finance*, Springer-Verlag (1998).
4. Knudsen T. S., Meister B. & Zervos M., On the Relationship of the Dynamic Programming Approach and the Contingent Claim Approach to Asset Valuation, *Finance & Stochastics*, **3**, pp. 433–449 (1999).
5. Trigeorgis L., *Real Options: Managerial Flexibility and Strategy in Resource Allocation*, MIT Press (1996).

AUTHOR INDEX